A Brief History of Numbers

A BRIEF HISTORY OF NUMBERS

LEO CORRY

UNIVERSITY PRESS

UNIVERSITY PRESS

Great Clarendon Street, Oxford, OX2 6DP,
United Kingdom

Oxford University Press is a department of the University of Oxford.
It furthers the University's objective of excellence in research, scholarship,
and education by publishing worldwide. Oxford is a registered trade mark of
Oxford University Press in the UK and in certain other countries

© Leo Corry 2015

The moral rights of the author have been asserted

First Edition published in 2015

Impression: 1

All rights reserved. No part of this publication may be reproduced, stored in
a retrieval system, or transmitted, in any form or by any means, without the
prior permission in writing of Oxford University Press, or as expressly permitted
by law, by licence or under terms agreed with the appropriate reprographics
rights organization. Enquiries concerning reproduction outside the scope of the
above should be sent to the Rights Department, Oxford University Press, at the
address above

You must not circulate this work in any other form
and you must impose this same condition on any acquirer

Published in the United States of America by Oxford University Press
198 Madison Avenue, New York, NY 10016, United States of America

British Library Cataloguing in Publication Data

Data available

Library of Congress Control Number: 2015930555

ISBN 978-0-19-870259-7

Printed in Great Britain by
Clays Ltd, St Ives plc

Links to third party websites are provided by Oxford in good faith and
for information only. Oxford disclaims any responsibility for the materials
contained in any third party website referenced in this work.

Dedicado a la memoria de la querida Tere

PREFACE

This book tells the story of the development of the idea of number since the days of the Pythagoreans and up until the turn of the twentieth century. The latter is more or less the time when currently prevailing conceptions about numbers reached their actual state, for all of their complexity (or perhaps we should rather say, for all of their simplicity). This is not the first book to tell a similar story, or, more specifically, to tell a story of roughly the same subject matter. Still, I believe that this book differs essentially from existing ones in content as well as in style. It certainly differs in scope and in the kind of historical material on which it is based.

For the sake of brevity, and given the somewhat informal character of this book, the historical account presented here will be, of necessity, selective. I have not attempted to be either exhaustive or fully balanced in telling this story. Choices had to be made, and I believe that my choices are fair, in terms of the scope and aims that I have pursued in writing the book. I think that the account is comprehensive and representative enough to provide readers with a fair view of the development of the concept of number. Hopefully, readers will also find my choices to be justified, coherent and illuminating.

The story told here focuses mainly on developments related to European mathematics (including ancient Greece). There is also a relatively lengthy chapter on the contributions of the medieval world of Islam. Because of considerations of space, I have left aside entire mathematical cultures such as those of the Far East (China, India, Japan, Korea) and Latin America, each of which came up with their own significant achievements and idiosyncratic conceptions.

This is a book about the historical development of important scientific ideas. Of course, these ideas were created and disseminated by actual persons, who lived and worked in specific historical circumstances. I devote some attention to these lives and circumstances, but only to the extent that they help us understand the ideas in their proper context. Accordingly, neither heroic duels at dawn nor tragic cases of suicide related to the failure in solving a certain problem appear here as part of my narrative. Not that such anecdotes are devoid of interest. But I thought that the intrinsic dynamic of the ideas is dramatic enough to warrant the continued attention of the reader.

In writing the various chapters, I have tried to reflect the most recent and updated, relevant historical scholarship, alongside some of the best classical one. This is a not a monographical text involving original historical research meant to be cited in subsequent works of fellow historians. Rather, it is a work of synthesis meant to provide a broad overview of what historians have written and to direct prospective readers to their works. I do not claim, however, to have reflected all the existing views on all the topics, nor all of the divergent and sometimes opposed interpretations that historians have suggested for the various episodes that I discuss. At the same time, given the style of the book, I did not want to fatigue the reader with a full scholarly apparatus to support each and every one of the claims I have made in the text. Direct references are only to texts that are cited as full paragraphs, or to texts that elaborate in greater detail on issues that I have only mentioned in passing but that, for reasons of space, I have left as brief remarks. Still, I have provided at the end of the book a somewhat detailed list of texts for further reading, organized more or less in accordance with the chapters of the book. This is my way to explicitly acknowledge the works on which I have directly relied for the various issues discussed. This is my way to indicate my great intellectual debt to each of these authors. Readers interested in broadening the scope of my account will find in those texts large amounts of additional illuminating material. The texts also provide direct references to the primary sources, which lend support to the historical claims that I put forward in this book. In fact, in consideration with space limitations, except in cases where I have cited directly, I do not include primary sources in the bibliographical list, in the assumption that interested readers can easily find detailed information either in the secondary sources or collections cited or by Internet search.

One typical reader that I have had in mind when writing the book is an undergraduate student in mathematics. As she is struggling to develop demanding technical skills in the courses that she is now studying, a book like this may help her make further sense, from a perspective that is different from that found in textbooks and in class, of the world of ideas that she is gradually becoming acquainted with. Her apprenticeship, I think, may gain in depth and richness by reading a book like this one. A second, typical kind of reader I have envisaged for the book is a teacher of mathematics. This book, I believe, may contribute to his efforts in bringing to the classroom a broader understanding of his own discipline, with an eye on the long-range historical processes that have been at play in shaping it.

I have also tried to write in such a way as to allow for a wider readership that would include intellectually curious high-school pupils with an inclination for mathematics, professional mathematicians, scientists and engineers, and other educated readers of various kinds. Hence I have tried to strike a natural balance between technicalities and broad historical accounts. In this respect, there are slight differences among the various chapters. Some are more technically demanding, some less so. In all cases, however, I have made my best efforts to write concisely and in a way that will be within the reach of readers of various backgrounds and will elicit their interest.

Still, I have certainly not tried to reach *every possible* readership. While writing, I have often had in mind a famous statement of Stephen Hawking in the preface to his *Brief History of Time* (from which I shamelessly took inspiration in choosing my own title). He declared that he had followed the advice of his publisher to the effect that he should not include too many formulae in the book, since each individual formula would reduce the number of potential readers by a half. Hawking wisely followed this advice and he included in the book only one formula, the inevitable $e = mc^2$. In retrospect, this was certainly a very good piece of advice, as the book turned into an unprecedented best-seller and it was also translated into an amazingly large number of languages.

I can only wish for such a success at the box-office with my book, but I have to confess that I have had to include more than one formula. Actually, the reader will find here *many* formulae and diagrams, though, in general, they are not particularly complicated. I think that anyone who thought that the topic of the book is appealing for her, in the first place, will not be deterred by what I have included. And I simply could not leave these formulae and diagrams out of my account, if I wanted to speak seriously about the history of numbers. Still, in places where I have thought that some readers would like to see further technical details, while others would find these details to be a hindrance to their understanding, I have relegated them to a separate appendix in the respective chapter. These appendices can be skipped without fear of loss of continuity in understanding the main text, but I strongly suggest that the reader should not do that and should rather devote the necessary effort where needed. I think that such effort will be rewarded. In some places, I have provided some detailed proofs or technical explanations in the main text, assuming that they will be of interest to most readers and that they are necessary for achieving the full picture that I am trying to convey.

Several friends and colleagues have been kind enough to read parts of the book and to offer me their useful critical comments. They have helped me improve the contents, the structure and the style of my text to a considerable extent. I thank them sincerely: Alain Bernard, Sonja Brentjes, Jean Christianidis, Michael Fried, Veronica Gavagna, Jeremy Gray, Niccolò Guicciardini, Albrecht Heeffer, Victor Katz, Jeffrey Oaks and Roy Wagner. I also acknowledge the wise advise of three anonymous referees.

It is a pleasure to thank Keith Mansfield and his team at OUP (Clare Charles and Daniel Taber) for their help and editorial support. My sincere thanks go to Mac Clarke for a superb job as copy editor of my text. Likewise, I want to thank Kaarkuzhali Gunasekaran and her team at Integra Software Services for their highly professional approach in typesetting this book.

Special thanks go to my dear friend Lior Segev for having introduced me into the basics of LaTeX. I had been able thus far, somehow, to avoid doing this, but, in retrospect, I am glad to have learned the craft now under his guidance.

I am always happy to acknowledge the continued and friendly support of Barbara and Bertram Cohn, as the proud incumbent of the Chair in History and Philosophy of Exact Sciences at Tel Aviv University that bears their name.

Finally, I thank my dear family for their unconditional support and for their blind trust that whatever I do, it must surely be important and praiseworthy. As always, I have done my best efforts to stand up to their high expectations.

Tel Aviv University, May 2015   Leo Corry

CONTENTS

1. **The System of Numbers: An Overview**   1
   - 1.1 From natural to real numbers   3
   - 1.2 Imaginary numbers   9
   - 1.3 Polynomials and transcendental numbers   11
   - 1.4 Cardinals and ordinals   15

2. **Writing Numbers—Now and Back Then**   17
   - 2.1 Writing numbers nowadays: positional and decimal   17
   - 2.2 Writing numbers back then: Egypt, Babylon and Greece   24

3. **Numbers and Magnitudes in the Greek Mathematical Tradition**   31
   - 3.1 Pythagorean numbers   32
   - 3.2 Ratios and proportions   35
   - 3.3 Incommensurability   39
   - 3.4 Eudoxus' theory of proportions   42
   - 3.5 Greek fractional numbers   45
   - 3.6 Comparisons, not measurements   47
   - 3.7 A unit length   50
   - Appendix 3.1 The incommensurability of $\sqrt{2}$. Ancient and modern proofs   52
   - Appendix 3.2 Eudoxus' theory of proportions in action   55
   - Appendix 3.3 Euclid and the area of the circle   59

4. **Construction Problems and Numerical Problems in the Greek Mathematical Tradition**   63
   - 4.1 The arithmetic books of the *Elements*   64
   - 4.2 Geometric algebra?   66
   - 4.3 Straightedge and compass   67
   - 4.4 Diophantus' numerical problems   71
   - 4.5 Diophantus' reciprocals and fractions   78
   - 4.6 More than three dimensions   80
   - Appendix 4.1 Diophantus' solution of Problem V.9 in *Arithmetica*   83

5. Numbers in the Tradition of Medieval Islam — 87
- 5.1 Islamicate science in historical perspective — 88
- 5.2 Al-Khwārizmī and numerical problems with squares — 90
- 5.3 Geometry and certainty — 94
- 5.4 *Al-jabr wa'l-muqābala* — 97
- 5.5 Al-Khwārizmī, numbers and fractions — 100
- 5.6 Abū Kāmil's numbers at the crossroads of two traditions — 103
- 5.7 Numbers, fractions and symbolic methods — 107
- 5.8 Al-Khayyām and numerical problems with cubes — 111
- 5.9 Gersonides and problems with numbers — 116
- Appendix 5.1 The quadratic equation. Derivation of the algebraic formula — 120
- Appendix 5.2 The cubic equation. Khayyam's geometric solution — 121

6. Numbers in Europe from the Twelfth to the Sixteenth Centuries — 125
- 6.1 Fibonacci and Hindu–Arabic numbers in Europe — 128
- 6.2 Abbacus and coss traditions in Europe — 129
- 6.3 Cardano's *Great Art of Algebra* — 138
- 6.4 Bombelli and the roots of negative numbers — 146
- 6.5 Euclid's *Elements* in the Renaissance — 149
- Appendix 6.1 Casting out nines — 150

7. Number and Equations at the Beginning of the Scientific Revolution — 155
- 7.1 Viète and the new art of analysis — 157
- 7.2 Stevin and decimal fractions — 163
- 7.3 Logarithms and the decimal system of numeration — 167
- Appendix 7.1 Napier's construction of logarithmic tables — 171

8. Number and Equations in the Works of Descartes, Newton and their Contemporaries — 175
- 8.1 Descartes' new approach to numbers and equations — 176
- 8.2 Wallis and the primacy of algebra — 182
- 8.3 Barrow and the opposition to the primacy of algebra — 187
- 8.4 Newton's *Universal Arithmetick* — 190
- Appendix 8.1 The quadratic equation. Descartes' geometric solution — 196
- Appendix 8.2 Between geometry and algebra in the seventeenth century: The case of Euclid's *Elements* — 198

9. New Definitions of Complex Numbers in the Early Nineteenth Century — 207
- 9.1 Numbers and ratios: giving up metaphysics — 208
- 9.2 Euler, Gauss and the ubiquity of complex numbers — 209
- 9.3 Geometric interpretations of the complex numbers — 212
- 9.4 Hamilton's formal definition of complex numbers — 215
- 9.5 Beyond complex numbers — 217
- 9.6 Hamilton's discovery of quaternions — 220

10. "What Are Numbers and What Should They Be?"
   Understanding Numbers in the Late Nineteenth Century 223
   10.1 What are numbers? 224
   10.2 Kummer's ideal numbers 225
   10.3 Fields of algebraic numbers 228
   10.4 What should numbers be? 231
   10.5 Numbers and the foundations of calculus 234
   10.6 Continuity and irrational numbers 237
   Appendix 10.1 Dedekind's theory of cuts and Eudoxus' theory of
      proportions 243
   Appendix 10.2 IVT and the fundamental theorem of calculus 245

11. Exact Definitions for the Natural Numbers: Dedekind,
    Peano and Frege 249
    11.1 The principle of mathematical induction 250
    11.2 Peano's postulates 251
    11.3 Dedekind's chains of natural numbers 257
    11.4 Frege's definition of cardinal numbers 259
    Appendix 11.1 The principle of induction and Peano's postulates 262

12. Numbers, Sets and Infinity. A Conceptual Breakthrough at
    the Turn of the Twentieth Century 265
    12.1 Dedekind, Cantor and the infinite 266
    12.2 Infinities of various sizes 269
    12.3 Cantor's transfinite ordinals 277
    12.4 Troubles in paradise 280
    Appendix 12.1 Proof that the set of algebraic numbers is countable 287

13. Epilogue: Numbers in Historical Perspective 291

References and Suggestions for Further Reading 295
Name Index 303
Subject Index 306

CHAPTER 1

The System of Numbers: An Overview

Mathematics and history, history and mathematics. One can hardly think of two fields of knowledge that are more different from each other—some would say outright opposed—in both their essence and their practice.

At its core, mathematical knowledge deals with *certain*, *necessary* and *universal* truths. True mathematical statements do not depend on contextual considerations, either in time or in geographical location. Generally speaking, established mathematical statements are considered to be beyond dispute or interpretation.

The discipline of history, on the contrary, deals with the *particular*, the *contingent* and the *idiosyncratic*. It deals with events that happened in a particular location at a particular point in time, and that happened in a certain way but could have happened otherwise. Historical statements are always partial, debatable and open to interpretation. Arguments put forward by historians keep changing with time. "Thinking historically" and "thinking mathematically" are clearly two different things.

But, if by "thinking historically" about whatever topic, we mean anything other than just establishing a chronology of facts, then an interesting question is whether we can think historically about the ways in which people have been "thinking mathematically" throughout history, and about the processes of change that have affected these ways of thinking. If mathematics deals with universal truths, how can we speak about mathematics from a historical perspective (other than establish the chronology of certain discoveries)? What is it that changes through time in a discipline that is, apparently, eternal?

This is precisely what this book is about: a brief historical account (which is not just a chronology) of how people have thought about numbers in changing historical circumstances. Not just *what* they knew about numbers, but also, and mainly, *how* they knew it and what did they think *about* what they knew. The kind of questions we will be pursuing here include the following:

- What were the basic concepts around which knowledge about numbers and about their properties was built in different historical contexts?

- How did it happen that these concepts changed through time?
- What were the main mathematical problems that required, in a certain historical context, the use of numbers of various kinds?
- How were numbers written in different cultures, and how did different mathematical notations either help or hinder further developments in the conceptions of numbers and in the techniques of calculation?
- What was the relationship, in the various mathematical cultures, between arithmetic and other, neighboring, disciplines (mainly geometry), and what were the philosophical conceptions of the mathematicians involved in arithmetical activity?
- How did practical considerations encourage (or sometimes discourage) the adoption of certain ideas pertaining to numbers?
- What roles did the institutions of knowledge that developed in the various cultures play in promoting or preventing the development of a given conception about numbers?

Numbers are important in our world. The world around us is saturated with them. Numbers appear routinely and increasingly not only in scientific and technological contexts, but also in the news and in commerce, as well as in many aspects of our private life. The vernacular of natural science—and particularly of physics—is mathematics, at the heart of which lies numbers. Also in the social sciences—economics and political science above all—the language of numbers is central to much of both its theorizing and its empirical work. In public life, numbers are not just ubiquitous as a tool for explanation, but also a necessary means for administration and control. The main tools used nowadays in bureaucratic systems all around the world rely on the processing of data and (sometimes manipulative) analysis of numbers. The centrality of digital computers in all aspects of life just emphasizes and makes even more manifest the truth of these basic facts.

Numbers are present in so many aspects of our day-to-day life that we take their presence for granted. But this situation is by no means an inevitable law of nature. Rather it is the result of a very specific, long, convoluted and multilayered historical process. A very important turning point in this process came in the seventeenth century, in the framework of what is typically known as the "scientific revolution." Developments in disciplines such as the science of mechanics during this period turned them into all-out mathematical branches, in opposition to what was the case in the Aristotelian tradition that had dominated European intellectual life since the late scholastic period around the fourteenth century. The dominant ideal of explanation of natural phenomena before the seventeenth century did not encourage (and indeed sometimes actively opposed) the search for mathematically formulated laws of nature. During the eighteenth and nineteenth centuries, the role of numbers became increasingly central in natural science as well as in other aspects of knowledge and day-to-day life.

These kinds of significant transformations that affected the *role* of numbers in society throughout history have attracted the attention of historians, and, indeed, many educated audiences are well aware of them to various degrees of detail. What is less recognizable at first sight, and has typically escaped the attention of those same audiences, is that even the very idea of *what is a number* and how it is used *within mathematics* has itself been at the focus of a more circumscribed yet no less long and complex process of debate, evolution and continuous modification. This book focuses on this process, a process that is historically important yet subtle and rather inconspicuous within the overall accepted conceptions about the development of science.

Of course, anyone who gives some thought to it will not fail to realize that all of our current knowledge about numbers and about their properties was achieved as the accumulated product of the efforts of generations of mathematicians who devoted great intellectual energies to it. But, more than with any other field of knowledge, this process tends to be conceived as essentially linear and straightforward. More than with any other field of knowledge, the role of the historian reconstructing this process tends to be seen as that of a chronologist whose main task is just to determine who did what for the first time and when.

Typically, the *process* leading to our present conceptions about numbers and to our knowledge of their properties is presumed to be clear, historically unproblematic and unsurprising, except perhaps for the dates and names to be attached to each step. The idea of a truly *historical process*—where individuals and groups are faced with dilemmas and need to make real choices between alternatives, where wrong alleys, detours and dead-ends are sometimes followed for long periods of time, where contrary views are held by parties who are equally knowledgeable within their subjects—seems to many to be foreign to the development of mathematics and in particular to the development of ideas related to numbers.

I intend to show that the history of numbers is an intriguing story that developed at various levels, in a surprisingly non-linear fashion, involving many unexpected moves, dead-ends and also, of course, highly ingenious ideas and far-reaching successes. The outcome of the story is the creation of a beautifully conceived world of numbers, as it crystallized by the turn of the twentieth century. Since then, some new ideas and refinements of the existing ones have been added, but our basic conception of the system of numbers and how it is built was essentially attained by that time, after centuries of important breakthroughs accompanied by hesitations, misunderstandings and uncertainties.

In order to make this account more perspicuous and effective, I have chosen to begin with an introductory chapter that is technical and non-historical. It is intended to offer a general overview of the system of numbers such as it is conceived nowadays. This includes a general description of the various kinds of numbers and the relationships between them, the accepted ways to write these numbers, and the assumptions underlying them. I have also included some truly basic results about numbers. Some prospective readers of this book will surely be acquainted with the material discussed in this chapter, but it is convenient to provide a common basic language with which to proceed to the historical account that will appear in subsequent ones.

1.1 From natural to real numbers

The very systematic conception of the world of numbers that is accepted nowadays envisages a carefully constructed hierarchy of various classes of increasing complexity that starts with the natural numbers and gradually adds new types—negative, rational, irrational—up to the real numbers. This view was attained only by the turn of the twentieth century. If we consider earlier periods, such as between the seventeenth and nineteenth centuries, in which so much important scientific progress was achieved in Europe, particularly in mathematics and in mathematical physics, it may come as a great surprise that the concept of number was still quite confused and lacked anything like a proper foundation. Remarkably, this did not substantially hinder the enormous progress in those disciplines where mathematics was now playing a central role.

This in itself is a non-trivial historical phenomenon that deserves closer attention. Indeed, there exists an interesting and not always properly emphasized gap between the actual, rather erratic, historical evolution of mathematical ideas, on the one hand, and, on the other hand, the subsequent, neat textbook presentation of these ideas as part of a perfectly structured body of knowledge. Students of mathematics typically learn in their university courses, say of calculus, a very tidy, comprehensive and well-arranged picture of the discipline, where each stage of the gradual presentation of results seems to arise naturally and smoothly from the previous one. Historically speaking, in contrast, these ideas evolved in a much more chaotic, unordered and unexpected way. As a matter of fact, sometimes they evolved in a succession that is the exact opposite to how they are conveniently presented in retrospect.

In the case of the calculus, for example, the concept of limit, typically learnt in the early lessons of any course in analysis, only arose at the beginning of the nineteenth century, after many of the basic techniques of calculation with derivatives and integrals and of solving differential equations had already been developed. And in turn, in order to provide a truly well-elaborated, formal definition of limit, it was necessary to come up with a much clearer idea of the foundations and basic structure of the system of real numbers, which became available only at the end of the nineteenth century. This, in turn, required a much better knowledge of the concept of set, which started to be developed only around this time and became well understood only in the first decade of the twentieth century.

The most basic idea related to the concept of number is that of counting. We associate this idea with the concept of *natural* numbers, namely those appearing in the series $1, 2, 3, 4, 5, \ldots$ This is the starting point of arithmetic both in the *cognitive* sense (i.e., the way in which any child begins to learn numbers) and in the *historical* sense (i.e., arithmetical knowledge, in whatever culture, starts from knowledge about the natural numbers). This may seem an obvious point, but it is far from being so. As already mentioned, these two lines of development – cognitive and historical – typically differ, especially when it comes to the more advanced aspects of mathematics. We will see many important examples of this in what follows. But here at very the basis of arithmetic, they essentially coincide.

It is customary to denote the collection of natural numbers by the letter $\mathbb{N}$, as follows:

$$\mathbb{N} = \{1, 2, 3, 4, \ldots\}.$$

An important issue that arises immediately in connection with the natural numbers is the special role accorded to the prime numbers within this collection. A prime number is usually defined as a natural number having only two exact divisors: itself and 1 (and hence 1 itself is typically not considered to be a prime number, but this is just a matter of convention). Such numbers attracted special attention from very early on, and they continue to be at the focus of the most advanced mathematical research nowadays.

Prime numbers provide the elementary building blocks of the entire system of natural numbers in a very precise and clearly definable way: every natural number can be written as a product of prime numbers, and this *in a unique way* (except for the order of the factors). The number 15, for instance, is the product of two primes, $15 = 3 \times 5$, and this is the only way to write it as such a product. The same goes for a different example: $13,500 = 3^3 \times 2^2 \times 5^3$. Again, there is no other way to write this number as a product of prime factors. This property of the system of natural numbers provides a powerful tool for proving many other theorems in the theory of numbers. Because of its importance, mathematicians usually dub it "the fundamental theorem of arithmetic."

Another interesting result related to prime numbers is that there are infinitely many of them. This was already known to mathematicians in ancient Greece (and we will speak about this later on), and you should notice that the result is far from self-evident. Indeed, since natural numbers arise from all kind of products involving combinations of primes and their powers, one might well imagine that a finite number of primes (perhaps a large number) would suffice to generate *all* natural numbers by different combinations and by taking ever larger powers of the same primes. But, as we will see, this is not the case.

Beginning from the idea of the natural numbers, we can systematically extend the concept of numbers and thus introduce additional types, which are required in order to be able to solve various types of algebraic equations. The first extension comes by way of the *negative* numbers, which are necessary if we want to be able to solve such a simple equation as $x + 8 = 4$. Here we are required to find a number that if added to 8 will yield the result 4. If we had just landed on Planet Earth without any previous knowledge of arithmetic and we had been taught to calculate only with natural numbers, our immediate reaction to the requirement implied by this equation would be that it has no possible answer. As we know from experience, this is what children are sometimes taught in the early grades of primary school. If presented with the operation "from 4 subtract 8," the expected answer could be: "impossible" or "no solution."

Also considered in historical perspective, the possibility of solving an equation by providing a negative number as the answer took very long to be accepted as possible or legitimate. We will talk about this at length. But here I want to begin from the opposite direction, and to simply state, or postulate, that -4 is "that number that solves the equation $x + 8 = 4$." I do not even question the legitimacy of postulating such a number. Negative numbers, accordingly, are numbers that, if added to any given number, yield another number that is smaller than the given one. This is not a very good mathematical definition, and we will see a better one later on, but for now it helps us expand our arsenal. The collection of positive natural numbers and negative ones, all taken together, is called the collection of *integer numbers*, or simply "integers." This collection is usually denoted by the letter $\mathbb{Z}$, as follows:

$$\mathbb{Z} = \{\ldots, -4, -3, -2, -1, 0, 1, 2, 3, 4, \ldots\}.$$

Notice that, without having said much, the number 0 has suddenly made an appearance here. This number also has an interesting history in itself, some of which will appear later in the book. At this point, we can define it, by analogy with the definition of negative numbers above, as a number that solves the equation $x + 4 = 4$, or as a number that, if added to any given number, yields the given number itself.

We can continue expanding the arsenal of necessary (or possible?) numbers with the help of equations that cannot be solved with integers. Consider, for instance, the equation $3x + 2 = 4$. Here we are required to find a number that when multiplied by 3 yields 2 (so that the result when added to 2 yields 4). Obviously, none of the integers that have been introduced thus far will be of use, because, for example, already $1 \times 3 = 3$, which is greater than 2. We can again give up and say "no answer." Alternatively, we can say that any equation of this kind must have a solution, which in this case is the fraction $\frac{2}{3}$, which indeed fulfills the requirement, since $3 \cdot (\frac{2}{3}) + 2 = 4$.

We can thus simply define a fraction as a division of two integers, and later on we will come up with a more mathematically satisfactory definition. The two integers entering into division to define a fraction may be both of the same sign (either positive or negative), in which case the fraction is positive, or they may be of opposite sign, in which case the fraction is negative. And of course, every integer can be seen as a fraction, in which the integer is divided by 1. On the contrary, fractions in which a number is divided by 0 are not accepted as legitimate. The collection of all fractions is called the system of *rational* numbers and is typically denoted by the letter $\mathbb{Q}$, as follows:

$$\mathbb{Q} = \left\{ \text{all numbers of the form } \frac{p}{q}, \text{ where } p \text{ and } q \text{ are integers and } q \neq 0 \right\}.$$

The fractions just introduced are "common fractions" $\frac{p}{q}$, but, as is well known, they can equivalently be written as "decimal fractions," such as $0.5 = \frac{1}{2}$. Here we come across yet another interesting story with historical roots that will be discussed in the forthcoming chapters: how did this non-trivial identification between common and decimal fractions come about?

Let us now move into yet another class of numbers, which is needed for solving an equation such as $x^2 = 2$. In this case, we are required to find a number such that, if multiplied by itself, yields 2. Notice that, on the face of it, there is no apparent reason to assume that this equation cannot be solved with the help of the rational numbers alone. We begin with the observation that $1^2 = 1$ and $2^2 = 4$. Hence, under the reasonable assumption that there is a good correspondence between arithmetic operations with rational numbers and their ordering (and, indeed, this is not just a gratuitous assumption, but we will not prove it here), we can easily reach the conclusion that the number we are looking for, and which solves the equation $x^2 = 2$, should be a number between 1 and 2. Thus far, there is no reason to expect that no rational number will do. If we multiply, for example, 1.5 by itself, we obtain 2.25, which is greater than 2. Continuing this process of approximation by trial and error, we may try with a smaller number, say 1.4. Multiplied by itself, it yields 1.96, just under 2. We then try a larger one, say 1.41 ($1.41 \times 1.41 = 1.9881$), and we continue in this way to approximate gradually to a number whose square is 2 (sometimes from above, sometimes from below):

1, 2, 1.5, 1.4, 1.41, 1.42, 1.414 and so on. We can of course write this sequence as a sequence of common, not decimal, fractions: $1, 2, \frac{3}{2}, \frac{7}{5}, \frac{141}{100}, \frac{71}{50}, \frac{707}{500}, \ldots$

This way of looking for the value of the desired number is perhaps tiresome, but there is, at this point, no reason to assume that we will not end up finding an exact value of a rational number that when multiplied by itself yields 2. This is a number that we can call "the square root of 2," and we indicate it by $\sqrt{2}$. But it so happened that at some point in history it was realized that no rational number exists whose square equals 2. How exactly this insight was attained and in what terms it was formulated, this is something we will discuss in Chapter 3. At this point, I just want to stress, without the slightest exaggeration, that this discovery represents one of the most seminal moments in the history of numbers, and we will see why.

We realize, then, that in order to solve the equation $x^2 = 2$, we need to consider the possible existence of numbers that cannot be expressed as fractions. Again, we postulate the existence of that number, and give it a name, $\sqrt{2}$, and we define it as a number such that if multiplied by itself yields 2. Moreover, we define the irrational numbers as those that cannot be written as fractions. Other irrationals include roots of other numbers, such as the root of 97, $\sqrt{97}$ and combinations thereof, such as $\sqrt[5]{2\sqrt{97}} - \frac{5}{3} + \sqrt[3]{2} + \sqrt{52}$. Notice, of course, that their irrationality is not embodied simply in their being written as a root, since $\sqrt{4}$, which is the number 2, is of course rational, and so is $\sqrt{\frac{9}{25}}$, which is the rational number $\frac{3}{5}$. In each case, we would need to find out whether the said roots can or cannot be written as fractions.

On the other hand, and perhaps more interestingly, not all irrational numbers arise as roots of rationals or combinations thereof. There are, in addition, irrational numbers of a very different kind, and indeed some of them are rather well known, such as the number π. There is no standard sign to indicate the collection of irrational numbers separately, but the system of numbers that includes both the rational and irrational numbers taken together is known as the system of *real* numbers and is usually denoted by the letter $\mathbb{R}$.

Notice that the natural numbers are also integers, that the integers are also rational, and that the rationals are also real numbers. This important hierarchy of classes of numbers can be diagrammatically represented as in Figure 1.1.

Thus far, I have introduced the various number systems by assuming, without any qualification, that if a certain kind of numbers is needed in order to be able to solve some kind of equation, this in itself justifies the acceptance of such numbers as legitimate mathematical entities. This has been a standard attitude in mathematics since the early twentieth century, and the only constraint imposed upon mathematical entities that are defined in this way is that they should not create any logical contradiction with the existing edifice of mathematics that we take as a starting point of our investigation. I have introduced the successive systems of numbers without actually checking that any of the extensions does not lead to contradictions with the previously existing ones, but, as we will see in later chapters, this can be done.

It should be noticed, however, that this very liberal approach to allowing legitimate mathematical citizenship to any idea that arises in our mind, provided only that it not creates a contradiction with the existing body of mathematics, is in itself the outcome of a convoluted and interesting historical process that deserves closer inspection. In earlier historical stages, there were acute discussions among mathematicians as to

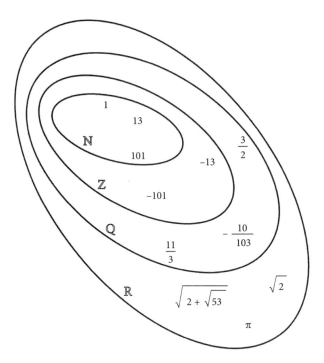

Figure 1.1 The various number systems, from the naturals to the real.

whether it makes mathematical sense of any kind, and whether it is convenient or philosophically sound, to speak of negative or of irrational numbers.

Ideas about the nature of numbers are intimately connected with the ways in which they are represented. One aspect of this concerns the symbolic dimension. In our culture, the typical symbolic representation of numbers is by means of the decimal notation, about which we will have much to say in terms of historical development. But another important aspect of representing numbers concerns the graphical or pictorial dimension. For us, the typical way to do this is with the help of a straight line, as in Figure 1.2. Numbers appear here spread over a straight line, the positive numbers growing indefinitely to the right-hand-side direction, and the negative ones equally indefinitely to the left. Over this same line, not just the integers but also all real numbers are laid, as in Figure 1.3.

The underlying idea behind this representation is that to each real number there corresponds one, and only one, point on the straight line, and vice versa. This important idea, needless to repeat, has an interesting history of its own. An important

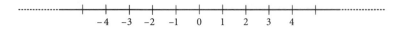

Figure 1.2 Graphical representation of numbers along a straight line.

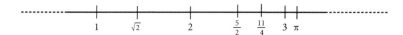

Figure 1.3 Graphical representation of rational and real numbers along a straight line.

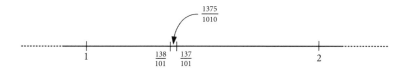

Figure 1.4 Rational numbers densely spread over a straight line.

property of the rational numbers that is also reflected in their representation on the lies is that, given any two such numbers, there is always another rational number lying between them. If we take, for instance, two fractions, $\frac{137}{101}$ and $\frac{138}{101}$, which are quite close to each other, we can take the number $\frac{1375}{1010}$, which is the arithmetic mean between the two, and hence lies between them. This can also be easily seen in terms of decimal fractions, whereby the ordering of the three results is evident: 1.35643564..., 1.36138613861..., 1.366336633.... Graphically, the situation corresponds to the diagram in Figure 1.4.

This situation can be more generally formulated by saying that, given any pair of rational numbers a, b, there is always another rational number lying between them. It is clear that $\frac{(a+b)}{2}$ lies between them, but, as a matter of fact, the case is that *infinitely many* rational numbers lie between a and b, since one could go on finding numbers between a and $\frac{(a+b)}{2}$, as well as between $\frac{(a+b)}{2}$ and b, and so on, indefinitely. This seems to indicate that we can fill points corresponding to the entire straight line just with the help of these densely packed rational numbers. Where do the remaining real numbers, all the irrational ones, enter the picture then, if this is really the case?

This is a rather non-trivial and highly interesting question that arose only in the last third of the nineteenth century, as it became clear that this property of "density" (namely, that between any two rational numbers there is always an infinite number of rational numbers) is different from another, much stronger, property, namely, "continuity." The German mathematician Richard Dedekind was the first to define this distinction very clearly and to show that while continuity appears with the real numbers, it does not appear with the rationals, which are "only" dense. This he did in a famous booklet of 1872, entitled *Stetigkeit und irrationale Zahlen* ("Continuity and Irrational Numbers"), in which the real numbers were rigorously defined for the first time. Notice how late in history such fundamental ideas regarding numbers were finally clarified! We will have much to say about this.

1.2 Imaginary numbers

The systems of natural, integer, rational and real numbers introduced above are the basic components of the world of numbers as we currently conceive it, but they are not the only ones. There are further notions about types of numbers that either divide these four types into subtypes or extend them into new realms. Solving, for example, the equation $x^2 = 2$, or $x^2 - 2 = 0$, requires, as we have seen, the existence of irrational numbers. Consider now a seemingly similar equation, namely, $x^2 + 1 = 0$. Solving this equation requires finding a number whose square equals -1. On the face of it, for the uninitiated, this sounds truly impossible, given that the product of two equal-signed real numbers is always positive, and hence the square of any number is positive, and hence it cannot be equal to -1. But we already know that mathematicians have a simple way to overcome this apparent difficulty by simply postulating the necessary number and hence establishing its existence.

The number in question is a mathematical entity typically denoted by the letter i, which is defined as a number with the property that $i^2 = -1$. That's all. We simply postulate the existence of a number that we call i, and such that $i = \sqrt{-1}$, and by setting $x = i$, the equation $x^2 + 1 = 0$ is automatically solved. To be precise, it is not enough to postulate the existence of such a number. One must also define a full arithmetic that conveniently incorporates this number, and into which all the existing arithmetic of the real numbers can be fully extended without giving rise to any kind of contradiction. This can be easily done, but in historical perspective—well, we can already guess—it took a long and intricate process until such a complete integration was attained (more on this in Chapter 9).

Throughout history, it turns out, there were mathematicians (often some of the best mathematicians of their era) whose idea of what is a number ruled out the possibility that there may exists numbers that are "less than nothing," or numbers that cannot be written as a fraction, or numbers that when multiplied by themselves yield negative numbers. An important part of the story that I will be telling in this book concerns the ways in which such ideas arose and developed and how, and under which mathematically circumstances, they had to be gradually modified until the current ideas about numbers consolidated. The case of $i = \sqrt{-1}$ is perhaps the most notorious one and we will devote considerable attention to it, but it is by no means unique. The letter i was originally chosen for denoting this somewhat mysterious entity because the number $\sqrt{-1}$ was initially called "imaginary."

With the help of imaginary numbers, we can create the "complex numbers," each of which comprises a "real" and an "imaginary" part. Thus, a complex number is usually written as $a + bi$, where both a and b are real numbers. The complex number $\sqrt{-1}$ can be written as $i = 0 + 1i$, which means that its "real" part is 0 and its "imaginary" part is 1. On the other hand, every real number is also a complex one, whose imaginary part is 0: the number 35, for example, is $35 + 0i$. The arithmetic of the complex numbers takes into account the arithmetic of the real ones, and also the basic defining property of i, namely, $i^2 = -1$. From here, it follows, among other things, that $i^3 = -i$ and $i^4 = 1$. The set of complex number is usually denoted by the letter $\mathbb{C}$, and it is conveniently represented in a graph over the plane as indicated in Figure 1.5.

Notice that I have chosen to represent in the figure a few specific numbers: $3.5 + 2i$, $-4 + i$, and $1 - \pi i$. Notice also that the horizontal axis represents the real numbers, and it remains identical to the graphical representation in terms of a line introduced in

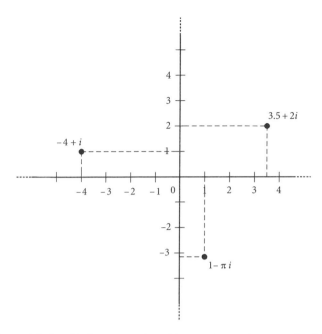

Figure 1.5 Graphical representation of the complex numbers over the plane.

the previous figures. The very need to address the question of possibly speaking about roots of negative numbers was an important historical source of original and, indeed, decisive ideas on the path to the modern conception of number. Complex numbers, to be sure, were not fully understood by mathematicians until the mid-nineteenth century. But in spite of their apparently artificial origins, it turned out very soon after their full incorporation into mathematics that they had important uses—and this was the real surprise—in *physics* and in electrical engineering. Indeed, important fields such as electrodynamics reached scientific maturity only after they were formulated in exact mathematical terms, thanks to the possibility of using complex numbers as part of this formulation.

1.3 Polynomials and transcendental numbers

I have presented thus far the most basic concepts that I would like every reader of this book to keep in mind while going through the following chapters. In the remainder of the chapter, I would like to introduce some further ideas that will add depth to the discussion to follow, and that will appeal to those readers with a somewhat more thoroughgoing interest in the topic (there's no need to worry about too deep mathematics being discussed here—it can be safely skipped by readers who may find it too demanding).

In the discussion above, I have systematically introduced new number systems wherever it was noticed that the existing ones did not provide solutions for certain equations. But one may well ask if this is the only way to introduce new classes of yet unrecognized numbers, namely, according to the need to solve certain equations. In order to answer this question, it is necessary to draw attention to the kind of equations considered thus far. They are all specific examples of a more general kind usually called polynomial equations, such as $x^2 + 2 = 0$, $5x^4 + x^{10} = -8$ and $5x^3 + 8x^{21} = x^{10} + 7$. We always have here one unknown number, x, that appears with various of its powers, $x, x^2, x^3, \ldots, x^n$, the latter being preceded by coefficients. Formally, all of these questions can be described as being of the form

$$a_n x^n + a_{n-1} x^{n-1} + a_{n-2} x^{n-2} + \ldots + a_2 x^2 + a_1 x + a_0 = 0.$$

The highest power of the unknown, the number n, is called the "degree" or "order" of the polynomial (and of the equation). Each power of the unknown, x^k, is multiplied by a coefficient that may be chosen, for instance, to be a rational number, a_k. One of the coefficients, a_0, does not multiply any of the powers of x in the equation. Also, some of the coefficients may be 0. The equations we have mentioned thus far have all been of degree 2 (usually called "quadratic equations"):

$$x^2 + 1 = 0, \quad x^2 + 3x - 10 = 0, \quad x^2 + 8 = 4, \quad x^2 - 2 = 0.$$

There are also irrational numbers that are solutions of polynomials of higher degrees. For example, after some tedious calculations, one can see that the number $1 + \sqrt[5]{2} + \sqrt[5]{4} + \sqrt[5]{8} - \sqrt[5]{16}$ is the solution of a polynomial equation of degree 5, namely, the equation

$$x^5 - 5x^4 + 30x^3 - 50x^2 + 55x - 21 = 0.$$

I also stated that the number π, which does not look like a root of a rational number or a combination of such roots, is indeed irrational, in the sense that it cannot be written as a fraction involving integers. But the real difference between π and the other irrational numbers is not just in their appearance (involving or not involving roots of rational numbers), but in a much deeper and highly significant respect, namely, that there is no polynomial equation with rational coefficients such that π is a solution of it!

The discovery of this intriguing property of π came as a big surprise at the time of its publication in 1882, by the German mathematician Ferdinand von Lindemann (1852–1939), who achieved immediate mathematical glory thanks to it. The proof is difficult and requires deep and broad mathematical knowledge in order to be understood. But what needs to be stressed here is that we are talking about a very "heavy" and, in a sense, unusual mathematical property. This is a property of "inexistence" or "impossibility," and proving that such properties hold may be, in most cases, very demanding. In this case, what Lindemann had to prove is that for *any* polynomial whatsoever, of *whatever* degree and with whatever rational coefficients, if we replace its unknown quantity x by π, the result will *never* be 0.

It is obvious that Lindemann could not prove inexistence by examining all the polynomials one can conceive of, since this is an infinite collection. The proof has to bring to the fore some *reasons of principle* that hold for all possible polynomials and that explain the said impossibility. This is, I repeat, heavyweight mathematics. From the point of view of our exposition of the various systems of numbers, what Lindemann's result indicates is that among the real numbers we can distinguish two types: numbers like 2, $\frac{3}{5}$, or $\sqrt{2}$ that can be realized as solutions to polynomial equations with rational coefficients, and numbers such as π that cannot be thus realized. Numbers of the first type are usually called *algebraic* numbers, and they comprise both rational and irrational numbers. We will denote here their collection by the letter $\mathbb{A}$. Numbers of the second type are called "transcendental numbers" and, of course, all of them are irrational numbers in the first place. Another well-known example of a transcendental number is e, the base of the natural logarithms. The proof that e is transcendental, which is also far from trivial, had already been presented in 1873 by the French mathematician Charles Hermite (1822–1901).

Research into transcendental numbers and their properties is nowadays a very active and demanding field of mathematical research. An example of a question that mathematicians may ask about this kind of numbers is: given any algebraic number, rational or irrational, a (not 1 or 0) and another algebraic but irrational number b, is it possible that the number a^b will be transcendental? This turns out to be a particularly difficult question to deal with, and there are many others similar to it on whose solution mathematicians are currently hard at work.

Another important and interesting fact related to the transcendental numbers is that among the real numbers, an "enormously larger" quantity of numbers are transcendental than they are algebraic. This may sound strange at first sight, since in both cases we are speaking about infinite collections. In what sense can an infinite collection be much larger than a second infinite one? As will be seen in Chapter 12, this statement can be given a very precise mathematical meaning, and, indeed, the proof of this somewhat surprising mathematical fact is not particularly difficult. It appears in Appendix 12.1.

Now, after having defined the complex and the transcendental numbers, we can further sharpen the picture of the hierarchy of the number systems, in the manner schematically represented in Figure 1.6.

A further question that may be asked at this stage is whether we can still find some polynomial equation that cannot be solved with the arsenal of numbers provided by the complex numbers, and all the other systems subsumed under them. This would require, as in all previous stages in our gradual construction, the introduction of yet another kind of numbers not yet defined. As it happens, however, it can be proved in very precise mathematical terms that no such further expansion is necessary once we have reached the complex numbers. That is, one can prove that, given a polynomial equation

$$a_n x^n + a_{n-1} x^{n-1} + a_{n-2} x^{n-2} + \ldots + a_2 x^2 + a_1 x + a_0 = 0,$$

such that all the coefficients a_k are complex numbers, there exists at least one complex number that is a solution to the equation. In other words, in the case of the complex numbers, a polynomial equation always has solutions from within the same domain of numbers that define the equation itself. But notice that in speaking here of "complex

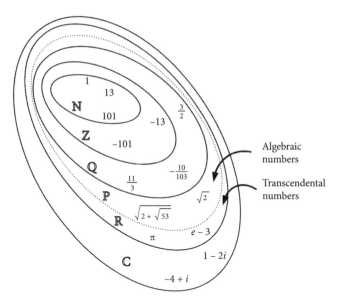

Figure 1.6 The various number systems, from the naturals to the complex (including the algebraic and the transcendental).

coefficients" and "complex solutions," I mean numbers that could also be natural, integers, rationals, or real, given that, as indicated in Figure 1.6, numbers of those kinds are also complex numbers. Hence, if we are given the equation $2x^2 - 5x + 1 = 0$, this is a polynomial equation of second degree with *complex* coefficients, since 2, −5 and 1 are complex numbers (well, they are also integers, of course).

We are thus certain, because of the theorem just mentioned and even before we begin to try and solve it, that this equation has *at least one* solution that is a complex number (which could, in particular also be natural, integer, or real). Because of this result, the complex numbers are said to be "algebraically closed," and this is a property that is not satisfied by any of the smaller systems discussed above: natural, integers, etc. The paramount importance that mathematicians attribute to this result is reflected in the name they give to it: "the fundamental theorem of algebra." Some algebraic knowledge of the kind learnt in high-school suffices to realize that the theorem has the following alternative formulation: any polynomial equation of degree n, with complex coefficients, has exactly n solutions, which are complex numbers.

The important insight embodied in the fundamental theorem of algebra was reached as part of the long and complex processes leading to the consolidation of the modern idea of number. Important advances, for instance, appeared in the work of René Descartes, about which we speak in some detail in Chapter 8. A complete proof, however, was provided for the first time by Carl Friedrich Gauss, one of the greatest mathematicians of all time. As a consequence of this theorem, we understand that the complex numbers bring to completion the gradual process of extension of the number system that we began with the basic idea of natural number and that is most clearly manifest in Figure 1.6. The historical process whereby all the relevant concepts were elaborated, I hasten to stress once again, was far from the orderly and systematic tour we have just taken in this chapter. The remaining chapters of the book will be devoted to an overview of that historical process.

1.4 Cardinals and ordinals

To conclude this chapter, I would like to mention two basic concepts associated with the idea of a natural number that are very important for understanding our story in the following chapters: "cardinals" and "ordinals." These two terms suggest two parallel meanings that we commonly associate with the idea of a natural number: (1) the idea of *plurality* of objects in a collection and (2) the idea of *position* within a given sequence. Take the number 4 as an example. On the one hand, it embodies the idea of that which is common to certain classes of collections, such as the following three:

{*Allegro, Adagio, Rondo, Presto*},
{*Lennon, McCartney, Harrison, Starr*},
{*Soprano, Alto, Tenor, Bass*}.

On the other hand, it also embodies the idea that we know how to order, in some natural and obvious manner, the following three collections:

{*Lennon, McCartney, Harrison*},
{*Lennon, McCartney, Harrison, Starr*},
{*Lennon, McCartney, Harrison, Starr, Epstein*}.

We can think of the number 4, then, as embodying the idea of a specific kind of plurality. This is the cardinal 4. At the same time, the ordinal 4 embodies the idea of a specific way to order entities, which is always preceded by the ordinal 3 and is always followed by the ordinal 5. I think that it is not difficult to realize that these two ideas are indeed different. What is more difficult is to describe an actual case in which they embody *different* mathematical situations. Indeed, if we take the collection {*Lennon, McCartney, Harrison, Starr*}, and if we think of different sequences that can be written with these elements, we will always obtain a sequence that, abstractly speaking, is associated with one and the same ordinal 4 (except perhaps for a change in the names of the elements in the sequence). We might take either *Lennon, McCartney, Harrison, Starr* or *Starr, Lennon, McCartney, Harrison*, or any other order involving these names. There is always one first element (after which come three additional ones), one second element (before which comes one and thereafter two more), a third one (before which come two and thereafter one more), and a fourth one (before which come three and none thereafter). It is always the same idea of order. What is then the point in speaking about two separate ideas, cardinal and ordinal, if they always represent the same situation? Can they somehow be understood separately? The answer to these questions is that there may indeed be situations where one cardinal may be associated with more than one ordinal, but this will happen only in the *infinite* case.

Dealing with infinite cardinals is a tricky business that has to be handled with extreme care, but I will give here a brief indication of some of the problems involved, so that the difference between ordinals and cardinals may at least be sensed. While any finite collection of natural numbers can be ordered, essentially, only in one way (except for changes in the names, as we have just seen), ordering an infinite collection may yield several possibilities. If we take the natural numbers, for instance, we have in the first place the standard ordering: $1, 2, 3, 4, 5, 6, 7, \ldots$ But we can easily think of the following alternative ordering: $1, 3, 5, 7, \ldots 2, 4, 6, 8, \ldots$ In this ordering, all even numbers are postulated to be "greater than" any given odd number (e.g., 80 is "greater than" 8,000,003). Notice that this is *really* a different ordering and not just a way of *renaming* the members of the sequence. How do we see this? Because in the standard ordering of the natural numbers, there is only one element for which no other number can be indicated as its immediate predecessor (i.e., 1), whereas in the alternative order just presented, there are *two* elements with that property, namely, 1 and 2. Indeed, in the alternative ordering, 2 is greater than any given odd number, but no specific odd number can be said to appear immediately before the number 2. By the same token, one may imagine yet another alternative ordering, such as $1, 4, 7, 10, \ldots 2, 5, 8, 11, \ldots 3, 6, 9, 12, \ldots$ Here all multiples of 3, for instance, are postulated to be larger than any specific number that is not a multiple of 3 (e.g., 81 is "greater than" 8,000,003). We find in this way three numbers, 1, 2 and 3, with no immediate predecessor. We have thus seen one collection with a given cardinal (which happens to be an infinite one, namely, the cardinal of the collection of all natural numbers) to which we associated three different ordinals, namely three essentially different ways of ordering the entire collection.

Admittedly, by allowing the consideration of infinite collections, we seem to be moving into a realm that looks artificial, in contrast with the simple idea of "natural numbers," with the help of which we will never count "up to infinity," as it were. And yet, these examples are very important at this stage, if only to illustrate the ways in which seemingly simple concepts related to numbers can require some sharpening and further consideration once we begin to ask innovative questions about them. The precise difference between ordinals and cardinals for the finite and infinite cases was first formulated in the last third of the nineteenth century by Georg Cantor, and we will have more to say about this in Chapters 11 and 12.

CHAPTER 2

Writing Numbers—Now and Back Then

Thinking about the essence of numbers and about ways to write them are intimately connected with each other and have been throughout history. This is an important point I will be stressing all through the book. In order to be able to speak properly about the historical development of approaches and techniques of number writing, I will devote the first part of this chapter to explaining how this is understood nowadays. In the second part of the chapter, I will briefly present some of the approaches followed for number writing in the ancient mathematical cultures of Egypt, Babylon and Greece.

2.1 Writing numbers nowadays: positional and decimal

The numbering system we use nowadays is decimal and positional. It is decimal because all numbers can be written with the help of just ten symbols, or digits: 0, 1, 2, 3, ..., 9. It is positional because the value we associate with each of these digits depends not just on the symbol itself, but also on its position within the number where it is used. For instance, in writing the number 222, we use the same digit three times, and in each appearance within one and the same number this digit represents a different value: the right-most 2 represents the value "two," the middle 2 represents the value "twenty," and the left-most 2 represents the value "two hundred."

We are used to thinking about numbers in terms of a very specific system of numeration and of a very specific way to write them. It is just out of habit that we take for granted the rather unusual fact that these three *different quantities*—two, twenty, two hundred—are denoted by the *same symbol*. In fact, we hardly tend to notice that we are able to distinguish among them just because their corresponding symbols appear in different positions within the number. If we thought of these three quantities separately as collections of pebbles, for example, we would immediately recognize (as

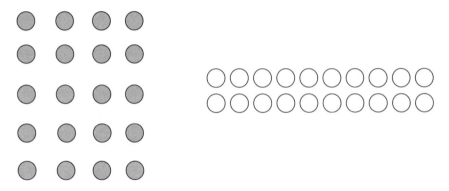

Figure 2.1 Two ways of representing the number 20 with pebbles.

exemplified in Figure 2.1) that we can separate the quantity represented from the way to represent it: the twenty pebbles can be arranged as two lines of ten pebbles each (2·10) or, alternatively, as five lines of four pebbles each, or indeed according to many other possible arrangements.

Whenever we write a number in our decimal–positional system, we write shorthand for a sum of powers of ten, each of which is multiplied by the value of a digit standing in the relevant place. In our example, the symbol 222 is shorthand for the following sum of powers:

$$222 = 2\cdot10^2 + 2\cdot10^1 + 2\cdot10^0.$$

If we take another example, say the number 7014, this is shorthand for the following sum of powers:

$$7014 = 7\cdot10^3 + 0\cdot10^2 + 1\cdot10^1 + 4\cdot10^0.$$

This second example stresses, of course, the importance of the use of the symbol 0 as part of the system. It indicates the *absence* of one of the powers in a specific sum (10^2 in this example). Without this idea, it would be difficult within the positional system to distinguish between numbers such as 7014, 71400, 7104, or 714. The rise and development of the idea of 0, and its incorporation into the decimal–positional system, as well as the realization of the idea of 0 as a number in its own right, are all historical developments of great importance and interest in our story. Historians of mathematics have devoted attention to them, and we will also touch upon them in various places throughout the book.

It is important to stress that the use of the positional system requires a choice of base. This is an arbitrary, or at least contingent, choice in the sense that there are no clear-cut mathematical reasons for necessarily preferring one over the other. This choice may be based on all kinds of cultural or practical considerations, which the historian may try to investigate. In principle, there is no reason why one would not work with a positional system that uses a base different from ten. The choice of ten as the preferred base is itself, once again, the outcome of a certain, specific historical process, which could have led into different directions. And it is particularly important to stress, moreover, that there exist numbering systems that are not positional and others that are not decimal. In the past, some mathematical cultures indeed adopted such systems.

Non-decimal numeration systems appeared in various cultures throughout history. Babylonian mathematics, for example, used sixty as the base of its system. The Mayan mathematical culture in Mexico, before the arrival of the Spanish conquistadores, used twenty as their base. But it is not necessary to go that far back into history to find a system that is not decimal. We find one nowadays in the world of digital computers, where numbers are internally stored and processed in binary representation, namely, with the help of only two symbols: 0 and 1. More precisely, in electronic computers, numbers are presented and processed as strings of tiny electric charges on some kind of medium, and the presence or absence of a charge is taken to represent either 1 or 0, respectively. This is the same positional principle used in our decimal system, but with base two, and any such string of 0's and 1's can be easily translated into decimal representation. So, for instance, the binary string 1001011 represents the following sum of binary powers:

$$1 \cdot 2^6 + 0 \cdot 2^5 + 0 \cdot 2^4 + 1 \cdot 2^3 + 0 \cdot 2^2 + 1 \cdot 2^1 + 1 \cdot 2^0.$$

This is equivalent to

$$1 \cdot 64 + 0 \cdot 32 + 0 \cdot 16 + 1 \cdot 8 + 0 \cdot 4 + 1 \cdot 2 + 1 \cdot 1 = 64 + 8 + 2 + 1.$$

Thus, the binary string 1001011 is equivalent to the decimal value 75.

For the purposes of interfacing with a human operator, the binary values handled by the computer may conveniently be translated and presented in their decimal representation. But, at the same time, it is often convenient to translate binary values into representations that use the base eight (octal representation) or, more often, base sixteen (hexadecimal). The hexadecimal representation requires the use of sixteen symbols, of course, and these are typically taken to be the ten standard digits 0, 1, 2, 3, ... 9, to which six letters are added, A, B, C, D, E, F, representing, respectively, the decimal values 10, 11, 12, 13, 14, 15. The hexadecimal value D1A, for example, represents the following sum of powers:

$$D \cdot 16^2 + 1 \cdot 16^1 + A \cdot 16^0.$$

If we translate the decimal values of A (10) and D (13), we obtain the decimal equivalent of the string, namely:

$$13 \cdot 256 + 1 \cdot 16 + 10 \cdot 1 = 3354.$$

The reason for adopting the base two as fundamental for representing numbers in computers is a practical one, and it is absolutely clear: the presence or absence of an electric charge is easily translated into a binary language. So, here, we have a cultural, historically conditioned reason for adopting a certain system of numeration. It might be claimed that if we overlooked what actually happened in history, and tried to choose the most convenient system of numeration based on some rational criterion, then the binary system would be the natural choice, and should be preferred over the decimal one even outside the world of computers. The reason for this would be that the binary system requires fewer symbols, two instead of ten, thus *facilitating* arithmetic activity for humans. But then one might equally claim that the fewer symbols a positional system of numeration uses, the longer and more confusing the strings of digits become, thus making arithmetic activity more *difficult* for humans. It would be more convenient, one might claim along this train of thought, to use the hexadecimal system with its sixteen symbols and shorter strings. Well, faced with this dilemma, we should perhaps be content with the base that we actually use, ten, because it strikes a nice balance between the two tendencies. In the artificially designed world of electronic computers, one may adduce all kind of technical considerations for influencing such a choice. In history, however, processes tend to be more complex and harder to control.

So, what can we say about the historical reasons for the choice of ten as a possible base for the numeration system? Two questions are actually involved here. First, on what grounds was the base ten chosen in a specific time and place, as a convenient and useful one? Second, how did it happen that, once it had been adopted and put to work in some specific mathematical culture, the decimal system became universally accepted and dominant as the preferred one? Concerning the second question, there is a long and complex story to be told. We will see some of its crucial stages. As for the first question, it is more a matter of speculation, since no evidence exists that can help us clarify it. Still, some cultural historians have tried to suggest possible explanations. One of the commonly accepted ones, which I just mention in passing and will not discuss in detail, relates the base ten to the number of fingers on our hands and to the assumption that primitive cultures counted with the help of their fingers.

Whether the choice of base ten derived from the number of our fingers or from any other source, this cannot at the same time be the explanation for the adoption of a positional system. Indeed, some decimal but non-positional systems were used in certain cultures in the past. An example of this appears in the ancient Hebrew numeration system that assigns a separate value to each letter of the alphabet. This idea appears also in other ancient cultures. The main feature of such a system is that the value represented by a symbol does not depend on its position, The number 222, to take again this example, is written in the Hebrew letter–value system, using *three different symbols*, as רכ"ב. Here ב stands for 2, כ stands for 20 and ר stands for 200. Notice, however, that, while not positional, there is an interesting component of decimality in this system. The first ten letters, א to י, are used for values 1 to 10, and then the following eight letters, כ to פ, are used for the eight round values 20 to 90. Thereafter, the four remaining letters in the Hebrew alphabet are used for values 100, 200, 300 and 400. Other numbers may be written as combinations of the above. Curiously a small element of positionality, also enters of necessity here, for denoting high values. So, for instance, while the number 758 is written as תשנח (i.e., from right to left, ת = 400, ש = 300, נ = 50, ח = 8), the number 5758 is written as התשנח, i.e., by adding, in front of תשנח (758), the symbol for 5, ה. So we find in this system an interesting mixture of ideas, but still something different from our accepted decimal–positional system. And in this system, one can readily see, there is no use for a zero, and indeed no such idea appears in it.

Another well-known non-positional system is the Roman system. Also here, the letters indicate values, not according to their position: I for 1, V for 5, X for 10, L for 50, C for 100, D for 500 and M for 1000. They can be combined according to rules for yielding other numbers, such as VIII for 8. Certain rules imply subtraction rather than addition: XL, for instance, stands for 40, since X is subtracted from a greater value, L, that comes immediately thereafter. The value 222, for instance, requires in this system a rather long string: CCXXII. Also here, a decimal orientation clearly underlies a non-positional system.

One of the interesting features of the decimal–positional system is that it works equally well for both integer and fractional values: the fraction $\frac{1}{2}$, for instance, can be written as 0.5. We attach a value to a decimal fraction according to the same principle we use for integers. Indeed, the decimal fraction 0.531, for instance, represents the following sum of decimal powers:

$$0.531 = 5 \cdot 10^{-1} + 3 \cdot 10^{-2} + 1 \cdot 10^{-3}.$$

Or, in other words,

$$0.531 = 5 \cdot \frac{1}{10} + 3 \cdot \frac{1}{100} + 1 \cdot \frac{1}{1000}.$$

Or, to take a different example,

$$725.531 = 7 \cdot 100 + 2 \cdot 10 + 5 \cdot 1 + 5 \cdot \frac{1}{10} + 3 \cdot \frac{1}{100} + 1 \cdot \frac{1}{1000}.$$

And here we have another fundamental property of the decimal–positional system that we usually take for granted, namely, the ability to recognize, on the basis of representation alone, what kind of number we are dealing with. An integer, for example, is a number whose decimal representation has no digits after the point. In the decimal representation of a rational number, to take another example, the digits after the point eventually repeat themselves in a certain order (such as in $\frac{138}{101}$ = 1.366336633 . . . or $\frac{1}{5}$ = 0.2). In the decimal representation of an irrational number, on the contrary, there is no such repetition (for instance, π = 3.14159 . . .).

The insight that integers and fractions can be written according to the same principles played a major historical role in allowing a broad, unified vision of number. Moreover, this insight teaches an important lesson about the developments we want to consider in this book, because the logic of numbers and the logic of history did not work in parallel here. Indeed, it was not the case that first a clear idea developed of fractions and integers being entities of a same generic type (numbers) and thereafter a notational system developed for conveniently expressing this idea. Rather, it was exactly the other way round: since it was realized that one and the same approach can be followed in writing fractions and integers, the idea was gradually (and not easily) developed that it may make sense to begin seeing fractions and numbers as essentially representing a common, general, idea. This important process will be discussed in some detail in Chapter 7.

But the historical processes connecting symbols and ideas about numbers have also worked in the opposite direction. Given the great power of the ideas embodied in the decimal system of numeration, one must exercise care in its use, since some deep-seated habits related to our continued use of this flexible notational system may sometimes mislead us. This is not just a matter of mistakes by individuals, but rather one of much broader and significant underlying historical processes. In some remarkable historical situations, mathematicians were repeatedly misled by naming, notation, or symbolism, and a breakthrough was reached precisely when a clear separation was envisioned between that which belongs to the symbols and that which belongs to a mathematical idea that is independent of the symbols used to represent it. This was the case, for instance, with the very slow and convoluted process leading to the legitimation of the idea of complex numbers, and we will have time to see some of the details of this in Chapter 9.

In the more immediate case of notational systems for numbers, for instance, it is important to notice that a given number may be represented in more than one way. The number 1 can also be written, in this view, as 0.9999... (with the clear convention that the ellipsis "..." indicates that the digit 9 in this decimal representation continues to appear *indefinitely*). There are various ways to see that these are two representations of one and the same number. For example, subtracting $1 - 0.9999\ldots$, yields $0.0000\ldots$, or, in other words, 0. Hence the two are one and the same number. The sophisticated interplay of ideas behind the decimal–positional system of numerations, then, may have pitfalls that one must be aware of.

The decimal system is central not just to number notation but also to systems of measurement and finance. The accepted unit of length, the meter, for instance, is typically divided or expanded according to decimal multiples to which we also add convenient prefixes. Thus, the centimeter is a one-hundredth part $\left(10^{-2}\right)$ of a meter, while a millimeter is a one-thousandth part $\left(10^{-3}\right)$ of it. On the other side, a kilometer indicates one thousand $\left(10^{3}\right)$ meters. We can find in books and encyclopedias (or on the Internet) additional prefixes, which are, however, much less common in day-to-day usage: for example, decimeter is a one-tenth of a meter and a decameter denotes ten meters.

The continued use of prefixes like "kilo" or "milli" is an interesting source of common mistakes. A square kilometer, for instance, is not an area of one thousand square meters, but rather the area of a square with side of one kilometer length $\left(10^{3} \times 10^{3}\right)$, and hence a square of one million $\left(10^{6}\right)$ square meters. More interesting is a kind of related difficulty common nowadays in the area of electronic data storage and communication. A "kilobyte" of data should indicate one thousand (10^{3}) bytes, but for historical reasons the term was adopted to indicate a different value. Indeed, as data are organized on an electronic memory in binary representation, rather than decimal terms, in the early stages of electronic computing it was found to be useful to use the term kilobyte to indicate 2^{10} (i.e., 1024) bytes. After all, this number is close enough to 1000, isn't it? With time, this ambiguity became well entrenched in the field, and, on the other hand, the amounts of data handled began to grow exponentially. In this way, the gap between the direct meaning of the terms used and what they were meant to represent became larger and larger. A megabyte (MB), for example, which literally means "one million bytes," is alternatively taken to indicate either 10^6 or 2^{20} $(1,048,576)$ bytes, and sometimes also $2^{10} \times 10^3$ $(1,024,000)$ bytes. A gigabyte (GB), which literally means "one hundred million bytes," is usually taken to indicate 10^9 bytes, but sometimes it is also used for 2^{30} $(1,703,741,824)$ bytes. This ambiguity does create problems and misunderstanding, and it is even misused in a manipulative way for marketing purposes. Thus, a hypothetical vendor might offer, say, hard disk drives of 100 GB, with an actual capacity of 100×10^9 bytes, and a specific salesmen would explain to the inattentive customer that he is purchasing a capacity of 100×2^{30} bytes. That's a big difference. As time passes by and the industry develops, the gap continues to grow between the possible interpretations, and it poses an interesting challenge to engineering organizations in charge of setting universally accepted industrial standards.

These examples, and many others that could be mentioned here, help emphasize the contingent character of the use of the decimal system in various contexts. On the face of it, it seems quite natural to assume that a mathematical culture that has adopted the decimal system for writing numbers will also use it, without any further considerations, for measuring and for subdivisions of currencies. History in this case was more complex than what commonsense would have us believe. The French Revolution, for instance, as part of the far-reaching reforms it imposed on many aspects of public life, devoted much energy and professional thought in order to reach a decision in 1795 about a sweeping adoption of the decimal metric system in most fields of measurement: length, weights and volumes (exceptions to this were measurements of time and angles—also for interesting historical reasons!). Some of this was already in use at the time, of course, but here came a compelling, thoroughgoing government decision accompanied by declarations that located it in the context of the universal principles of the revolution.

This was also the time when the accepted prefixes were introduced: kilo, hecta, deca, centi, milli, etc. While this was taking place in France, the British, on their side, went on with their rather cumbersome measurement styles, whereby each different unit is differently divided into its own subunits. A yard, for example, comprises three feet, while a foot comprises twelve inches. A pound, on the other hand, subdivides into sixteen ounces, whereas there are fourteen pound in a stone. This was the case also with the British currency up until 1971. As late as that, the British pound sterling was divided into twenty shillings, each of which subdivided into twelve pence. In the United States, on the other hand, currency was divided decimally from very early on, but they continued with the non-decimal approach in many other aspects of life, even after the British had abandoned it. Again, the reasons for all these differences are historically and culturally determined. There is plenty of evidence, for instance, that the British decision not to go the French way was substantially influenced by nationalistic considerations. There are very interesting historical processes behind the seemingly simple issue of measurements systems in different countries and cultures, and all of this will remain (unfortunately) beyond the scope of this book. Still, in later chapters, we will see some of the important stages in the processes that led to the development, adoption and dissemination of ideas related to the decimal system.

2.2 Writing numbers back then: Egypt, Babylon and Greece

After having presented a more or less systematic view of the basic concepts that we will be using in the historical account to come in the following chapters, I move now to more specifically historical issues, and I want to focus on a more detailed description of the numeration systems used in three ancient mathematical cultures: Egypt, Babylon and Greece. Of course, there were other ancient mathematical cultures, such as in China, India, Japan, Korea, in some South- and Central-American indigenous cultures, and in the ancient Hebrew culture. In all of these, we can find interesting ideas, also concerning notation, but I have chosen these three because they are more directly connected to our own Western mathematical tradition.

I begin, then, with Egypt. Hieroglyphic writing was developed about 3000 BCE, and it included signs for writing numbers. These signs embodied a numeration system that was decimal and non-positional, and they represented the various powers of ten as follows:

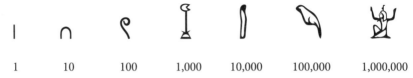

Now, any given number can be written as a combination of these symbols, as shown in the following example, which reproduces signs carved on a stone found at the Karnak temple, and which nowadays is exhibited at the Louvre in Paris. They are estimated to have been written around 1500 BCE:

It is important to keep in mind that the Egyptian culture we are discussing here spanned a period of more than 2000 years, and hence it is obvious that the hieroglyphic signs underwent many transformations. The signs displayed here are just meant to be representative. Still, the basic ideas of the system remained constant. Hieroglyphs were relatively easy to carve on stone, and hence they are commonly found on the walls of temples and graves, or on decorative utensils like vases. They were not fit, however, for the use of scribes, who wrote documents on all kinds of papyri with the help of straws. Various kinds of cursive-like writing were therefore developed for such purposes, such as hieratic and demotic writing, each of which had its own signs for writing numbers. The hieratic system for number writing, for example, is not a direct transliteration of the hieroglyphic system, and it is closer to the Hebrew approach of assigning values to letters, with one symbol for each value between 1 and 9, additional symbols for each of the values 10, 20, ..., 90, and then additional symbols for the values 100, 200, ..., 900. Any number up to 9999 could be written as a concatenation of the relevant symbols. The number 5234, to take one concrete example, is written as ⌐𝟑𝟰, which is a combination of the signs ⌐ = 4, 𝟑 = 30, 𝟰 = 200 and 𝐼𝑓 = 5000. As the system is non-positional, the particular order in which the symbols are written is irrelevant. Moreover, in such a non-positional system, there is no direct need for the idea of zero (or of a symbol for it). Still, while no sign for zero appears in Egyptian texts on arithmetic topics, it does appear in some administrative texts possibly dating from the reign of Senruset I (1956–1911 BCE).[1]

It is indeed easier to write hieratic symbols than hieroglyphs on a papyrus, but this does not mean that it is easier to perform arithmetic with them. The number of symbols to be learnt in the hieratic system, to begin with, is ten times as large as that in the hieroglyphic one. More interesting is the comparison when it comes to operating with the two systems of signs. If we want to add 253 and 746 in the hieroglyphic system, to take one example, we simply put together the signs corresponding to the two numbers, as follows:

But then, since we have obtained eleven signs for the tens and nine for the hundreds, we can substitute all of this by one sign of ten and one of thousand, and we thus obtain the following:

The hieratic, non-positional system cannot of course allow for such an easy system of adding and subtracting.

There is another remarkable idea related to the Egyptian notation for numbers that must be mentioned here, as providing an illuminating example of the close interrelation between the development of a mathematical idea and of the appropriate symbolic language to deal with it. Retrospectively seen, we can describe this idea as the introduction of so-called unit fractions, i.e., fractions whose numerator is 1. In the hieroglyphic system, this can be done very simply by appending to any number the

---

[1] Dates for individuals in the ancient and medieval periods are often approximate. All dates are CE unless noted as BCE.

special sign ⌢ (and in the hieratic system we obtain the same by adding a dot over the number). Two examples of hieroglyphic notation for unit fractions are

$$\frac{∩∩∩}{||} \frac{⌢}{??} \qquad \frac{⌢}{\frac{|||}{||}}$$

We would read these two numbers nowadays as $\frac{1}{232}$ and $\frac{1}{5}$, respectively. Also in retrospect, we know nowadays that any common fraction can be written as a sum of unit fractions, and even in more than one way. So, for instance, $\frac{1}{5} + \frac{1}{5} = \frac{2}{5}$, or $\frac{1}{3} + \frac{1}{15} = \frac{2}{5}$.

Now, the name "unit fractions" may be misleading in this context. The ability to think of the reciprocals of the natural numbers (the half, the third, the fifth, etc.) and the availability of a convenient notation to indicate them can arise independently of a more elaborate idea of fractions, of which "unit fractions" are a particular case. And indeed, when we look at the extant evidence (appearing in a considerable amount of papyri that have survived), we see that the mathematical culture of ancient Egypt lacked a *general* idea of fractions. At the same time, the idea of taking a "part" of a whole arose naturally in the framework of the solutions to the specific problems that they dealt with.

Assume, for example, that we are required to divide five loaves of bread among eight slaves. In present-day terms, we would say that each slave has to receive $\frac{5}{8}$ of a loaf. We may divide each loaf of bread into eight pieces and then give to each slave five of the pieces thus obtained. Each slave would indeed receive the stipulated $\frac{5}{8}$ of a loaf. In the unit fraction system, however, we find a much more natural way to approach this practical problem: we can simply cut four of the loaves into halves, and the remaining loaf into eight pieces. In this way, each slave will receive one half and one eighth of a loaf, namely, $\frac{1}{2} + \frac{1}{8}$. Of course, in purely arithmetical terms $\frac{1}{2} + \frac{1}{8} = \frac{5}{8}$. But in the real-world situation involved, we have two different ways to approach it, one being more natural and directly appropriate for the problem than the other.

And here is another example: suppose we are asked to compare two fractions, say $\frac{3}{4}$ and $\frac{4}{5}$, and decide which is larger. According to our view and using the kind of notation we are used to, we will find it more natural to write both numbers in their decimal expressions, 0.75 and 0.8, which are easier to compare. We may as well write them with a common denominator, say $\frac{15}{20}$ and $\frac{16}{20}$, and then easily compare them. But in the case of unit fractions, the comparison is direct and self-evident: $\frac{1}{2} + \frac{1}{4} + \frac{1}{20}$ is obviously greater than $\frac{1}{2} + \frac{1}{4}$. The Egyptian system of writing what we see as unit fractions, then, is not a partial or deficient version of the broader system of fractions known to us, but an autonomously developed and highly consistent idea that is naturally associated with the mathematical–historical context in which it developed. I should add that in some Egyptian texts we find special signs for $\frac{2}{3}$ and, less often, also for $\frac{3}{4}$. We will see in Chapter 3 that this important idea also arises in the context of Greek mathematics.

A nice historical turn of events that I want to mention here is that the term "Egyptian fraction" is used nowadays in modern number theory to denote a sum of different unit fractions. There are many interesting problems and open conjectures related to the possible representation of common fractions as Egyptian fractions. For example, the number 1806 has the interesting property that it is the product of the primes 2, 3, 7 and 43, and at the same time it satisfies the following property involving an Egyptian fraction: $1 = \frac{1}{2} + \frac{1}{3} + \frac{1}{4} + \frac{1}{43} + \frac{1}{1806}$. Numbers with this property are called "primary pseudo-perfect numbers" and they began to be systematically investigated only in the year 2000, which is about four thousand years after the ancient Egyptians introduced their own "unit fractions."

We move now to the Babylonian system of number writing. Around 1600 BCE, the Babylonians developed a considerable body of arithmetical knowledge, which has reached us as part of the many cuneiform texts written on clay tablets that were preserved for centuries. The arithmetic cuneiform texts comprise long lists of problems and calculations in which numbers are represented with the help of a sexagesimal positional system (i.e., base sixty). The system, however, did not use sixty symbols, but only two, the combinations of which produced all other numerals. Thus, in spite of the base sixty, the entire system displays clear evidence of an underlying decimal approach. Indeed, two basic symbols help in representing all numbers between 1 and 59 by way of groupings by tens (denoted as ◁) and by units (denoted as ˥). Here are some examples:

Numbers over 59 were represented as strings of such combinations, in conjunction with a positional interpretation based on powers of 60. So, for instance, the string

would represent the number $1 \cdot 60^3 + 30 \cdot 60^2 + 51 \cdot 60 + 23$, that is to say (in decimal terms) 327 083. One can immediately notice the most significant shortcoming of this system, namely the absence of the idea of zero, and of course of a symbol to represent it. Without zero, as already said, no positional system can really work smoothly. Indeed, without the possibility of representing an empty place in the sequence of powers of the base, every string of symbols cannot be interpreted unambiguously. Take, for example, the following string of two symbols:

It can represent any one of the values $30 \cdot 60 + 23$, or $30 \cdot 60^2 + 0 \cdot 60 + 23$, or $30 \cdot 60^2 + 23 \cdot 60$, or many others. From the texts that remain on extant clay tablets, we learn that the Babylonians typically established a clear context for a problem from which the actual value of the natural number used could be easily understood, without any danger of ambiguity. At the same time, however, the Babylonians extended their system so that fractions could also be written in it, and here the possible ambiguities arising from a lack of zero became even more apparent. Fractions were written in a way similar to our decimal fractions, but with base sixty. In this system, for instance, the string

could represent not only $30 \cdot 60^2 + 23 \cdot 60 + 56$ or $30 \cdot 60^3 + 0 \cdot 60^2 + 23 \cdot 60 + 56$, but also $30 \cdot 60^2 + 0 \cdot 60 + 23 + 56 \cdot \frac{1}{60}$ or $30 \cdot 60 + 0 + 23 \cdot \frac{1}{60} + 56 \cdot \frac{1}{60^2}$ and so on.

The question of the origins of base ten for our decimal system has already been mentioned. A similar question, as well as similarly speculative answers, arises in the case of base sixty adopted in the Babylonian system. Historians have sometimes pointed out that earlier cultures, such as the Sumerians and the Akkadians used the same base sixty, though within a much less developed system of arithmetic knowledge. The Greek mathematician Theon of Smyrna, who lived between about 70 and 135, stressed in relation to the Babylonian system that sixty is divisible by a large number of factors (1, 2, 3, 4, 5, 6, 10, 12, 15, 20, 30), which makes it a convenient choice of base in many respects. Mathematically speaking, Theon was certainly right in stressing this point, and the fact is that the base sixty was and remains the preferred one in geometry and in astronomy to this day.

All through history, to be sure, there were all kinds of efforts to change this situation and to adopt the decimal system in astronomy as well. We will see a very remarkable example of this in Chapter 7, when we discuss the mathematical contributions of the seventeenth-century Flemish mathematician Simon Stevin. Stevin was very successful in finally consolidating and promoting the decimal system in various fields of knowledge, and he devoted his efforts to do the same for astronomy. But he was not able to convince astronomers to follow his ideas, and base sixty remained the standard one in this discipline. Still, in spite of the strong plausibility and internal logic behind Theon's argument as a retrospective explanation for the adoption of base sixty, it is hard to believe that any numbering system was designed and adopted in an ancient mathematical culture on the grounds of a rational and well-elaborated conception alone, as if formulated in the framework of a professional committee. Like many other issues discussed in this book, historical factors were at play in various ways behind the scenes of the adoption of a certain mathematical idea or notation. At any rate, there is of course no direct evidence to either confirm or refute a historical claim based on Theon's remark.

Last but not least, we look briefly at the systems of numeration used in the Greek mathematical culture, whose concepts of number continued to be dominant until at least the seventeenth century, and which we will discuss in Chapters 3 and 4. A main point I will stress in the forthcoming chapters is that ancient Greek mathematics was far from being a monolithic or even homogeneous enterprise, and it actually comprised diverging, and sometimes opposite, views. This is especially the case when it comes to the issue of notation. The various islands and regions where this ancient culture with all of its manifold manifestations developed stressed their pride of independence and autonomy in all aspects of life. This may certainly have influenced the development of local systems pertaining to numbering, coins, measuring and weighing. So, in this last part of the chapter, I can only describe briefly some of the more commonly used systems of numeration in the ancient Greek mathematical culture.

It is important to mention first of all a system that is very similar to the ancient Hebrew one described above. This is a decimal, non-positional system in which the first letters of the alphabet are used to represent the values 1 to 9, the next ten to represent the values 10 to 90, and then another nine for the values 100 to 900. The signs used included some letters that are not currently in use in the modern Greek language. The various numbers are created, as we already know, by using appropriate combinations: 345, for instance, is written as $\tau\mu\varepsilon$ ($\tau = 300$, $\mu = 40$, $\epsilon = 5$), whereas 534 is $\phi\lambda\delta$ ($\phi = 500$, $\lambda = 30$, $\delta = 4$). In some texts, when these letters were used to represent numbers, some indication, such as a bar or a dot, was added over the numerals (e.g., $\overline{\phi\lambda\delta}$) to distinguish them from words or letters used to label a figure. Also, for numbers over 1000, the letter ι was added as an indication to change the standard value assigned to each letter. While α indicates 1, for example, the combination 'α or $_\iota\alpha$ indicates 1000. With time, additional ideas were introduced for indicating even larger numbers. So, from the Greek word $\mu\nu\rho\iota\acute{\alpha}\varsigma$ (myrias), meaning ten thousand or, indeed, a myriad, the letter M was adopted to indicate numbers multiplied by this amount. The number 40,000, to take an example, could be written as $\overset{\delta}{M}$. The numbers thus appearing over the letter M could actually be large, including hundred or thousands. The number 5,340.000, for example, could be written as $\overset{\phi\lambda\delta}{M}$. This rather cumbersome system, however, was changed later on by writing the number before, rather than on top of, the letter M. We find interesting uses of large numbers in astronomy books, such as by Aristarchus of Samos (310–230 BCE), famous for his suggestion of a heliocentric system. In his book, he wrote a number as large as 71,755,875 with the help of the basic values (using the symbols E = 5, O = 70, P = 100, Ω = 800, Z = 700), and arranging them in the following way: $_\iota$ZPOEM$_\iota$EΩOE. Of course, it is not by coincidence that such large numbers appear precisely in an astronomical treatise. The great mathematician Apollonius of Perga (ca. 262–ca. 190 BCE), who also made important contributions in astronomy, went even further when he suggested writing an additional sign over the M, in order to indicate multiples of 10,000. The number 587,571,750,269, for instance, he wrote as $\overset{\beta}{M_\iota}$EΩOE$_{\chi\alpha\iota}$ $\overset{\alpha}{M_\iota}$ZPOE$_{\chi\alpha\iota}$$\Sigma\,\Xi\Theta$. (Here the combination $\chi\alpha\iota$ is the Greek word for "plus," and the additional symbols used are Σ = 200, Ξ = 60 and Θ = 9).

Also Archimedes of Syracuse (ca. 287–ca. 212 BCE) devoted some thought to developing ways for writing large numbers, and his contributions to this issue are know to us via a rather curious treatise, *Sand Reckoner*. Rather than focusing on tens of thousands (10^4) as a possible starting point, Archimedes adopted its square (10^8), and on this base he introduced a special quasi-positional approach. He put the strength of this method to the test by way of a thought experiment of sorts in which he counted the amount of grains of sand needed to fill up the universe (meaning by this the "universe" as it was then conceived, of course). This would be, so he hinted, the "largest thinkable number." His calculations yield a number in the range of 10^{51} to 10^{64}, a number that in his system was easily written. Beyond the success of the system in fulfilling its aim, an interesting curiosity to be mentioned in this regard is that, according to current knowledge, if we set to do a similar calculation of the amount of grains of sand needed to fill up the solar system, based on the idea of the average radius of Pluto's orbit, we would obtain a number that is quite close to 10^{51}.

A last important source to be mentioned here is a famous treatise on astronomy, the *Almagest*, written about 130 by Claudius Ptolemy (85–165). This was the canonical book of the discipline up to well into the seventeenth century. Since Ptolemy adopted the Babylonian sexagesimal system as the main way for presenting astronomical data and computations, this has remained the accepted one in the discipline to this day. He modified the Babylonian system, though, in a crucial respect, namely, by introducing the symbol *O* to indicate an empty place in the power representation of the numbers. It is remarkable, however, that in spite of the enormous impact and influence of his book, only a few astronomers that followed him adopted this particular aspect of his work. Nor was it particularly stressed by other mathematicians, contemporary or subsequent to him, and the full adoption of the zero will have to wait to make its entrance from a different source, as will be seen later on.

CHAPTER 3

Numbers and Magnitudes in the Greek Mathematical Tradition

We are ready to begin our exploration of the historical development of the concept of number. The scientific culture of ancient Greece is an ideal starting point for doing this. Not because it is either the earliest or the only ancient culture to have developed arithmetical knowledge worth the name—I have already mentioned the Babylonian and Egyptian mathematical cultures, which developed much earlier. Neither was this the only ancient culture to have developed conceptions of number with significant mathematical and historical interest. But the Greek mathematical tradition is, after all, the real source from which (with important additions from the Islamic world, as we will see) European mathematics developed through the late Middle Ages, the Renaissance and the seventeenth century. Understanding the way in which the Greeks thought about, and worked with, numbers is a fundamental part of our story, and it is the focus of this and the next chapter. Subsequently, we will see how mathematicians of later generations and in several cultures exploited the full *potential* inherent in the Greek idea of number. At the same time, we will see how these same mathematicians worked hard to liberate themselves from some of the *limitations* that affected the Greek conceptions from their inception.

"Greek mathematics" is an umbrella term used to designate a broad range of different traditions, and it should not be thought of as referring to a monolithic body of knowledge with rigidly defined methodologies and exclusive normative constraints. Indeed, this term refers to a period that spans more than 800 years, beginning in the sixth century BCE, with Thales of Miletus (the first mathematician known by name) and going up to the third or fourth century CE (depending on how we define what is "a Greek mathematician"). It is habitual to refer to Hypatia of Alexandria (ca. 370–415) as the last Greek mathematician (incidentally, Hypatia was also the first woman mathematician in history to be known by name; her persona was brought to public attention in the 2009 Hollywood film *Agora*, starring Rachel Weisz). Also in terms of geography, the area of activity of Greek mathematicians spread from Asia Minor to the East, to Syracuse in Sicily to the West, and to Alexandria in Egypt to the South.

This historical and geographical heterogeneity poses many problems for the historian, especially when trying to reconstruct something as flimsy and elusive as "the concept of number" in Greek mathematics. But there are, in addition, challenging methodological problems related to the nature of the extant evidence. Evidence about this many-sided mathematical culture has come down to us in a rather fragmentary manner. Some important texts survived only via Arabic translations, to begin with. But even the material that has been preserved in the Greek language reached us mostly via much later transcriptions dating from the ninth century onward, written in a modernized Byzantine script. These transcriptions often followed norms of notation that embodied novel ideas and new systems of symbols and hence differed in important ways from those available at the time of writing of the original texts. Even the technical aspects of the way in which these later manuscripts were prepared, the ways that propositions and proofs were structured within the text, and the ways that diagrams were drafted were quite different from the original. Innovations such as these are far from being neutral to mathematical content. A historian trying to make sense of a mathematical text rewritten hundreds of years after the original one, while using different symbols and different notational conventions, has a really hard job in trying to sort out how the original really looked and what ideas were added by the successive rewriters/commentators/interpreters.

It is also important to stress that most of our knowledge of Greek mathematics, via extant texts, is part of a tradition that can be called "academic mathematics" or "scholarly mathematics." Most of the arithmetic that we will be discussing here is part of this tradition. Parallel to this, however, there was another tradition, usually called "logistics," that arose mainly in the framework of commerce and day-to-day life. Incidentally, the extant *direct* evidence of written texts is mostly of this latter kind of mathematics. It appears in relatively extensive, but always fragmentary, papyrus texts. They were written by the Greco-Roman administration in Egypt from the fourth century BCE onwards. So, very little of this kind of more practical mathematics appears in Byzantine manuscripts, whereas, at the same time, very little of the scholarly mathematics has remained in original papyrus texts.

In spite of all these methodological difficulties, I will nevertheless speak here of "Greek mathematics" in a generalized sense, while focusing on the main representative traits of this mathematical culture, particularly in what concerns the concept of number in its various manifestations. When talking about the concept of number in Greek mathematics, I will be referring mostly to the scholarly tradition as known to us via the extant texts associated with the big names of Euclid, Archimedes, Apollonius and Diophantus, and also, though to a lesser extent, to the astronomical tradition associated with Ptolemy.

3.1 Pythagorean numbers

I begin by referring to the work of Pythagoras of Samos (ca. 596–ca. 475 BCE) and the Pythagoreans, who were members of a school that was active in the city of Crotona, in what is today the south of Italy. Many legends and myths came to be associated with their name, and it is sometimes difficult to separate such legends from historical truth. Nevertheless, there is no doubt about the seminal importance of their mathematical contributions. The best known of these is the famous theorem about right-angled triangles that bears the name of Pythagoras: the square described upon the hypotenuse of a right-angled triangle is equal to the sum of the squares described upon the other two sides. In spite of its name, the theorem was also known to other ancient mathematical cultures. Our interest in the Pythagoreans, at any rate, lies in a different aspect of their work, namely, the focused attention and intense research they devoted to natural numbers and their properties.

The interest of the Pythagoreans in numbers goes way beyond the purely arithmetical. For them, number was the universal principle that underlies the cosmos and allows it to be understood. As part of a unique blend of a rational approach to understanding nature with numerology and other mystical practices, the Pythagoreans saw the natural numbers as a clearly discernible, stable element that hides behind the apparent chaos of day-to-day experience and helps to make sense of it. Relations among numbers explain, in their view, such disparate phenomena as the properties of geometric solids, the relative motions of celestial bodies and their possible configurations on heaven, and the production of musical harmonies.

The musical theory of the Pythagoreans, for example, seems to have arisen from the insight that vibrating strings create harmonic sounds when their lengths stand to each other in relations that can be expressed as exact ratios of natural numbers. An octave, to give an example, which is the interval between, say, C and the next C on the scale is obtained with two strings, whose lengths have the relation of 1 to 2 (i.e., one of them is twice the length of the second). The fourth (the interval between, say, C and F) is obtained with two strings with lengths having the relation of 3 to 4, whereas the fifth (the interval between, say, C and G) is obtained with two strings with lengths having the relation of 2 to 3. The theory of harmonies that remained valid, without significant changes, in the Western musical world at least until the seventeenth century was not very distant from this Pythagorean theory.

Additional manifestations of the same views of numbers and their significance, though more in the mystical direction, attributed to each number some kind of individual personality: masculine or feminine, perfect or imperfect, beautiful or ugly. Some of these traits can be expressed in purely arithmetic terms. A "perfect" number, for example, is one that equals the sum of its proper factors, as in the case of 6 (3 + 2 + 1 = 6) or 28 (14 + 7 + 4 + 2 + 1 = 28). There are also "amicable" numbers such as 220 and 284, for which the sum of the proper factors of the first equals the second, and vice versa:

$$284 = 1 + 2 + 4 + 5 + 10 + 11 + 20 + 22 + 44 + 55 + 110$$

$$220 = 1 + 2 + 4 + 71 + 142.$$

The Pythagoreans' preoccupation with numbers and their properties translated into a very high degree of arithmetic proficiency and also led to the development of very useful techniques and conceptual tools. First of all, they used to represent numbers with the help of pebbles arranged in various kinds of configurations. On the basis of the properties of the configurations alone, they derived general theorems about the properties of various kinds of numbers. A basic property of even numbers, for instance, is that they can be represented by two identical rows of pebbles. Odd numbers, in turn, if arranged in two rows, will always leave one separate pebble. From these elementary kinds of properties, the Pythagoreans deduced (as shown in Figure 3.1) that sums of even numbers are always even, and that when we add an even number of odd numbers, we obtain even numbers.

Configurations of pebbles could also correspond to geometric forms, or "figurate" numbers, as shown in Figure 3.2 for triangular, square, or oblong numbers (i.e., numbers of the form $n - (n + 1)$).

Again on the basis of the configurations alone, the Pythagoreans deduced general results about these kinds of numbers. For example, a square number equals a sum of two consecutive triangular numbers. Another example: an oblong number equals a triangular number added to itself, as shown in Figure 3.3. The centrality of figurate numbers in Pythagorean arithmetic is also connected to a defining feature of their basic conceptions, namely, that numbers are collections of units and that each of the units that comprise the numbers are themselves indivisible entities. Of course, when Pythagoreans spoke about numbers, they meant what for us are the natural numbers alone.

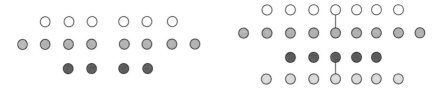

Figure 3.1 A sum of even numbers is always even (because it is also represented by two identical collections of pebbles); a sum of an even number of odd numbers is always even (because there is an even number of pebbles left after arranging the odd numbers in two rows, and taken together they add to an even number).

Figure 3.2 Triangular numbers, square numbers and oblong numbers.

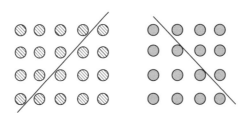

Figure 3.3 Relations among triangular, square and oblong numbers.

3.2 Ratios and proportions

Another key concept of Pythagorean arithmetic is that of "proportion," namely a situation in which four numbers, when arranged as pairs, determine the same ratios. For example, the pairs 2, 3 and 4, 6 are said to be in proportion. From the point of view of our current arithmetical conceptions, we understand this statement simply as involving the equality of two fractions: $\frac{2}{3} = \frac{4}{6}$. The Pythagoreans saw this quite differently. For them, the meaning was closer to something like this:

Whereas the unit taken twice measures the number 2 and taken thrice it measures the number 3, so does the number 2 when taken twice measures 4 and when taken thrice it measures 6.

More generally, in the Pythagorean tradition,

Four numbers are in proportion if as the first and the second are measured by a certain magnitude, so are the third and fourth measured by some other magnitude.

For simplicity, I will follow the convention of indicating a proportion between four quantities as $a{:}b :: c{:}d$. This notation was introduced in the Renaissance, and it did not exist in ancient Greece. In classical Greek mathematical texts, proportions were always referred to in a completely rhetorical manner, more or less as follows:

The ratio of a to b is the same as the ratio of c to d.

The difference between the Pythagorean concept of proportion and our current concept of equality of fractions is not just that the latter is written in symbols and the former is expressed just with words. Rather, it concerns the very way in which numbers were conceived and used. This is an important point that requires some elaboration. When we say nowadays that we conceive the proportion 2:3 :: 4:6 simply as an equality between fractions, $\frac{2}{3} = \frac{4}{6}$, we may have in mind two different, but equivalent, ideas: (a) that the two sides of the equation represent the same numerical value, namely, that of the decimal fraction 0.666 . . . , albeit written in two different ways as common fractions; (b) that we can simplify the fraction on the right-hand side of the equation, by dividing numerator and denominator by 2, and hence obtain the fraction on the left. For the Pythagoreans, *both* (a) and (b) would have lacked any clear meaning.

A ratio between two numbers, such as 2 and 3, did not represent for them either a number or a "numerical value," in the same sense that 2 or 3 did, or as $\frac{2}{3}$ does for us. A ratio was for the Pythagoreans, and thereafter for Greek mathematicians in general, a completely different kind of beast. A ratio was a way to *compare* numbers, and ratios could be compared with each other, but not *operated upon* as numbers were. Pythagoreans added, subtracted and also multiplied numbers. On the other hand, they just *compared* ratios, such as the ratio of 2 to 3 with that of 4 to 6. They did not *operate* with these ratios as they did with numbers. This difference in attitudes toward numbers and ratios had far-reaching consequences for the Greek ideas about numbers, as we will see.

An early important application of the concept of ratio and proportion appears in the definition of various kinds of numerical means and the study of their properties. Numerical means appear in earlier mathematical cultures, such as the Babylonian and the

Egyptian. The Pythagoreans may have learnt about them in their journeys. Given two numbers, a "mean" is an intermediate, third number, that satisfies some additional, well-defined condition. The most immediate case is, of course, that of the "arithmetic mean," in which the condition defining the intermediate number is that its distances to the two given numbers are the same. In modern terms, given two numbers a and b, the arithmetic mean is a number c such that $c - a = b - c$ or $c = \frac{a+b}{2}$.

The "geometric mean" is the one that more naturally adapts to the Pythagorean concept of proportion. Here, the defining condition is that the ratio of the first number to the intermediate one is the same as the ratio of the intermediate to the second. If we denote the geometric mean by d, then the condition can be symbolically expressed as $a{:}d :: d{:}b$. Using the modern language of fractions, we can say that d satisfies the equality $\frac{a}{d} = \frac{d}{b}$, and hence that, given two numbers a and b, the value of d is given by $d = \sqrt{a \cdot b}$. I stress again that this is *not* the way that the Pythagoreans saw the geometric mean.

In the Pythagorean theory of musical harmonies, the arithmetic and geometric means appeared prominently in conjunction with a third, the so-called sub-contrary or "harmonic mean." In this case, the intermediate number is defined by the following condition:

By whatever part of itself the larger number exceeds the intermediate one, by the same part of itself is the smaller one exceeded by the intermediate one.

Symbolically, this is a number h with the property that $(b - h){:}b :: (h - a){:}a$. Simple algebraic manipulation (which the Pythagoreans could not perform, since they lacked such a symbolic language) translates this property into the following: $h = \frac{2ab}{a+b}$. These three means—arithmetic, geometric and harmonic—characterize basic musical intervals. The arithmetic mean of an octave (1:2), for example, embodies the perfect fifth (i.e., $\frac{1+2}{2} = \frac{3}{2}$), while its harmonic mean embodies the perfect fourth (i.e., $\frac{2 \cdot 1 \cdot 2}{2+1} = \frac{4}{3}$).

I have emphasized the rhetorical rendering of the definitions in order to stress the difference between the way in which the Pythagoreans conceived these ideas and the way in which we can understand them (and even visualize them, we could say) from the vantage point afforded by the flexible symbolic language of modern algebra. And the difference is indeed historically significant. For one thing, even coming up with the idea of the various means and the relationships among them is extremely more difficult when conceived rhetorically as the Pythagoreans did. For another thing, while the Pythagorean concept of proportion could indeed allow them to *define* the geometric mean, it could not help in *calculating* it.

When we think of the geometric mean in terms of its algebraic formulation as an equality of fractions, $\frac{a}{d} = \frac{d}{b}$, we can very easily manipulate it and obtain the expression $d = \sqrt{a \cdot b}$. As in the case of the harmonic mean, this possibility was simply unavailable to the Greeks. The same is true of the interesting relation holding between the three means discussed above, $c{:}d :: d{:}h$, or, in modern terms: $d = \sqrt{c \cdot h}$. When seen from the perspective of the purely rhetorical definitions available to the Pythagoreans, this is much more than just a formal derivation, and it is indeed remarkable that they were aware of these relationships. Even more remarkable is the fact that using this same, purely rhetorical approach, the Pythagoreans studied up to ten different types of numerical means. These types are defined in the following table (all indicated with the letter *c*), for any pair of numbers *a*, *b*, with *b* > *a*. Even when written symbolically, as I write them here, this summary indicates how remarkable was the Pythagorean insight concerning numerical means:

(1) *a*:*a* :: *c*–*a*:*b*–*c*   (2) *a*:*c* :: *c*–*a*:*b*–*c*
(3) *a*:*b* :: *c*–*a*:*b*–*c*   (4) *b*:*a* :: *c*–*a*:*b* –*c*
(5) *c*:*a* :: *c*–*a*:*b*–*c*   (6) *b*:*c* :: *c*–*a*:*b*–*c*
(7) *b*:*a* :: *b*–*a*:*c*–*a*   (8) *b*:*a* :: *b*–*a*:*b*–*c*
(9) *c*:*a* :: *b*–*a*:*c*–*a*   (10) *c*:*a* :: *b*–*a*:*b*–*c*.

You can easily check by straightforward manipulation that relations (1), (2) and (3) represent, in fact, the arithmetic, geometric and harmonic means, respectively. But keep in mind that for the Pythagoreans it was much more difficult to check that this is the case. In order to express the arithmetical statement embodied in, say, proportion (3), they would have to say more or less the following:

Given two numbers, find a third number such that the ratio between the difference of the second to the first to the difference of the third to the second is the same as the ratio of the first to the third.

You immediately realize that formulating all these proportions and studying the relationships among them becomes a rather involved task in the absence of the kind of algebraic tools that are available to us nowadays. And we will see that this difficulty grew larger as the mathematical ideas that the ancient Greeks developed became more complex and sophisticated. Increasingly complex ideas and an adequate, flexible symbolic language to express them do not always develop at the same pace in the history of mathematics, as we will see throughout the book.

Based on the idea of proportion, the Pythagoreans approached yet another important concept that has accompanied mathematics since its very early stages, and to which we should also devote some attention here. This is the idea of the "golden ratio" (the Greeks, by the way, called this "mean and extreme ratio," and the term "golden ratio," which I will use here for simplicity, appeared much later and was widely adopted only sometime in the nineteenth century). The golden ratio has fascinated artists and mystics throughout history. Popular writers of various kinds have repeatedly associated it—typically with more enthusiasm than real justification—with the aesthetic appeal of famous buildings such as the Great Pyramid, the Parthenon, or Notre Dame, with the acoustic secret of the Stradivarius violins, or with the dimensions stipulated by God for Noah's Ark and for the Ark of the Covenant. The golden ratio pops up frequently in several biological contexts as well.

None of these real or imagined extra-mathematical manifestations of the golden ratio are the reason for mentioning it here. Rather, what is of interest for our story is the way in which, for the Pythagoreans, it got its exact expression *only* in terms of proportions, and in purely rhetorical terms at that. They defined it as follows:

Two magnitudes determine a golden ratio if the ratio of the larger to the smaller is the same as the ratio of their sum to the larger of the two.

Expressed in anachronistic symbols and in relation to Figure 3.4: if the segment *AB* is cut at point *P*, then the two resulting segments are in a golden ratio if the proportion

Figure 3.4 The golden ratio.

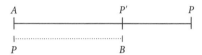

Figure 3.5 The recursive character of the golden ratio.

AB:*AP* :: *AP*:*PB* holds. In this case, *P* is called the golden section of *AB*. A nice property of the golden ratio, which can be noticed right away, is that if *AP* is cut at point *P′*, so that the segment *AP′* equals *PB* (as in Figure 3.5), then *P′* is the golden section of *AP*. Obviously, this recursive property can be applied indefinitely, and this has interesting mathematical consequences.

The main reason for bringing up the golden ratio here is its connection with yet another fundamental Pythagorean mathematical enquiry, namely that of the five-pointed star, which they called the "pentagram." Some ancient sources refer to the pentagram as a sign of recognition among the members of the Pythagorean sect. This has contributed to strengthening the emphasis laid by some writers on the mystical aspects of the Pythagoreans' interest in the golden ratio. But there are plenty of purely mathematical reasons to explain why the Pythagoreans devoted efforts to investigating this figure without having to decide if their mystical inclinations played any role in this. Among other things, the Pythagoreans addressed the task of constructing the pentagram with "straightedge and compass" alone. Such constructions, which played a truly central role in Greek mathematics in general, are constructions that assume two basic abilities, namely, (1) the ability to draw a circle of given radius and with any given point as its center; (2) the ability to draw a straight line between any two given points and to extend it indefinitely. In other words, this straightedge is not a "marked ruler," with distances marked on it, but rather just a ruler that allows one to connect, with a straight line, any two given points.

The Pythagoreans realized from very early on that constructing a pentagram requires the construction of an isosceles triangle in which the angles at the bases are the double of the remaining one. In modern geometric terminology, we can formulate this task, with reference to Figure 3.6, as follows:

To construct a triangle *QCP* with two basis angles β of 72° and the third angle of 36°.

It turns out that to complete this construction we need to be able to cut a given segment into its golden ratio. Thus, given a segment *AB*, constructing the pentagram involves the task of finding the point *P* that is the golden section of *AB*.[1]

---

[1] The details of this construction appear in the classical English edition of Euclid's *Elements*, first published in 1908, by the distinguished scholar Sir Thomas L. Heath (1861–1940). See there Proposition IV.10.

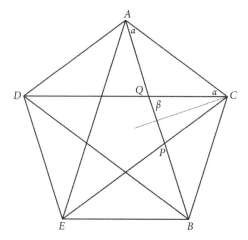

Figure 3.6 A regular pentagram having a given segment *AB* as its side. The point *P* cuts *AB* into two segments, *AP* and *PB*, forming a golden ratio.

While studying the properties of this, and of other associated geometric figures, the Pythagoreans made a surprising discovery that was to have far-reaching consequences for the further development of the concept of number. This is the discovery of "incommensurable" magnitudes, or magnitudes that lack a common measure among them. We need to devote focused attention now to these magnitudes and how they were discovered.

3.3 Incommensurability

What are these "incommensurable" magnitudes and why are they so important in our story? In order to explain, in the first place, what was at stake in the Pythagoreans' study of lengths and their relations, let me begin with the following, simple, and indeed seemingly trivial observation: any given length is measurable by another length that is smaller than it. Given a segment *AB*, as in the example displayed in Figure 3.7, it is always possible to find a smaller length *r* that "measures" *AB* an exact number of times. Also, if we are given a second segment *AC*, we can easily find some other length, say *s*, that measures *AC* an exact number of times.

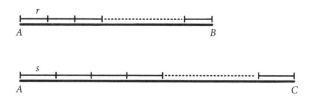

Figure 3.7 The length *r* measures the segment *AB*, and the length *s* measures the segment *AC*.

Notice that the units r and s were chosen independently and we don't know if they are equal or different, or which of them is larger than the other. We may, however, try to find now a *common* unit of measurement t that measures both of them. On the face of it, there seems to be no immediate reason to assume that such a unit t would not exist. For the Pythagoreans, moreover, in the framework of their views on number as the fundamental organizing principle of the universe—and certainly of all that pertains to mathematics—it would by all means be natural to assume that there exists such a common unit of measurement, small as it may be. As a matter of fact, so deeply ingrained was this assumption that they did not even take the step of formulating it explicitly. But, as in many other cases in the history of mathematics, deeply ingrained and implicitly assumed principles need to be called into question, and sometimes deeply revised, when they suddenly collide with new results and completely unexpected mathematical situations. This is precisely what happened in this case: contrary to their expectations, the Pythagoreans came to realize, and moreover to *prove*, that in certain situations no such common unit of measurement can exist.

The existence of such pairs of magnitudes with no common measure, that is to say, "incommensurable magnitudes," was really bad news for the Pythagoreans. To make things worse, one of the most prominent examples of incommensurability appeared in a perfectly simple and apparently transparent geometric situation, namely, in the ratio between the side and the diagonal of the square (Figure 3.8).

We can understand this simple situation in modern mathematical terms as follows: if we assign the value 1 to the side AB, then, using the Pythagorean theorem for right-angled triangles, the length of the diagonal AC turns out to be $\sqrt{2}$. The surprising discovery of the Pythagoreans was that some magnitudes, such as the one that might

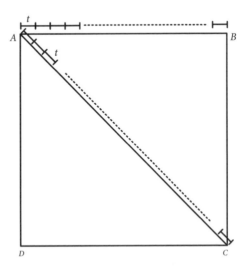

Figure 3.8 Figuring out the ratio of the diagonal to the side of a square. The Pythagoreans had assumed, as a matter of course, the existence of a small segment t that exactly measures any two given segments.

express the lengths of the side and the diagonal in a square, cannot be expressed as the ratio of two natural numbers. In other words, they discovered the existence of what we would call nowadays irrational numbers. The Pythagoreans would phrase the situation more or less as follows:

If AB is the side of the square and AC is its diagonal, one cannot find two integers n and m such that the ratio of AB to AC is the same as the ratio of n to m.

Now, if the simplest imaginable geometric situation cannot be described in terms of simple relations between natural numbers, how could the Pythagoreans claim for numbers (that is, *natural numbers*) the status of the ultimate, universal explanatory and organizational principle of the entire cosmos?

Legends of various kinds arose around this discovery. The Neoplatonist philosopher Iamblichus (245–325) reported that the member of the sect who first divulged the terrible secret "to those who were unworthy to receive it" perished by drowning in the sea. The main suspect was Hippasus of Metapontum, who lived in the fifth century BCE. We cannot be sure, however, if the gods themselves considered at the time that divulging this secret was impious to the extreme of imposing capital punishment on the culprit.

From the point of view of our story, what was at stake was not just a question of metaphysical beliefs. Rather, this was an issue of utmost *technical* as well as *foundational* importance for Pythagorean mathematics. Proportions were for them a main tool for developing and expressing important results, as we have seen, and their definition of proportion implicitly assumed that all quantities are pairwise commensurable. The existence of incommensurable magnitudes meant that if the arithmetic knowledge that they had worked so hard to create was to be preserved and further developed, finding a new coherent definition of ratio had become imperative.

In terms of hard historical evidence, little can be said about the circumstances in which the Pythagoreans came to *realize* that incommensurable magnitudes actually exist. Indirect evidence, however, indicates that this may have happened around the year 430 BCE, and also that it may have happened in connection with the Pythagoreans' investigations of the pentagram. We do know for certain, however, that they had a well-formulated proof of the incommensurability of the side and diagonal of a square, even though the details of this proof continue to be a matter of debate among historians (details on this debate appear in Appendix 3.1). We also know that the discovery of incommensurable magnitudes had important consequences for the development of the idea of number.

A first consequence was a thorough and systematic adoption of the fundamental distinction between (discrete) numbers (*arithmos*) and (continuous) magnitudes (*megethos*). The latter, comprising lengths, areas and volumes, can be indefinitely subdivided. Numbers, on the contrary, are always multitudes or collections of units, the unit itself being an indivisible entity. These numbers, moreover, can be conceived of as always counting some *concrete* kind of entity. The distinguished scholar Jacob Klein (1899–1978), famous as an interpreter of Plato's philosophical legacy, wisely pointed out that in the early Greek conception of numbers, the entities counted by the *arithmos*,

... however different they may be, are taken as uniform when counted; they are, for example, either apples, or apples and pears which are counted as fruit, or apples, pears and plates which are counted as 'objects.' .... Thus the *arithmos* indicates in each case *a definite number of definite things.*[2]

This basic distinction between (discrete) numbers and (continuous) magnitudes became highly influential throughout the Greek mathematical tradition, and in all those traditions influenced by it. It was transmitted to the Islamic world and later on to Europe, and it continued to be in the background of all discussions about the nature of number until well into the seventeenth century and beyond. The modern idea of number, as we will see, could never have consolidated as it eventually did as long as this distinction continued to be dominant. For one thing, irrational numbers would be, according to the Greek distinction, continuous magnitudes that cannot be conceived of as instances of the same general idea underlying the notion of the discrete natural numbers.

A second consequence of the discovery of incommensurable magnitudes concerned the already mentioned *technical* challenge of redefining the concepts of ratio and proportion. A new definition was needed that could account for incommensurable pairs of magnitudes as well. This challenge was successfully met with a new theory commonly attributed to Eudoxus of Cnidus (408–355 BCE), which became widely known via Book V of the famous treatise by Euclid of Alexandria (fl. 300 BCE), the *Elements*. The *Elements*, written about the year 300 BCE and comprising 13 books, presented in a systematic and exhaustive way the basic toolbox of elementary Greek mathematics (as well as some more advanced topics). In the forthcoming chapters, we will speak widely about Euclid's treatise, which, up until the seventeenth century, continued to provide a fundamental pillar for the development of mathematical ideas and techniques for problem-solving. The new Eudoxian theory of proportions, in particular, took center stage in the system of knowledge embodied in the *Elements*.

3.4 Eudoxus' theory of proportions

In Book V of the *Elements*, a ratio is defined as "a sort of relation in respect of size between two magnitudes of the same kind." This doesn't sound very useful or mathematically precise, to be honest. As with many of Euclid's definitions, however, we should not always try to see here the kind of formal definition that we typically expect to find in a modern mathematical textbook. If we define, for example, a straight line as the "shortest distance between two points," this is for us like the Chekhovian gun that is hanging on the wall in the first act of a play and is sure to be fired at someone in the third act. At some place down the road in our mathematical textbook, we are sure to use the definition of straight line given at the beginning, and to use it in its precise formulation to prove a crucial result. Not so with Euclid's definitions. Euclid says in Book I of the *Elements*, for example, that "a point is that which has no parts," but this defining property is not mentioned again in the book and it is never used as a crucial step in any of his proofs. A different case is that of the circle, which he defines as

---

[2] (Klein 1968, p. 46).

follows: "A circle is a plane figure contained by one line such that all the straight lines falling upon it from one point among those lying within the figure equal one another." The property that defines here the said point, which is of course the center of the circle, is indeed used crucially in some proofs. But other definitions, like that of ratio, are quite general comments about the ideas involved, and they are not always intended for direct use in proving results. Such definitions, in fact, belong more properly to the discourse about mathematics than to the technical mathematical content needed to develop the theories and the proofs.

The crucial term in Euclid's definition of ratio, at any rate, is that the two magnitudes compared are "of the same kind." This is what really counts in what he does with ratios. Euclid explained that ratios of two magnitudes exist only if they are "capable, when multiplied, of exceeding one another." This means that, for example, two *lengths* are magnitudes of the same kind, since we can repeatedly add the shorter of them to itself, as much as needed, until it surpasses the longer one. The same holds for two *areas*, two *volumes*, or, with some differences, for two *angles*. Consequently, we can define the ratio of two lengths, the ratio of two areas, or the ratio of two volumes. On the contrary, no matter how many times we add a line segment to itself, it will never "surpass" any given area, and hence no ratio can be defined between a length and an area.

Once we know that we can form ratios of two magnitudes of the same kind, the next step is to explain how to compare any two such given ratios. If we are given two magnitudes A and B of the same kind (e.g., two areas), they define a first ratio, which, just for convenience, we anachronistically denote here by $A:B$. Keep in mind, however, that Eudoxus and Euclid (and all other Greek mathematicians as well) did not have this symbolic privilege, and that they did all their reasoning on ratios in full rhetorical manner. Likewise, if we are given two additional magnitudes C and D of the same kind (e.g., two lengths) they define a second ratio, which we denote by $C:D$. Notice that C is of the same kind as D, but the two need not be of the same kind as A and B. Eudoxus' theory defines the conditions under which the ratio $A:B$ is the same as the ratio $C:D$. Now, if any two magnitudes were commensurable (as was just assumed before the discovery of incommensurables), then the comparison would be an easy matter. Indeed, if $A:B :: m:n$ and $C:D :: j:k$, then the four magnitudes are in proportion, $A:B :: C:D$, if the integers that define the ratios are themselves in proportion, i.e., if $m:n :: j:k$. It would be quite easy to check, along with the Pythagoreans, that these two ratios *of integers* are proportional.

But the challenge that Eudoxus had to address was precisely that we cannot always count on the help of such ratios of integers for making a decision. He suggested a truly ingenious way to overcome the difficulty involved here. His insights bring him quite close to some of the main ideas with the help of which, at the end of the nineteenth century, Richard Dedekind satisfactorily defined the real numbers for the first time. Some mathematicians and historians have even claimed that Eudoxus' and Dedekind's ideas are equivalent. We will nevertheless try to see the differences in the long run, when we discuss this in detail in Appendix 10.1. Eudoxus' achievement is very impressive, no doubt, by any standard, and it was indeed highly influential in centuries to come. Let me explain it here.

Eudoxus' approach is more easily understood in modern symbolic formulation, as follows: Given two ratios of pairwise comparable magnitudes A, B and C, D, they are said to establish a proportion $A:B :: C:D$ if and only if, given any two integers n and m, one of the following conditions hold:

(1) If $mA > nB$, then $mC > nD$,
(2) If $mA = nB$, then $mC = nD$,
(3) If $mA < nB$, then $mC < nD$.

The original, purely rhetorical, definition appearing in the *Elements* is much harder to follow, and it was certainly much harder to conceive, to formulate in precise terms, and to *use* at the time of the Greeks. At the same time, unlike Euclid's definition of ratio or of point, this one is really intended for use in the proofs. It is important that we actually present (at least once) the original definition as it appears in Book V of the *Elements*:[3]

Magnitudes are said to be in the same ratio, the first to the second and the third to the fourth, if any equimultiples whatever be taken of the first and third, and any equimultiples whatever of the second and fourth, the former equimultiples alike exceed, are alike equal to, or alike fall short of, the latter equimultiples taken in corresponding order.

Notice that, like the Pythagoreans, Euclid also writes that the ratios in question are "the same," and not that one ratio is "equal to" the second. Once again, I want to emphasize that this difference in wording is far from superficial. For Euclid, two numbers could be "equal" (if they were the same multitudes of whatever unit) and, in a different sense, two triangles could also be "equal" (if they had the same area). But "ratios" were neither numbers nor magnitudes. Comparing them meant checking not whether they were equal in one of the accepted senses of the word, but rather whether or not they were "the same."

In order to turn Eudoxus' rather cumbersome definition of proportion into a mathematical tool of "practical" value, Book V of the *Elements* was devoted precisely to providing a considerable number of significant results involving proportions and their use. Later generations of mathematicians, up until the seventeenth century, continued to use all these results very efficiently as tools to prove ever new and more complex mathematical results, and also to solve open geometric problems. At the same time, this well-elaborated and useful theory of proportions was also instrumental in perpetuating the conceptual separation between numbers and ratios.

In order to provide a clearer idea of how the concept of proportion was actually used in specific situations by the Greeks and their followers, I want to present now two examples of propositions from Book V of the *Elements*. I begin with proposition V.16, which in the Heath edition reads as follows:

V.16: If four magnitudes be proportional, they will also be proportional alternately.

Symbolically (and, I stress again, anachronistically) expressed:

Given four magnitudes of the same kind, a, b, c, d, if $a:b :: c:d$, then $a:c :: b:d$.

---

[3] All quotes from Euclid's *Elements* are taken from the Heath edition.

From a modern point of view, this looks to be a trivial result: the proportion can be seen as an equality of fractions, $\frac{a}{b} = \frac{c}{d}$, from which, by a simple side-transference, we get a second equality $\frac{a}{c} = \frac{b}{d}$. But Euclid, as we already know, could not rely on any kind of symbolic manipulation of this kind, and his ratios were not fractions. Rather, he had to apply the long and rhetorically cumbersome definition of proportion to the two pairs a, b and c, d, and then show that the two pairs a, c and b, d also satisfy the property stipulated in the definition. This is not a difficult task, but it does require a somewhat lengthy and verbose argumentation that one would rather avoid repeating every time that this situation arises. Hence, Euclid wanted to provide it as part of the basic toolbox he was laying down in his treatise. The case with V.18 is similar:

V.18: If magnitudes be proportional *separando*, they will also be proportional *componendo*.

Symbolically expressed:

If $a:b :: c:d$, then $a+b:b :: c+d:d$.

Once again, we might be tempted to think of a very simple algebraic proof for this identity, as follows:

Since $\frac{a}{b} = \frac{c}{d}$, therefore $\frac{a}{b} + 1 = \frac{c}{d} + 1$, and hence $\frac{a+b}{b} = \frac{c+d}{d}$.

Not so with Euclid, for whom the ratio was not a fraction that can be operated upon. The full Euclidean proofs of V.16 and V.18 appear in Appendix 3.2. You can see there in detail how the Eudoxian definition could actually be used to prove, without the help of symbolic manipulation, these and similar results.

3.5 Greek fractional numbers

Ratios and proportions, then, were a crucial idea lying at the heart of classical Greek mathematics. While I have tried to make clear that ratios were not fractions, I did not mean to say that "fractional quantities" do not appear in any way in Greek mathematical culture. They do appear, mainly in calculations found in commercial or astronomical texts. They differ from the ratios and proportions just discussed, but they also differ from our idea of common fractions $\frac{p}{q}$. The basic conception behind these Greek fractional quantities runs parallel to the one behind numbers as "collections of units." These collections of units comprise the couple, the trio, the quartet, etc., and so, in the same way, one can imagine the parallel reciprocals, or "parts of a unit": the half, the third, the fourth, etc. At the same time, it is also easy to extend, in a natural way, any existing notation for numbers into a similar notation for parts, without thereby having to assume any prior concept of a common fraction. So, for example, if γ represents "the trio," then its reciprocal, "the third," might be denoted, for instance, with the help of a sign such as ´, $\acute{\gamma}$ (or likewise with the help of any other similar indicator, such as a superposed dot).

As in the case of Egyptian mathematics, these reciprocals are not really "unit fractions." They are not particular cases of a more general idea of common fraction, in which for some reason we decided to allow for only one kind of numerator, namely, 1.

In Greek commercial texts where these reciprocals appear, they are often presented in an ordered sequence. The sequence usually begins (apparently for some historical reason) with two special terms, $\acute{β}$ and $∠$, representing, respectively $\frac{2}{3}$ and $\frac{1}{2}$. It then continues with all the reciprocals as follows: $\acute{β}, ∠, \acute{γ}, \acute{δ}, \acute{ε}, \ldots$.

In those texts where they appear, these Greek "parts" found a natural place among results that involve "inexact" divisions. A typical expression of the kind where they appeared is the following:

$$\ldots \text{of the } ιβ \text{ the } \acute{ιζ} \text{ is } ∠ίβίζλδυάξη.$$

A straightforward translation of this into modern notation, using the idea of common fraction, would be

$$\frac{12}{17} = \frac{1}{2} + \frac{1}{12} + \frac{1}{17} + \frac{1}{34} + \frac{1}{51} + \frac{1}{68}.$$

But the original text does not suggest the idea of a separately given fraction, $\frac{12}{17}$. Neither does it suggest that we are asked to represent or to calculate the value of that given fraction by way of arithmetic addition of other fractions having 1 as their numerator. The mathematical situation involved here is somewhat different. While for us $\frac{12}{17}$ represents the product of 12 times $\frac{1}{17}$, the expression "of the $ιβ$ the $\acute{ιζ}$" is an indication that we have gathered together 12 instances of that particular kind of part, the reciprocal of 17, $\acute{ιζ}$. The difference is subtle but important. Let me add some remarks that may help clarifying it.

Phrases such as "of the m the nth," which are found repeatedly in extant papyrus texts, appear often abbreviated by superposition, as in $\overset{n}{m}$. But even when this kind of abbreviation became available to Greek mathematicians, they did not use it in the framework of a well-developed arithmetic of fractions. In current notation, the arithmetic of fractions embodies the important (and for us seemingly obvious) insight that

$$\frac{m}{n} + \frac{p}{q} = \frac{m \cdot q + p \cdot n}{n \cdot q}.$$

If, in using the notation $\overset{n}{m}$, the Greeks had somehow operated with numbers "of the m the nth" as we operate with fractions, then we would find some identity similar to the following (I use here +, just for illustration, but of course they could have used any other symbol):

$$\overset{n}{m} + \overset{q}{p} \quad \text{yields} \quad \overset{n \cdot q}{m \cdot q + p \cdot n}.$$

But we do not find anything similar to this in the extant evidence.

Thus, a more suggestive way to understand how the Greeks saw these fractional quantities and the abbreviated way of writing them is to compare with similar instances that were common at the time. I have already mentioned the Greek approach to writing large numbers, in which myriads were written by superposition of symbols, as in $\overset{\phi\lambda\delta}{M}$. There are also examples in commercial texts where numbers were superposed over an accepted symbol that indicates a weight or a currency unit. Think about these examples, together with $\overset{n}{m}$ taken to represent "of the m the nth," and recall Jacob Klein's characterization of *arithmos* mentioned in Section 3.3. One gathers together or counts, in each case, a definite number of units, or currency units, or myriads, or weights, or halves (or thirds, or fourths, etc.). For example, "of the $\iota\beta$ the $\iota\zeta$" means that we counted, or gathered together 12 of the reciprocals of 17. So, it is always a collection, or a multitude, and not the product of an integer by a reciprocal, or by a Myriad, or by a unit of weight. Seen in this light, the early Greek use of fractional quantities—mainly in the context of commercial and astronomical texts—represents an early stage of a process whereby common fractions would eventually be organically incorporated into the overall system of numbers. Significant developments would still be needed, however, before this portion of the general picture could be completed by developing a full arithmetic of these fractions.

3.6 Comparisons, not measurements

Comparison of numbers, comparison of figures and comparison of ratios, then, were the main tools on which the bulk of theoretical Greek mathematical knowledge was built from its early stages. Accordingly, comparison, rather than measurement, of lengths, areas, or volumes was the main platform on which the great masters of classical Greek mathematics produced their impressive works. And this comparison was always of magnitudes of the same types. This had far-reaching consequences for the continued development of the idea of number, and it is important to see why.

Let us take the simple case of the area X of a circle. For us, the most adequate way to express what we know about this area is by means of a formula that indicates that it is the product, $X = \pi r^2$, of the square of the radius r by a certain constant (which we call nowadays π). For the Greeks, however, the typical, relevant way to express what they knew was by comparison of areas, expressed in the familiar language of ratios and proportions. In the case of the area of the circle, we find this expressed in Proposition XII.2 of Euclid's *Elements* as follows:

XII.2: Circles are to one another as the squares on their diameters.

In modern terms, the meaning of this is that if we have two circles of areas X, x and diameters Z, z, respectively (Figure 3.9), then we have the proportion $X{:}x :: Z^2{:}z^2$. The proof of this proposition is very interesting, and it is also historically important because it involves approximating a circle by a finite polygon and then "passing to the infinite." The details are somewhat beyond the scope of this book, but I present them in Appendix 3.3 because they may be of interest to anyone curious about how the Greeks managed an infinite approximation in situations like this. What is of direct interest for our discussion here is that, as I have already stressed, there is nothing like a formula and only comparisons of areas via ratios and proportions. Neither did Euclid measure or assign lengths to the diameter of the circle or to the sides of the polygon in his proof.

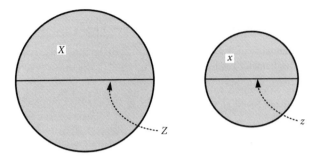

Figure 3.9 *Elements*, XII.2: the ratio of the areas of the circles is the same as the ratio of the squares of their diameters.

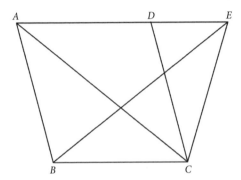

Figure 3.10 *Elements*, I.41: the area of the parallelogram *ABCD* is twice the area of the triangle *BCE*.

Another example taken from the same classical treatise concerns the even simpler case of the area of a triangle. We do not find in Euclid's text a statement about this area being half the product of the base times the height, as we would express nowadays in a formula: $T = \frac{b \cdot h}{2}$. What we do find is a proposition stating that if a parallelogram and a triangle are built on the same base and between two given parallel straight lines, then the area of the parallelogram is twice that of the triangle (Figure 3.10). So, again, areas are compared, not measured.

More complex cases appear in the always fascinating works of Archimedes, but still the same kind of formulation is used there as well. Where we would say today that the area of the spherical surface of a sphere with radius r is $4\pi r^2$, Archimedes stated that "the area of the spherical surface is four times the area of a great circle in that sphere." Similarly, Archimedes does not tell us that the volume of the cone is given by the formula $V = \frac{1}{3}\pi hr^2$. Rather, he says that it is "one-third the volume of a cylinder having the same basis and the same height."

If we want to get the broader picture, it is important to emphasize that a mathematician like Archimedes dealt in some cases with questions of areas and volumes in ways others than the comparisons that I have just mentioned. For example, in his famous treatise *On the Measurement of the Circle*, he presented an interesting approximation of what for us is the value of π. His own formulation was the following:[4]

MC.3: The circumference of any circle is three times the diameter and exceeds it by less than one-seventh of the diameter and by more than ten-seventyoneths.

This is typically considered to be the first serious estimation of the value of π. In modern terms, it can be taken to mean the following:

$$3\frac{1}{7} < \pi < 3\frac{10}{71}.$$

But Archimedes' own formulation and proof of this statement was, nevertheless, closer to the use of ratios and comparison of areas just explained than this symbolic formulation might make us believe. I will give just some details of this result, to illustrate what is really involved here.

Archimedes' short treatise opens with a proposition that would seem, on first sight, to solve the difficult problem of squaring the circle, i.e., finding a square whose area equals that of a given circle. This classical problem, which occupied the best minds in Greek mathematics, is, incidentally, precisely a problem of comparison of areas in the sense explained above (in this case, the task is to show that two figures, one given and another one required to be constructed, are *equal* [in area, of course]). The opening proposition, illustrated in Figure 3.11, states that

MC.1: A circle is equal to a right-angled triangle one of whose sides is equal to the radius and the other to the circumference of the circle.

This seems to solve the problem of squaring the circle, because we are given the circle, and we come up with a triangle whose area equals that of the circle (and, of course, if we have this triangle, then we can easily construct a square of equal area, thus apparently solving the problem). It *does not* actually solve the problem, however. Archimedes did not say how we can construct (with straightedge and compass) the side of the triangle that is equal to the circumference. Well, actually, he *could not* have said it, because, as we now know (this was proved in the nineteenth century), it is impossible to do so. And yet, Archimedes did show the way to making progress with the problem, because the focus had moved at this point to the question of how to construct the said side of length c.

Figure 3.11 *Measurement of the circle*: The circle with radius r and circumference c equals the right-angled triangle with corresponding sides.

---

[4] Quoted in (Dijksterhuis 1987, p. 223).

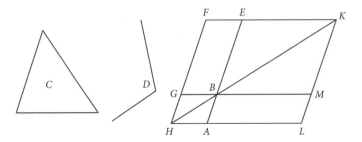

Figure 3.12 *Elements*, I.44: triangle *C*, angle *D* and segment *AB* are given. The figure on the right is constructed to represent a parallelogram *BMLA* equal in area to *C*, its side being equal to *AB* and the angle *ABM* being equal to *D*.

Constructing a *line* with a certain property seems to be, on the face of it, much easier than constructing a figure, but in the end it turns out that this was not really the case. Archimedes did not solve the problem with his geometric proposition, but still, after having stated it, he moved on to calculate in a very sophisticated manner the value of the ratio between the diameter and the circumference of a circle. He did this by approximating the circle with inscribed and circumscribed 96-gons. For reasons of space, I will not go here into the details of Archimedes' calculational procedure.[5] I only want to emphasize that this procedure embodies a significant, and rather unique, example in classical Greek geometry of working out the *actual value* of a measurement of a magnitude, rather than just comparing two magnitudes of similar kind. Unlike in this example, comparing, rather than measuring (i.e., associating a numerical value to a magnitude), was the typical attitude found in classical Greek geometry.

3.7 A unit length

There is yet another important aspect to this way of thinking about magnitudes and their sizes. In the scholarly mathematical culture of ancient Greece, there was no comparison of—much less operations with—magnitudes of *different* kinds. The Greeks added, subtracted and compared *only magnitudes of the same kind*. They did not think these situations in terms of an abstract, general concept of number that was assigned to magnitudes and operated upon. Such an idea turned out to be very elusive, and it took hundreds of years of developments and progress in geometry and arithmetic for it to arise and consolidate, and indeed to impose itself almost, one might say, against the will of the mathematicians involved. An important turning point in this evolution will appear in the work of Descartes in the seventeenth century, as we will see in Chapter 8. But since the connecting line between Euclid and Descartes and the change implied in Descartes' work are so fundamental for our story, I would like to conclude this chapter by referring to an important result that appears in the *Elements*, namely Proposition I.44, and by saying some words about it in preparation for a later discussion.

[5] See (Berggren et al (eds.) 2004, pp. 7–14; Dijksterhuis 1987, pp. 222–229).

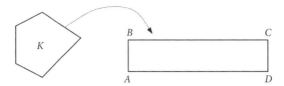

Figure 3.13 The area of a polygon K is transformed into a rectangle ABCD, with side AB having unit length.

Proposition I.44 of the *Elements* presents a task of geometric construction which is defined as follows (Figure 3.12):

I.44: To a given [segment of a] straight line to apply, in a given rectilinear angle, a parallelogram equal to a given triangle.

The details of the construction and the proof are not complicated, but they are not so important for us here. What is important is to quote the editorial notes that Heath added to this proposition in his edition of the *Elements*. This proposition represents— he wrote—"one of the most impressive in all geometry," and this is because it allows "the transformation of a parallelogram of any shape into another with the same angle and of equal area but with one side of any given length, e.g., a *unit* length." Based on this result, Heath reached the conclusion that in Euclid's view, given any polygon, one can construct a rectangle of the same size and with one of its sides having the length "one." Figure 3.13 depicts the geometric situation that Heath had in mind.

Why is this situation so important in the eyes of Heath? He is definitely thinking about the use of numbers in a geometric context in a way that, if indeed found there, would attribute to the Greeks an approach that is totally different from the one I have been trying to explain thus far. Indeed, let us assume that the area of the polygon K is 10. If the length of AB is indeed the unit length, and if the area of the rectangle ABCD is 10, as implied by this interpretation, then the length of AD would be 10. In other words, with the help of this proposition as Heath interprets it, we can find a length that is equal to any given area. The area equals 10 (area units?) and also the length equals 10 (length units?). We would have here, if this were the case, a general and abstract understanding of number that is equally applicable to measure magnitudes of various kinds. But this is exactly what, in my explanation above, is untypical of what we find in Greek classical geometry.

On what grounds, then, is Heath able to translate a proposition of the *Elements* so that it will end up embodying an approach that I have been claiming does not appear there? Or does it appear, after all? My point is that Heath is interpreting this proposition with the help of concepts that are foreign to the Greek mathematical culture. Some of them were developed much later, and some of them existed but were not used by the Greeks as their core approach to scholarly mathematics. Of course, from a purely mathematical point of view, the translation seems plausible. But we are looking here for *historical* accuracy and not for *mathematical* plausibility. We want to know what was Euclid's concept of number and how he used (or didn't use) it in this specific context. In the kind of geometry with which Euclid, Archimedes, Apollonius and their contemporaries were involved, the classical conception was that numerical values were not attached to lengths or to areas so that calculations could be performed with them.

Heath wants to see here the rectangle as a geometric translation of the arithmetic operation of multiplying two numbers, and in other parts of his editorial notes to the *Elements* he also suggests that the ratio of two lengths should be interpreted as a division of two magnitudes. But a close inspection of the original Greek mathematical texts shows that this interpretation is foreign to their spirit, and particularly to the idea of number as understood and used in them. The meaning of these conflicting interpretations will became increasingly clearer as we make progress in our story, and particularly when we arrive in the seventeenth century and take a closer look at the work of Descartes.

Appendix 3.1   The incommensurability of $\sqrt{2}$. Ancient and modern proofs

How did the Pythagoreans *reach the insight* that there exist incommensurable magnitudes? Not much is known that can help in giving serious historical answers to this question. And yet, historians have come forward with interesting conjectures with varying degrees of plausibility. The only solid piece of available evidence deals with the more specific question of how the Pythagoreans possibly *proved* that the diagonal of the square is incommensurable with its side. But we should keep in mind that the discovery of the possibility of the existence of incommensurable magnitudes and the specific proof in question are two completely (though related) matters, particularly because this is a proof by contradiction. It is evident that mathematicians do not discover new mathematical results by way of proofs by contradiction. Such proofs require that we realize or assume in advance the statement that is to be proved, so that we may start the proof by assuming its negation.

Before we attempt to reconstruct the proof that the Pythagoreans possibly knew and on which we have some evidence, let me begin by presenting here the standard proof for the irrationality of $\sqrt{2}$ that is accepted and typically taught nowadays. This is a short and very beautiful proof, which can be formulated as follows.

Assume $\sqrt{2}$ to be a rational number, or, in other words, assume $\sqrt{2} = p/q$, where p and q are integers having no common divisors (if they have, simply divide both by that common divisor, until you get a simplified fraction with no common factors for the numerator and denominator). In particular, we may assume that p and q are not both even numbers. Now square both sides of the equation, and move the denominator to the left-hand side, and you get $p^2 = 2q^2$. Hence, p^2 is an even number, and it follows that p is also an even number (because if p were odd, $p = 2k + 1$, then $p^2 = (2k + 1)^2 = 4k^2 + 4k + 1$, which is clearly an odd number). Hence, we can write p as $p = 2k$. But since $p^2 = 2q^2$, we obtain $4k^2 = 2q^2$, or $2k^2 = q^2$, and this means that q^2 is an even number. Hence, again, q itself would be an even number. But then we would have a contradiction, since we have assumed that p and q cannot be both even numbers, and hence our original assumption is false. It follows, finally, that $\sqrt{2}$ is an irrational number.

Now back to the Pythagoreans. The historical evidence we have about their proof appears in a text by Aristotle (*Prior Analytics*, I.23), which reads as follows:

For all who effect an argument per impossible infer syllogistically what is false, and prove the original conclusion hypothetically when something impossible results from the assumption of its contradictory; e.g., that the diagonal of the square is incommensurate with the side, because odd numbers are equal to evens if it is supposed to be commensurate. One infers syllogistically that odd numbers come out equal to evens, and one proves hypothetically the incommensurability of the diagonal, since a falsehood results through contradicting this.

As part of a general discussion about legitimate ways of proof and of logical inference, Aristotle mentioned the Pythagorean proof of incommensurability as a representative example of a proof by contradiction. And, in this case, what is assumed, according to his account, is that the diagonal and the side are commensurable, whereas the contradiction that then obtains is that "odd numbers come out to even." How should we understand this Aristotelian account? Heath gives the following, rather straightforward interpretation in his edition of the *Elements*:[6]

Suppose *AC*, the diagonal of a square, to be commensurable with *AB*, its side.

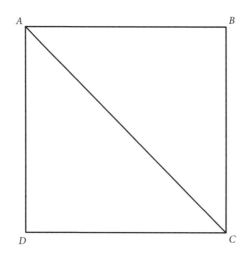

Let $\alpha:\beta$ be their ratio expressed in the smallest numbers.
Then $\alpha > \beta$ and therefore necessarily $\alpha/\beta > 1$.
Now $AC^2:AB^2 = \alpha^2:\beta^2$,
and since $AC^2 = 2AB^2$ [because of the Pythagorean theorem],
hence $\alpha^2 = 2\beta^2$.
Therefore α^2 is even, and therefore α is even.
Since $\alpha:\beta$ is in its lower terms, it follows that β must be odd.
Put $\quad\quad\quad \alpha = 2\gamma$;
therefore $\quad\quad 4\gamma^2 = 2\beta^2$,
or $\quad\quad\quad\, \beta^2 = 2\gamma^2$,

[6] (Heath 1956, Vol. 3, p. 2).

so that $β^2$, and therefore $β$, must be even. But is $β$ also odd: which is impossible. This proof only enables us to prove the incommensurability of the diagonal of a square with its side, or of $\sqrt{2}$ with unity.

Heath's interpretation is mathematically plausible, givenAristotle's account. Here and there, the existence of a common measure is assumed, and a contradiction is reached whereby an even number is shown to be odd as well. Heath's interpretation is also easily seen to correspond to the currently accepted proof of the irrationality of $\sqrt{2}$. But again, speaking as historians, mathematical plausibility is important, but not necessarily valid as historical evidence. The extant Greek sources do not show any evidence of symbolic manipulation of the kind assumed by Heath. The purely algebraic statement that from $α = 2γ$ we can deduce $4γ^2 = 2β^2$, for example, is not something that we find in the text. As a matter of fact, Euclid did not have *any* symbolic language that allowed abstract manipulation of symbols. Moreover, there is a clear historical difference (in terms of the concepts involved) between the claim that the diagonal is incommensurable with the side and the claim that $\sqrt{2}$ is an irrational number. As I have already emphasized several times, in the Greek mathematical culture, the ratio between diagonal and side was by no means a number but was just that, a ratio.

Heath's line of historical argumentation belongs to a broader historiographical tradition, usually called "geometric algebra." This tradition postulates the view that the Greeks did develop some algebraic kind of *thought* and techniques, but that they did not support it by the kind of symbolic algebraic *notation* known to us. They had the basic idea, it is claimed, but they did not have the right notation. I discuss this interpretive tradition and the problems related with it in some detail in Section 4.2 and in Appendix 8.2. Here I would like only to hint at the difficulties involved in such an approach by presenting an alternative to Heath's interpretation of this passage of Aristotle. This alternative not only fits mathematically the contents of the extant Aristotelian text, but also seems much more historically plausible than Heath's, for the simple reason that, unlike Heath, it relies *only* on ideas that are known to have been developed and currently used by the Pythagoreans in their work. Thus, this interpretation preserves the Pythagorean spirit, and at the same time it does not assume any kind of mathematical ability or idea that is foreign to what the Greek texts known to us explicitly display. For this purpose, let us consider a diagram (Figure 3.14) that appears in the Platonic dialogue *Menon* (in relation to a different issue) and that was certainly not foreign to the spirit of Greek mathematics. Let us assume now that segments *DH* and *DB* are commensurable. This means that there is some small segment that measures each of these an exact number of times. Let us assume that the said segment enters exactly n times in *DB* and exactly m times in *DH*. We can surely assume that n and m have no common factors, since otherwise we can take an even smaller common measure.

By looking at the diagram, we notice that the squares *DBHI* and *AGFE* represent square numbers (in the sense of the Pythagoreans figurate number described above). In addition, *AGFE* represents a square number that is twice the number represented by *DBHI*. Hence *AGFE* represents an even number, and hence *AG*, the side of *AGFE*, is itself an even number. Notice that this last inference is also based on the typical kind of arguments that the Pythagorean favored, using only figurate numbers and reasoning on the basis of the configurations (as pebbles, as we saw above) that define them: if a

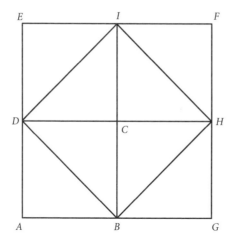

Figure 3.14 A diagram from Plato's *Menon*.

square number is even, so is its side. But then we deduce that the square *AGFE*, whose side is an even number, can be divided into four. Hence the square *ABCD*, which (as the diagram shows) is a quarter of *AGFE*, does itself represent a number. But again by purely figurate considerations on the diagram (i.e., counting the number of equal triangles on each), *DBHI* is a square number that is the double of the square *ABCD*, and hence *DBHI* represents a square number that is even. Hence its side *DB* represents itself an even number. But this contradicts the previous conclusion that *DH* is an even number, while *DH* and *DB* are measured by numbers n and m that are not both even.

We have thus seen two different ways to interpret an ancient mathematical text: Heath's interpretation based on the use of algebraic symbolic manipulation that is not explicitly found in the text, and an alternative one that relies on mathematical tools that are not explicitly in the cited text either, but that are known to have been used at the time.

Appendix 3.2 Eudoxus' theory of proportions in action

In order to give the reader a more direct feeling for what the Euclid–Eudoxus theory of proportions is all about, how it was used, and what were its limitations, I present here some details on the proofs of two propositions from Book V of the *Elements*, V.16 and V.18.

As stated above, Euclid's original definition of proportions as sameness of ratios is the following:

Magnitudes are said to be in the same ratio, the first to the second and the third to the fourth, if any equimultiples whatever be taken of the first and third, and any equimultiples whatever of the second and fourth, the former equimultiples alike exceed, are alike equal to, or alike fall short of, the latter equimultiples taken in corresponding order.

For simplicity, I have rendered this in modern algebraic terms by saying that given two ratios of pairwise comparable magnitudes A, B and C, D, they are said to establish a proportion, $A{:}B :: C{:}D$, if and only if, given any two integers n and m, one of the following conditions holds:

(1) If $mA > nB$, then $mC > nD$,
(2) If $mA = nB$, then $mC = nD$,
(3) If $mA < nB$, then $mC < nD$.

Still stated in the same, modern, terms, Proposition V.16 states that

If $a{:}b = c{:}d$, then $a{:}c = b{:}d$.

Of course, thinking in terms of modern symbolic algebra and its simple rules of manipulation, this statement does not require a separate formulation, let alone a separate demonstration. In its own historical context, however, it certainly does involve a meaningful mathematical idea that requires a somewhat elaborate proof. In the Heath edition, the proposition is formulated as follows:

V.16: If four magnitudes be proportional, they will also be proportional alternately.

Let A, B, C, D be four proportional magnitudes, so that as A is to B, so is C to D. I say that they will also be alternately, that is, as A is to C, so is B to D.

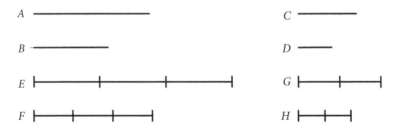

For of A, B let equimultiples E, F be taken and of C, D other, chance equimultiples G, H. Then, since E is the same equimultiple of A that F is of B, and parts have the same ratio as the same multiples of them, therefore, as A is to B so is E to F. But as A is to B, so is C to D; therefore also as C is to D, so is E to F. Again, since G, H are equimultiples of C, D, therefore, as C is to D, so is G to H. But, as C is to D, so is E to F; therefore also, as E is to F, so is G to H. But, if four magnitudes be proportional, and the first be greater than the third, the second will also be greater than the fourth; if equal, equal, and if less, less. Therefore, if E is in excess of G, F is also in excess of H, if equal, equal, if less, less. Now E, F are equimultiples of A, B and G, H other, chance, equimultiple of C, D; therefore as A is to C, so is B to D.

If we translate the basic idea of this proof into modern symbolic terms, we get the following. In order to prove that $a{:}c = b{:}d$, we need to prove that for given any pair of integers n, m, if $ma >=< nc$, then $mb >=< nd$. We begin, therefore, by writing down the magnitudes ma, mb and nc, nd, and then we go through the following three steps:

1. $a:b = ma:mb$ and $c:d = nc:nd$;
2. from $a:b = c:d$, there follows $ma:mb = nc:nd$;
3. from $ma:mb = nc:nd$, there follows $ma >=< nc, mb >=< nd$.

Notice that each of these steps concerns separate propositions (V.15, V.11 and V.14, respectively) that were previously proved rhetorically (in the style of the proof just quoted and not without effort) in the same book of the *Elements*.

Let us take a look now at Proposition V.18. This proposition has to be seen in conjunction with V.17, since they both deal with the following, similar mathematical situations: if we are given a proportion $a:b = c:d$, and if we modify both ratios in the same way, is the proportion thus preserved? "Modification" means, in this case, that instead of comparing the original ratios $a:b$ and $c:d$, we compare either the ratios $(a+b):b$ and $(c+d):d$ or the ratios $(a-b):b$ and $(c-d):d$. And Propositions V.17 and V.18 state that, indeed, also the pairs of modified ratios continue to yield a proportion.

These and some additional ways of transforming a given proportion appeared repeatedly in dealing with geometric problems in the Greek mathematical culture and also as late as the seventeenth century, including in such influential books as Newton's *Principia*. The historical importance of theorems such as V.17 or V.18 can hardly be exaggerated. The Latin names of the modifications dealt with in the two propositions (*separando* and *componendo*, respectively) were preserved in the English editions of the *Elements*, such as Heath's. I also use them here. I present in detail only the proof of V.18, but it depends on that of V.17 (which the interested reader will easily find in the *Elements*), and therefore I begin by formulating both propositions:

V.17: If four magnitudes be proportional *componendo*, they will also be proportional *separando*.

In modern symbolic formulation, V.17 states that if $a:b = c:d$, then $(a - b):b = (c - d):d$. V.18, on the other hand, states that if $a:b = c:d$, then $(a + b):b = (c + d):d$. The original statement and proof of V.18 (including the diagram) in Heath's version reads as follows:

V.18: If four magnitudes be proportional *separando*, they will also be proportional *componendo*.

Let AE, EB, CF, FD be magnitudes proportional *separando*, so that, as AE is to EB, so is CF to FD; I say that they will also be proportional *conponendo*, that is, as AB is to BE, so is CD to FD.

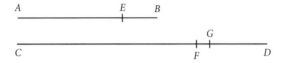

For, if CD be not to DF as AB to BE, then, as AB is to BE, so will CD be either to some magnitude less than DF or to a greater. First let it be in that ratio to a less magnitude DG. Then, since as AB is to BE, so is CD to DG, they are magnitudes proportional *componendo*; so that they will also be proportional *separando*. Therefore, as AE is to EB, so is CG to GD. But also, by hypothesis, as *AE* is to *EB*, so is *CF* to *FD*. But the first *CG* is greater than the third *CF*; therefore the second *GD* is also greater than the fourth *FD*. But it is also less; which is impossible. Therefore, as *AB* is to *BE*, so is not *CD* to a less magnitude than *FD*. Similarly we can prove that neither is it in that ratio to a greater; it is therefore the ratio to *FD* itself.

This proof introduces a form of argumentation, "double *reductio ad absurdum*," that was fundamental in the Greek mathematical culture. In order to prove that two magnitudes are equal, we reach two separate contradictions after assuming (1) that the first magnitude is *larger* than the second, and then (2) that it is *smaller* than the second. This kind of argumentation applied equally to sameness of ratios and was an important tool within the Eudoxian theory as well. Indeed, in order to prove that four magnitudes are in proportion, $a:b = c:d$, the Greeks would derive a "double *reductio ad absurdum*," for example, from the assumption not that the four said magnitudes are in proportion, but rather that $a:b = c:e$ are in proportion, where (1) $e < d$ and (2) $e > d$. This strategy proved particularly important when dealing with areas and volumes of curved shapes, namely, in situations involving what we could call nowadays "passage to the infinite" (an example of which appears in Appendix 3.3).

So, if we render Euclid's proof of V.18 in modern symbolic terms, we obtain the following:

1 Assuming that the proportion in question does not hold (i.e., assuming that $(a + b):b \neq (c + d):d$), implies that it will hold when the fourth element is either greater or smaller than d (i.e., it implies assuming $(a + b):b = (c + d):(d \pm x)$).
2 If, for instance, $(a + b):b = (c + d):(d - x)$, then, by V.17, $a:b = (c + x):(d - x)$ (recall that V.17 states that if $p:q = r:s$, then $(p - q):q = (r - s):s$).
3 But $a:b = c:d$, and hence $c:d = (c + x):(d - x)$.
4 But $c < (c + x)$, while $d > (d - x)$, and hence the proportion in (3) is impossible.
5 Hence, the assumption in (1) is false and $(a + b):b = (c + d):d$ indeed holds (and here Euclid also states that we may repeat the argument if in step (2) we take $(a + b):b = (c + d):(d + x)$).

Having understood the basic argument, the reader may still feel that the translation to algebraic terms does not make the argument more immediately transparent. Why couldn't Euclid just have used an argument such as $a/b = c/d \Rightarrow a/b + 1 = c/d + 1 \Rightarrow (a + b)/b = (c+d)/d$? Well, the point is precisely that his argument was *not* devised by first thinking in algebraic, or quasi-algebraic, terms, and then expressing the idea in a non-algebraic, purely rhetorical way for lack, say, of the proper symbolism. Rather, this proof stresses that the Euclidean/Eudoxian way to deal with proportions involved, indeed, an idiosyncratic and certainly non-algebraic perspective.

Another important point to notice here is the following: in step (1), Euclid is implicitly assuming a rather non-trivial fact, namely, the necessary existence of a fourth proportional magnitude for any given three a, b, c. Indeed, he is assuming that if d does not complete the proportion with the other three, then there must be some $d \pm x$ that *does* complete it. Now, when we look at Book IX of the *Elements*, which is entirely devoted to arithmetical propositions, we realize that Euclid was aware of the need to discuss more explicitly the possible existence of the fourth proportional, and he did so for the case of *numbers*. Given the three numbers 4, 5 and 2, for example, it is obvious that no integer n exists such that 4:5 :: 2:n. But in Book V, and specifically in Proposition V.18, Euclid was dealing with *continuous magnitudes*, and in the proof he made the assumption without qualifications or any further comments.

This assumption started to be criticized in the sixteenth century by commentators and editors of the *Elements*. The prominent Jesuit mathematician Christopher Clavius, to take an important example about whom we will speak in greater detail in Section 6.5, included the assumption of the existence of a fourth proportional among the general axioms of the *Elements*. A more interesting criticism came from yet another Jesuit priest, Giovanni Gerolamo Saccheri (1667–1733), who in his very original commentaries on the *Elements* came up with ideas that would eventually prove important on the way to the creation of non-Euclidean geometries in the nineteenth century. Sacchieri questioned the legitimacy of Euclid's assumption about the existence of a fourth proportional, and claimed that it has to be proved separately. But moreover, and more importantly, he also showed how this could be done in the special case in which all the magnitudes in the proportion are straight line segments. His proof relied only on propositions that do appear in the *Elements* prior to V.18. Somewhat later, Sacchieri was also able to prove the claim for the more complicated case in which the magnitudes are polygons.

Appendix 3.3 Euclid and the area of the circle

Proposition XII.2 of Euclid's *Elements* describes the relationship between areas of circles and the squares of their diameters. Not by way of a formula, as we do it nowadays, but by way of comparison of ratios, in the best style of classical Greek geometry. If we denote, as in Figure 3.9, the areas of the two circles by X, x and their diameters by Z, z, then the proposition states that the following proportion holds: X:x :: Z^2:z^2.

The proof is based on applying two ingenious ideas developed by the Greeks for carrying out complex proofs. First is the method typically known as "the method of exhaustion," helpful in dealing with areas of curvilinear figures by means of approximation with rectilinear ones. Second is the method of "double *reductio ad absurdum*," which I have mentioned in Appendix 3.2. But the most fundamental principle on which the proof is based is the so-called Principle of Archimedes, which Euclid proved earlier in his treatise as Proposition X.1. In Euclid's formulation, it is meant as a criterion to turn the more or less general idea of the continuity of magnitudes into a practical tool for use in proofs. It is formulated as follows:

X.1: Two unequal magnitudes being set out, if from the greater there be subtracted a magnitude greater than its half, and from that which is left a magnitude greater than its half, and if this process is repeated continually, there will be left some magnitude which will be less than the lesser magnitude set out.

We will see exactly how this idea is put to use in the proof. But there is yet another result, XII.1, that is necessary for the proof. It is the following:

XII.1: Similar polygons inscribed in circles are to one another as the squares of the diameters.

Notice that XII.2 is exactly the same result, but formulated for "circles" instead of "similar polygons":

XII.2: Circles are to one another as the squares of the diameters.

The proof for polygons is straightforward (I skip the details). The real challenge is how to move from the polygon to the "polygon with infinite sides," that is, the circle. Euclid did not have at his disposal the tools of the infinitesimal calculus that started to be developed in the seventeenth century. These tools are meant precisely for handling situations like this, involving a "passage to the infinite". Let us see, then, an outline of Euclid's argument for handling that passage with the tools that were available to him (I make it easier for the reader by introducing some symbols that do not appear in the original).

The diagram accompanying the proposition in the Heath edition is depicted in Figure 3.15. What needs to be proved is the validity of the proportion $X{:}x :: Z^2{:}z^2$. The proof is by double *reductio ad absurdum* in the following sense: we will assume that the said proportion does not hold, but rather that $X{:}S :: Z^2{:}z^2$. We will then assume that (1) $S < x$ and reach a contradiction and then that (2) $S > x$ and reach a second contradiction. The sameness of the ratios is thus the only alternative left.

The process of "exhaustion" consists in successively inscribing polygons of $2k$ sides inside the circles: square, then octagon, and so on. The polygons inscribed in X we denote by Y_n, with the number of sides of Y_{n+1} being twice that of Y_n (e.g., Y_2 is, in Euclid's diagram, the octagon *ARDQCPBO*). The polygons inscribed in x we denote by y_n (e.g., y_2 is, in Euclid's diagram, the octagon *ENHMGLFK*). And, by slight abuse of language, the letters, X, x, Y_n, y_n denote both the figures and their areas. The question now is to what extent do we come closer to the circle in each of these steps of approximation with polygons? This can be seen with reasonable precision if we zoom into the left-upper corner of the smaller circle in Euclid's Figure 3.15. What we see is depicted in Figure 3.16. The shaded region in the upper part of the diagram is that part of the circle that remains outside the polygon y_n (whose side is *FE*). It clearly takes more than half of the rectangle. The segments *FK* and *KE*, in the lower part of the diagram, are the sides of the next polygon y_{n+1}. The diagram shows that in passing from y_n to y_{n+1}, we have subtracted a triangle, which is half the rectangle and hence *more than half* of the shaded region. So, the overall process is as follows: we start with a circle and subtract

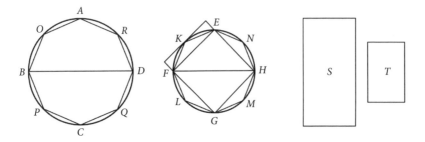

Figure 3.15 *Elements*, XII.2: Circles are to one another as the squares on their diameters.

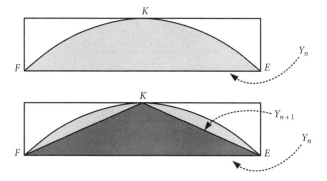

Figure 3.16 *Elements*, XII.2: Zooming into the process of exhaustion.

from it a square, which is obviously greater than half the circle; then from that which is left, we subtract an area magnitude greater than its half, and we repeat this process continually, going from y_n to y_{n+1}. Proposition X.1 warrants that after we repeat this process a sufficient number of times (in our words, for n large enough), we will reach the polygon y_n so that the difference $x - y_n$ between the area of the circle and that of the polygon is less than any area chosen in advance.

All that is left now is to take the area $x - S$ as the bounding area chosen in advance. If so, after taking n large enough, we obtain $x - y_n < x - S$. It follows that $S < y_n$. At this point, the proof is not yet complete, but the trick is already done. We had chosen S to be some area smaller than x; we now see that no matter what we chose as S in the proportion that was assumed, $X{:}S :: Z^2{:}z^2$, if we continue to exhaust the circle with inscribed polygons (or, in our language, if we chose some n large enough), then we will always be able to have a polygon lying between x and the putative S.

In order to complete the proof, we now invoke Proposition XII.1 and apply it to the polygons Y_n and y_n, to obtain: $Y_n{:}y_n :: Z^2{:}z^2$. Since we are assuming that $X{:}S :: Z^2{:}z^2$, we can conclude that

$$Y_n{:}y_n :: X{:}S.$$

And since we have shown that $S < y_n$, the proportion just obtained implies that $X < Y_n$. But we have constructed Y_n as inscribed within X, and hence we have attained the desired first contradiction. The alternative assumption, namely, that $S < x$, leads in a similar way to a second contradiction, as desired.

It should be noticed that "exhaustion procedures" such as exemplified here remained a main tool for calculation of areas and volumes well into the seventeenth century, before the creation of the infinitesimal calculus. Notice also that the name "Principle of Archimedes" is interestingly anachronistic when used with reference to the *Elements*, as Archimedes was active several decades after Euclid. It is the case, however, that in his own works, Archimedes did rely on a similar idea with astounding success for solving difficult problems.

CHAPTER 4

Construction Problems and Numerical Problems in the Greek Mathematical Tradition

Whereas the bulk of the contents of Euclid's *Elements* is devoted to purely geometrical topics, Books VII–IX deal with purely arithmetical ones. This is not to say that these are textbooks devoted to teaching methods for writing numbers or for calculating with numbers. Those topics were not treated at all as part of the *Elements*. What Euclid's books on arithmetic contain are general propositions about properties of natural numbers, about the relationship between a given number and its factors, and, in particular, about the properties of prime and composite numbers. Basically, then, the arithmetical books of the *Elements* comprise what we nowadays call "Elementary Number Theory." The propositions are always accompanied by rigorous proofs similar in many respects to those appearing in the geometric books (including diagrams), and yet an important difference exists that needs to be stressed: unlike in the geometric books, Euclid does not open his treatment of arithmetic by introducing a system of postulates that would serve as the foundation for arithmetical knowledge. This is a rather noteworthy fact that has always attracted the attention of historians, and we should also devote some thought to it.

It is well known that one of the most important contributions of Greek mathematics in general, and of Euclid's books in particular, was the way in which they transformed mathematics into a strongly deductive science, with propositions rigorously derived from definitions and postulates that are clearly stipulated at the outset. While this is certainly the case for Euclid's geometry, it is much less so for his arithmetic. Not because the results are not presented in a strictly deductive way, but because they are not derived from basic postulates that are taken to be self-evident. Indeed, a definite system of such postulates for arithmetic appeared only in the last third of the nineteenth century. This is a fundamental point that should always be kept in mind when examining the historical development of the concept of number.

4.1 The arithmetic books of the *Elements*

Book VII of the *Elements* begins only with definitions but not with postulates. Like the Pythagoreans, Euclid defined a number as "a collection of units." A divisor of a number is another number that "measures" it; a prime number is defined as a number that is "measured" only by the unit, and so on. Euclid's inquiry into the properties of numbers begins from these definitions. Among other things, Euclid developed a theory of proportions for numbers that draws directly on the Pythagorean tradition and that is based on the very simple definition of ratios between numbers that I mentioned in Chapter 3. Symbolically, this can be expressed as follows: four numbers a, b, c, d are proportional if a measures b in the same way that c measures d. Thus, for instance, 4:6 :: 6:9, because a certain number (in this case 2) measures 4 twice and 6 three times, whereas another number (in this case 3) measures 6 twice and 9 three times.

The discovery of incommensurable magnitudes implied, as we saw, that such a definition cannot work for magnitudes in general, and this is what led to a more sophisticated definition of proportion, such as that provided by Eudoxus' theory. But in the case of numbers, the Pythagorean definition indeed suffices, and Euclid preferred to use this simpler definition in his arithmetical books. Still, also in this case, a "ratio" of two natural numbers is not in itself a number, and it continues to be a different kind of a mathematical entity. Euclid never added, subtracted, or multiplied ratios, even when these were ratios of numbers. Interestingly, the status of the unit continues to be slightly ambivalent with Euclid, as it was with the Pythagoreans. "One" is the unit, and therefore it is not a collection of units, and hence not a number according to the definition. On some occasions, however, "one" is treated as a number whenever this helps in stating a general proposition or in proving them. An overview of some of the arithmetical propositions of Euclid will be of great help in understanding the concept of number as it was seen in the classical Greek mathematical culture.

Among the arithmetical propositions proved in Euclid's books, one of the most famous is IX.20 which states that there are infinitely many prime numbers. Since the concept of "infinity" typically remained beyond the scope of legitimate mathematical practice in the Greek mathematical culture, Euclid worded this result in the following way: "prime numbers are more than any assigned multitude of prime numbers." Euclid's proof of this important result is both famous and simple, and it is worth presenting here, since it points to another interesting aspect of his treatment of number. It goes as follows:

Let a, b, c be the assigned prime numbers. Take the number d to be the product of the three assigned primes, and consider the number e, which is equal to d plus the unit (notice here that this is one of those cases in which the unit is used as a number). Now, upon measuring e by any of the assigned primes $a, b,$ or c, we will always obtain a remainder of 1. Hence none of the three assigned prime numbers a, b, c measures e. And since each given natural number is either itself prime or may be measured by a prime number (and this is a result previously proved by Euclid), e must be prime (since it is not measured by any prime), and hence the collection of three primes a, b, c does not exhaust the entire collection of prime numbers as initially assumed. So, the prime numbers are more than *this* collection, whose existence we assumed in advance, and hence we can conclude that prime numbers are more than *any* assigned multitude of prime numbers.

Euclid's proof—consistently (and rightly) acclaimed as exemplary in the history of mathematics—is perfectly acceptable from the point of view of today's standards, except for one detail of style, which today we would possibly write in a different way. What I mean is the statement that the collection of prime numbers a, b, c legitimately represents the idea of "any assigned collection." Today we would prefer to refer to "any assigned collection of prime numbers" as $a_1, a_2, a_3, \ldots, a_n$. In this way, we are able to symbolically represent the idea of a completely general collection of natural numbers with the help of the combined use of a general index n and the three dots that clearly indicate an indefinite amount of numbers in the collection. Euclid indicated that he meant "any collection," but in fact he took a rather specific one, namely, one comprising three numbers.

Logically speaking, doubt might arise as to the general validity of Euclid's argument. Perhaps, one might insist, the proof would not work with any arbitrary collection, because the inner logic of the proof presented by Euclid may contain a step that depends on the fact that the collection has exactly three numbers, and no more than that. Of course, this is not really the case with this specific proof of Euclid, but I stress this seemingly trivial point in order to emphasize once again just how much the existence of the right notation and symbolism may become crucial to our ability not only to express but sometimes even to think correctly about the mathematical situation involved in a particular question. The numbers involved in Euclid's proofs have proper names a, b, c that help us referring to them, but notice that Euclid did not operate on these letters (with operations such as addition, multiplication, or division) in the same way we can with our modern algebraic notation for the numbers $a_1, a_2, a_3, \ldots a_n$. Moreover, in Euclid's text, the number d is not defined as the "product" of the three, but rather as "the lowest number among those which are simultaneously measured by these three numbers" (we call this the "lowest common multiple"). Euclid's letters, then, are only *labels* rather than algebraic symbols that allow for formal manipulation. This is a crucial difference that will be overcome only as a consequence of a long and complex process described throughout the various chapters of this book.

Another interesting proposition appearing in the arithmetical books of Euclid is VII.30. It states that if a prime number p divides a product of two numbers $a \cdot b$, then either p divides a or p divides b. For example, the prime number 7 divides 42, which is a product $42 = 2 \cdot 21$, and hence it must divide either 2 or 21 (which indeed it does: $21 = 3 \cdot 7$). But 42 can also be written as $42 = 14 \cdot 3$, and again 7 must divide one of the factors, in this case 14. With time, this general property became so intimately identified with the concept of prime number itself that it was sometimes perceived as a possible alternative definition. However, as the concept of number continued to evolve, new abstract and more general systems were defined in the nineteenth century where the two meanings of a prime number (namely, on the one hand its being divisible only by itself, and on the other hand the property embodied in Proposition VII.30) were no longer coextensive. It was only then, as we will see, that these two properties were clearly separated (see Section 10.2).

4.2 Geometric algebra?

The issue of notation and its intimate connection with the idea of number are crucial throughout this book. They are also closely connected with the rise and development of the general idea of an algebraic equation. The full idea of an equation is absent from the Greek mathematical culture, but even in the very initial stages of this culture, it is possible to identify some significant seeds of the processes that would lead, around the seventeenth century, to its definite consolidation. These seeds are not found in the arithmetical books of Euclid, but rather in the important tradition of geometric problem-solving. This tradition comprised the development of construction methods for finding a magnitude (length, area, or volume) that would satisfy certain properties defined in advance. Our idea of an equation implies the task of finding certain numbers that satisfy a given condition, and the way from the equation to its solution passes through the formal manipulation of symbols and numbers according to the rules of algebra. Greek construction problems involved the search for a magnitude, and not a number. The tools that were at their disposal were not those of symbolic manipulation, but rather those of Greek geometry (Euclid-style or more advanced) and of Eudoxus' theory of proportions. Let us examine a concrete example of this, namely the one related to the "golden section."

As explained in relation with Figure 3.4, the golden section of the segment AB is given at point P, defined by the proportion $AB{:}AP :: AP{:}PB$. The point P can be found via the following construction problem, solved in Book II of the *Elements* (Figure 4.1):

Given a line AB, to cut it at point P, so that the square built on the segment AP will equal in area a rectangle whose sides are the given segment AB and the segment PB, which is the remainder left after taking off the section.

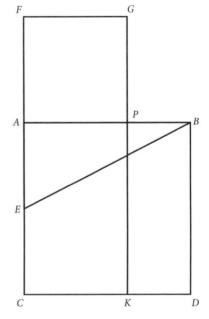

Figure 4.1 Euclid's *Elements*, Proposition II.11: Given the segment AB, construct a square $ABCD$. Bisect AC at E, and extend AC as far as F, so that EF equals EB. Complete the square $AFGP$ and extend GP as far as K. A previous result of Book II, II.6, implies that rectangle $CFGK$ equals square $ABDC$, from whence it follows easily that square $AFGP$ equals rectangle $PBDK$. P, of course, is the golden section of AB.

Faced with this kind of problem, one may be tempted to translate the geometric situation into algebraic symbolism. Indeed, if we write in the diagram $AB = a$, $AP = x$, then we may easily translate the requested condition "the area of $AFGP$ equals the area of $PBDK$" (or its equivalent, $AB:AP :: AP:PB$), as follows:

$$\frac{a}{x} = \frac{x}{(a-x)} \quad \text{or, alternatively,} \quad a \cdot (a - x) = x^2.$$

If we go along with the same kind of translation, the conclusion follows easily, that the length of the segment we found, AP, is actually a solution to a quadratic equation. Furthermore, since this construction is necessary for constructing the Pythagorean pentagram, we could move from the level of mathematical equivalence to that of historical plausibility and we might suggest, based on the algebraic translation above, that the Pythagoreans had a full solution to the quadratic equation.

I have already commented briefly on this kind of historiographical approach in Appendix 3.1, in the context of the discussion about the possible Greek proof of the incommensurability of $\sqrt{2}$. I say more about it in Appendix 8.2. The approach is commonly called by historians "geometric algebra." It is meant to help us make sense of what the Greeks were trying to do, by assuming that they had the right algebraic ideas, but lacked the symbolic language and hence formulated their results in geometric garb. A first serious problem with this approach is the separation it makes between the "ideas" and the way to write them. A second is that it is unavoidably anachronistic: in order to perform the translation into algebra, and to claim that this is what the Greek mathematicians really had in mind, we need to assume that the they possessed all those concepts and ideas related to numbers, equations, symbolic manipulation, etc. But these are precisely the kinds of concepts and ideas whose historical elaboration we are actually trying to understand here as a long, slow, convoluted and non-linear process. If, on the contrary, we continue to look at the construction of the golden ratio as a *purely geometric* question, with no algebraic underpinnings, then we may better understand the Greek views, with all their potential and limitations, as they themselves understood it.

4.3 Straightedge and compass

The construction of a golden section is a foremost example of "straightedge and compass" constructions. The Greeks made considerable efforts to find, wherever possible, solutions of this kind for their geometric problems of construction. At the same time, however, they did not spare any energy or ingenuity in trying to find alternative ways of solution, where such straightedge and compass constructions seemed not to be at hand. Let us examine an interesting example that illustrates what these alternative ways could look like. This example also illustrates additional interesting aspects of the Greek use of numbers.

Among all geometric construction problems that occupied the attention of the Greeks, three are especially prominent: (1) to divide a given angle into three equal parts, (2) to construct a cube whose volume is double than that of a given cube, and

(3) to construct a square with area equal to that of a given circle. These three problems became very famous, in the first place, for the simple reason that the Greeks did not succeed in solving them by means of straightedge and compass. And for a good reason! It was only as late as the nineteenth century that it was proved in precise mathematical terms that in fact it is *not possible* to solve these problems using methods of straightedge and compass. I have already mentioned thus, in Section 3.6, relation to Archimedes' *On the Measurement of the Circle*. The first two of these problems can be solved with the help of more powerful methods, such as the use of conic sections, as we will see now. The third, which is the famous "squaring of the circle" problem, cannot be solved even with the help of such methods. It turned out, after more than two thousand years of mathematical research, that the reason for this impossibility is the fact that π is a transcendental number. Again, a truly interesting mathematical topic that I cannot explain in detail here, and an extremely forceful example of how convoluted and surprising can be the history of mathematical ideas.[1]

In the case of the duplication of the cube, the first step toward a solution came quite early, in the hands of Hippocrates of Khios (who lived from about 470 to about 410 BCE, and is not to be confused with Hippocrates of Kos, whose name we associate with the physician's oath). Hippocrates found out that this problem is equivalent to another one, seemingly much simpler, namely the problem of constructing two "geometric means." Recall that if we are given two lengths a and b, their geometric mean is an intermediate length x such that $a:x :: x:b$. In Hippocrates' problem, for the same two lengths a and b, we are asked to find *two* intermediate lengths x and y such that the following proportions are satisfied:

$$a:x \ :: \ x:y \ :: \ y:b. \tag{4.1}$$

Think for a minute about this situation in modern algebraic terms: if we have a cube of volume 1 (i.e., one whose sides are all of length 1), and if we want to construct a cube of volume equal to 2, then what we need is to construct a side with length $\sqrt[3]{2}$. And indeed, if in the proportion (4.1) we set $a = 1$ and $b = 2$, we obtain $x = \sqrt[3]{2}$, as required.

Hippocrates did not have these algebraic tools at his disposal, and so his way into the solution was far more complicated. But his insight turned out to be of seminal importance, since it transformed a problem of finding a *solid body* according to a given condition (a cube that is the double of a given cube) into a problem of finding a *line* according to another given condition (double geometric mean). The solution to the latter seems much more reachable, and several Greek mathematicians undertook to solve the classical duplication problem by finding the said two geometric means x and y.

One particularly interesting suggestion was that of Menaechmus (ca. 380–ca. 320 BCE), who was one of Eudoxus' students. Menaechmus showed that the two values x and y could be found with the help of conic sections, namely, by intersecting a parabola and a hyperbola. Once again, the point of view of modern analytic geometry helps us understand why the solutions he proposed lead to a determination of

---

[1] See (Berggren et al (eds.) 2004, pp. 194–206, 226–230).

the two desired lengths. Indeed, if we take the proportions in (4.1) and write them as modern equations, we obtain the following system:

$$a/x = x/y \quad \text{or} \quad x^2 = ay,$$
$$x/y = y/2a \quad \text{or} \quad y^2 = 2ax,$$
$$a/x = y/2a \quad \text{or} \quad 2a^2 = xy.$$

Thinking in terms of analytic geometry, solving these equations is tantamount to finding the intersections of the geometric figures that the equations represent, namely, two parabolas and one hyperbola, respectively. To make things simpler, we may assume to be dealing with the case $a = 1$. The system of three equations we need to solve then is this:

$$x^2 = y, \quad y^2 = 2x, \quad 2 = xy.$$

From the first two of these equations, for example, we obtain a solution in the intersection of two parabolas, as in Figure 4.2. When the system is solved algebraically, we obtain as the point of intersection $x = \sqrt[3]{2}, y = \sqrt[3]{4}$. And, indeed, the volume of a cube with a side of $\sqrt[3]{2}$ is 2, as requested in the original duplication problem.

Menaechmus and his fellow Greek mathematicians could not count on these tools of analytic geometry, of course. All of their constructions and arguments supporting their proofs were purely geometric, and they used no symbolic algebra as I have done here. To be sure, even their definitions of the parabola and hyperbola were given in terms of geometric properties defined with the help of proportions. Equally geometric and not algebraic were the proofs about the properties of these sections.

Because of considerations of space, however, I must refrain from giving here the details of Menaechmus' purely geometric proofs.[2] It suffices for our purposes to

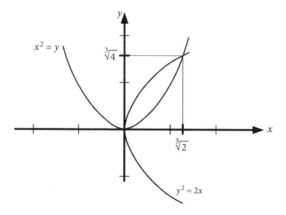

Figure 4.2 The solution to the problem of doubling the cube by intersecting conic sections.

[2] See (Fauvel and Gray (eds.) 1988, pp. 82–85).

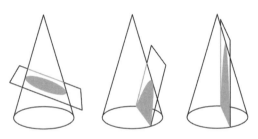

Figure 4.3 Conic sections: ellipse, parabola and hyperbola.

emphasize that his beautiful and surprising solution obviously departed from the requirement of construction with straightedge and compass. Indeed, "conic sections," as their name suggests, arise when a cone is intersected with planes in different angles (Figure 4.3), and cannot be produced with straightedge and compass on a plane. So, it is important to keep in mind that while straightedge and compass constructions were always *actively sought* by the Greeks in solving their geometric problems, they were by no means an *exclusive* standard of solution.

Only relatively late in history did these straightedge and compass methods received an exclusive, canonical status that was not intended when they were originally introduced. This change of status is quite interesting as part of our story here. In order to understand how it happened, we need to realize that the early Greek mathematical tradition passed on to later generations not only through the original works of Hippocrates, Euclid, Archimedes and the like, but also through rewritings, summaries and commentaries written in late antiquity. In other words, some of the most important texts that conveyed to later generations the basic ideas of classical Greek mathematics were written between three and six hundred years after the original works where these ideas had appeared. Authors active in this later historical period contributed many new and interesting mathematical ideas of their own and at the same time reinterpreted the works of their predecessors, while looking at them from the perspective afforded by their current knowledge. In the case of straightedge and compass constructions, it was the work of Pappus of Alexandria, who lived in the fourth century CE, that played an important role of reinterpretation.

Pappus' best-known work, the *Synagogue* or *Mathematical Collection*, comprises eight volumes that offer a broad summary of many topics of Greek mathematics. It is remarkable that the bulk of it has survived, and the earliest extant transcription, dating from the tenth century, is preserved in the Vatican Library. This compendium was highly influential in the Renaissance and later on, and many normative views about what is good mathematics and how it should be further practiced took their inspiration from this book. A passage of Pappus' text, which was frequently quoted, discusses various ways followed by "the ancients" (as he called them) in solving construction problems, and it was here that the preferred status of straightedge and compass methods was upgraded to mandatory.

Pappus introduced a distinction between three different kinds of solution. First are the "plane methods," based on the use of straightedge and compass. Second, there are "solid methods," which involve the use of conic sections ("solid" because they originate in a solid body, the cone). Finally, there are the "linear methods," involving "other" kinds of lines, such as spirals. Pappus went on to assert that "the ancients" strictly forbade a solution of a plane problem using solid methods. This assertion continued to be quoted from the Renaissance on, and it helped establishing the opinion that the Greeks did not seriously consider constructions beyond those based on straightedge and compass. But this opinion is historically inaccurate. Solutions to geometric construction problems of all three kinds mentioned by Pappus were actively pursued before him, and continued to appear and develop after him. These constructions came to play an important role leading to a full development of the idea of an algebraic equation and, because of that, to the development of the modern concept of number.

4.4 Diophantus' numerical problems

No name is more closely associated in the history of mathematics with the rise and development of the idea of equation than that of Diophantus of Alexandria (ca. 201–ca. 285). He lived in the third century CE, and he is no doubt one of the most interesting figures of late antiquity mathematics. In his famous book *Arithmetica*, he presented a large number of original techniques for solving problems in which the value of one or more unknown numbers is sought, given that they satisfy certain relations. These techniques were couched in an original and ingenious symbolic language of "abbreviated designations." Diophantus mobilized the full power of this language on behalf of the solutions he devised, by formulating them in terms of "equation-like" expressions such as the following:

$$\Delta^Y \bar{\delta} \overset{\circ}{M} \bar{\iota\varsigma} \wedge \zeta \bar{\iota\varsigma} \ \iota\sigma \ \overset{\circ}{M} \bar{\iota\varsigma} \wedge \Delta^Y \bar{\alpha}.$$

Retrospectively read from the vantage point of modern algebraic language, this equation-like expression can be interpreted in the following terms:

$$4x^2 - 16x + 16 = 16 - x^2.$$

It is for this reason that Diophantus is often called "the father of algebra," and it indeed makes historical sense to honor him with this respectable title. We will see below that he shares the honor with others, but this kind of shared fatherhood is not uncommon in the history of mathematics, especially when it comes to a term as ambiguous as "algebra." At the level of historical merits, moreover, readers surely know that nowadays, when we are given an equation and we are asked to find only *rational* numbers that satisfy it, then we say that this is a "Diophantine equation." Also, this makes a lot of historical sense, given that Diophantus' solutions were all, indeed, rational (actually only *positive* rational, as we will see). Of course, it was not the case that Diophantus could find all kinds of solutions to his problems, but decided, for whatever reasons, to limit himself to the positive, rational ones. It was, in fact, the other way round: Diophantus' concept of number was such that positive rational solutions were the only ones that he thought of. In this and in the next section, I discuss Diophantus' ideas about numbers and the way he used them in his problems. The details may become intricate for some readers, who will want to skip this section. But I think that diving into these details will prove worth the effort. Diophantus' work is of great inherent mathematical interest and originality. In addition, his ideas were greatly influential after they were rediscovered in fifteenth-century Europe.

The problem that opens the *Arithmetica* is a typical example of what we find in this book: "to write a number as a sum of two numbers with a given difference." The problem, as we see, was formulated in very general terms, but the first step typically followed by Diophantus in solving this, and indeed any problem, was to focus on a specific instance. He chose the number 100 in this case as the one that we are requested to write as the sum of two other numbers. In addition, he also chose a specifically given difference, 40. Only then did he proceed to work out the solution. Thinking about this problem in terms of current algebraic ideas, the solution would involve calling the smaller of the two numbers x and the larger number $x + 40$. We would then translate the condition into a symbolically written equation, $x + x + 40 = 100$, and from here we would deduce, by means of formal manipulation, that $x = 30$. Some elements of such an approach are also present, at least incipiently, in Diophantus' own way to the solution. At the same time, however, there are also noteworthy differences. Let us see some additional details in order to understand these differences.

The difficulties in understanding Diophantus' work are not only technical mathematical, but also methodological historiographic. The manuscripts that have reached us, and through which we know the *Arithmetica*, were written much later than the original Diophantine texts. This naturally leads to difficulties related to language and to style of manuscript preparation (a general methodological difficulty already mentioned in Chapter 3). Furthermore, of the 13 books that comprise the *Arithmetica* only 6 are extant in Greek, and until quite recently all the rest were believed to be lost. A manuscript discovered in 1968 in Meshed, Iran, however, contains an Arabic translation, usually attributed to Quṣṭā ibn Lūqā (820–912), of books IV–VII. The belated availability of these texts led historians to reconsider many long-held beliefs about Diophantus and his work. These historians had to face the limitations implied by the fact that they were engaging with this part of the *Arithmetica* only through the mediation of a translation produced in a much later, and completely different, mathematical culture. So, when speaking of Diophantus' ideas, we must continually keep these difficulties in mind, but we may nevertheless try to say something of historical value about them.

Let us focus on a well-known example of the kind of problems addressed by Diophantus, Problem 8 of Book II: "to divide a given square number into two square numbers." Before commenting on Diophantus' solution, it is worth mentioning that this problem may sound familiar to many readers. Indeed, around 1630, Pierre de Fermat (1601–1665) was immersed in studying Diophantus' *Arithmetica*, and upon arriving at this problem, he famously wrote a few lines in the margins of his copy of the book:

It is impossible for a cube to be written as a sum of cubes or a fourth power to be written as the sum of two fourth powers or, in general, for any number which is a power greater than the second to be written as a sum of two like powers. I have a truly marvelous demonstration of this proposition which this margin is too narrow to contain.

Little could Fermat imagine the enormous attention that these lines would attract in the centuries to come. In 1670, his son Samuel published an annotated edition of Diophantus' *Arithmetica* containing many remarks, conjectures and open problems, similar to the one just quoted. Fermat had added them while reading the Greek master. In the following decades, his comments received intense attention and soon led to important results and generalizations. But attempts to find the putative proof announced by Fermat in the specific marginal remark on Problem II.8 were unsuccessful for more than 350 years. The impossibility statement thus became known as "Fermat's Last Theorem," and it was fully proved only in 1994 by Andrew Wiles.

How did Diophantus go about solving this problem? In order to try and understand his approach, it may be helpful to begin by presenting his solution in modern algebraic terms. This would read roughly as follows:[3]

The first square is x^2 and the second is $16 - x^2$, and necessarily $16 - x^2$ is a square.

I produce a square from whatever number minus as many units as there are in the square of 16. This is $2x - 4$ whose square is $4x^2 - 16x + 16$, which will equal $16 - x^2$. Add now the negative factor to both sides and subtract equals from equals. We obtain $5x^2 = 16x$ and from here $x = 16/5$. Therefore, one of the numbers is 256/25, the other is 144/25 and the sum is 400/25 or 16, and all the numbers are squares.

In these terms, the solution looks more or less straightforward. Notice, however, that the *general problem* appearing in the original formulation was transformed here into *a specific instance* of it by (1) stating that the square number to consider is 16 and (2) indicating that one of the squares into which 16 is divided is of the form $2x - 4$.

Now, in order to take a closer look at the way that Diophantus himself conceived of this problem, we need first to make acquaintance with some of his basic terms as well as with the special symbols that he introduced. Let us begin with the particular instance into which the general problem is reduced in the first step. Rather than the symbolic formulation in terms of x^2, the following is much closer to Diophantus' original:

Let it be proposed to divide the square number 16 into two square numbers.

The "square numbers" that Diophantus has in mind here are not just the squares of some numbers, but "square numbers." This is not exactly the same thing, and we already know this. In the Pythagorean tradition of figurate numbers discussed in Chapter 3 (see Figure 3.2), we found "square numbers" alongside "triangular numbers" and "oblong numbers." Recall that the Pythagoreans investigated properties of figurate numbers, alongside other classes such as prime, even, or perfect numbers. In his *Arithmetica*, Diophantus was not proving general statements, but rather solving problems. But the basic idea of considering classes of numbers and relying on their general properties is not very different from that of the Pythagoreans. Alongside the "square numbers," which Diophantus calls τετραγώνοι ("*tetragônoi*"), he also

---

[3] Two classical editions of Diophantus' work are listed among the references for further reading at the end of this book. (Tannery 1893–95) comprises Greek and Latin versions of the text. (Ver Eecke 1959) comprises a French version. This quotation is taken from (Tannery 1893–95, pp. 91–93).

considered additional classes such as cube numbers, κύβοι ("kyboi"). In a different treatise, *On Polygonal Numbers*, of which only fragments are extant, he also discussed more general cases of figurate numbers in the Pythagorean tradition. So, a fundamental point in understanding Diophantus is that in his problems he is not just looking for a general, unknown number, but rather he addresses an arithmetical situation defined for numbers belonging to a specific class characterized by a property (such as "square numbers").

In stating and solving his problems, Diophantus typically followed the following steps:

1. Indicate the *entire class of numbers* on which the problem focuses (in this case, the class of "square numbers").
2. Indicate a task to be performed on any number of that class (in this case, to divide the number into two other square numbers).
3. Move into the more concrete realm of *specific numbers*, by choosing a particular case of a "square number" (in this case, the specific number is 16).
4. Make the problem even more specific by adding an additional condition not specified in the general formulation of the problem (in this case, one of the squares into which 16 is divided is of the form $2x - 4$).
5. With the help of "abbreviated designations," give "proper names," as it were, to the chosen number and to other specific numbers that arise as part of solving the problem.
6. Finally, having given these names, set up an equation-like expression that will lead to the answer.

The process leading from the equation-like expression to the final answer is also of interest, but before discussing it in some detail, some explanations are needed about the steps just described. Of particular interest are the "abbreviated designations" that constitute the core of Diophantus' highly praised symbolic language.

The most important of the "abbreviated designations" is, of course, that of the unknown number, ἀριθμός ("*arithmos*"). Diophantus denoted it by the special sign ς. In general, the unknown number appearing in the abbreviated expression is *not* the one mentioned in the general formulation of the problem. Rather, it is something that we would call today an "auxiliary variable." Other abbreviations include Δ^Y, used to denote the square of the unknown, δύναμις ("*dynamis*"), or $\overset{\circ}{M}$, used to denote the units, μονάς ("*monades*"). Using these abbreviations in the example of Problem II.8, Diophantus would write

Let us put the first square number to be $\Delta^Y \bar{\alpha}$.

In modern notation, we would identify Δ^Y, with the square number we are searching for, x^2, preceded by a coefficient 1, $\bar{\alpha}$. Accordingly, the second square number requested by the problem would be, in modern terms, $16 - x^2$. Diophantus gave a proper name also to this second "square number" with the help of yet another of his ingenious symbols, namely, ⋏. He wrote

The other square number will be $\overset{\circ}{M} \bar{\iota}\bar{\varsigma} \, ⋏ \, \Delta^Y \bar{\alpha}$.

Diophantus used standard Greek numerical notation to indicate that of the units, $\overset{\circ}{M}$, we have sixteen, $\bar{\iota}\bar{\varsigma}$, or that of the squares, Δ^Y we have one, $\bar{\alpha}$. It is interesting to notice that "one" appears here unambiguously and consistently as a number. I mentioned above the ambiguity involved in Euclid's arithmetic books, where the unit is sometimes not counted among the numbers properly speaking. This ambiguity cannot even arise in the context of a book devoted specifically to solving arithmetical problems, where the number 1 may be among the values of the numbers that we are looking after.

But now I turn to a subtle point, which is crucial for understanding Diophantus' concept of number. I used above the term "square" with *two* different meanings. A number appearing in the *general enunciation* of the problem as a "square number" is called a *tetragônon*. This refers to a class of numbers that we can identify as satisfying a *certain property*, as I have already explained. On the other hand, there is the square to which the expression $\Delta^Y \bar{\alpha}$ refers. This is a *dynamis*, and it refers to the square of a *specific* value (i.e., "the square of the unknown"). We have then *tetragônon* and *dynamis*, both of which we call "squares," but which are used differently.

Diophantus' expressions could become increasingly complex as he introduced, alongside the three basic species just mentioned, species representing higher powers as well as their reciprocals. Thus, $\kappa\acute{\upsilon}\beta o\varsigma$ ("kybos"), denoted by K^Y, was the term used in problems involving "cubic numbers." Other species and their corresponding symbols were as follows:

Square-square	*dynamodynamis*	$\Delta^Y \Delta$
Square-cube	*dynamocubos*	ΔK^Y
Cube-cube	*cubocubos*	$K^Y K$

In the fourth book of the *Arithmetica*, Diophantus also made use of species equivalent to x^8 and x^9 (but, for some reason, not x^7). This part of the treatise is extant only in the Arabic translation of Quṣṭā ibn Lūqā, but one can infer, speculatively, the signs Diophantus may have used in the original. What we do know from the Greek original is that Diophantus also defined the *reciprocals* of the various species in the introduction (and then used them in the treatise). He denoted these reciprocals by adding the letter χ as a superscript to each of the symbols. Thus, the reciprocal of the unknown, *arithmos*, was the $\alpha\rho\iota\theta\mu o\sigma\tau\acute{o}\nu$ ("*arithmoston*"), denoted by ς^χ. The reciprocal of the square of the unknown, *dynamis*, was the $\delta\upsilon\nu\alpha\mu o\sigma\tau\acute{o}\nu$ (or "*dynamoston*"), denoted by $\Delta^{Y\chi}$. And, in the same way, this worked with the higher species as well.

Diophantus also explained in great detail how to multiply each species by any other. The *dynamis* multiplied by the *kybos* yields the *dynamokybos* (ΔK^Y) (in modern terms, $x^2 \cdot x^3 = x^5$). The *arithmoston* multiplied by the *kybos*, in turn, yields the *dynamis* (K^Y) (in modern terms, $\frac{1}{x} \cdot x^3 = x^2$). But what looks so transparent in modern symbolic terms was much less so in Diophantus' abbreviated designations. In spite of these well-developed rules for operating with specific powers and their symbols, Diophantus did not formulate an abstract, general arithmetic of powers and had to specify each case separately. The reason, of course, is that his notation was not perspicuous enough to clearly bring to the fore such an underlying arithmetic.

This issue of symbolically representing powers of the unknown and formulating a full arithmetic of powers was to prevail well into the seventeenth century. Many systems of notation for powers were developed in various mathematical cultures, particularly in the Renaissance, and all of them had the same limitation. An important step leading to a more flexible and general conception of number appeared precisely when it finally became clear that the best way to indicate powers of a number or of an unknown is with the help of just the same sequence of numbers itself, 1, 2, 3 .... In retrospect, it is quite remarkable that this insight was so hard to reach. We will see more of this in the following chapters. At this point, I will just add that Diophantus was able to state quite explicitly, as part of his handling of symbolic expressions and their products, a kind of law of "sign multiplication" for negative expressions: "that which is missing multiplied by that which is missing yields that which is extant."

But how should we understand all this way of handling symbolic expressions of powers of the unknown, and their reciprocals? An algebraic expression such as "$4x^2 - 16x + 16$" is for us a polynomial whose various factors are connected via *operations* of addition and subtraction. We take it to symbolically stand for a single number, "$4x^2 - 16x + 16$." A symbolic expression of the kind used by Diophantus, on the contrary, is best seen as a *list* or, even better, as an *inventory* of various different "species" of quantities that have been gathered together to form a collection of objects, as it were. The expression $\Delta^Y \bar{\delta} \overset{\circ}{M} \bar{\iota\varsigma} \pitchfork \bar{\zeta\iota\varsigma}$ tells us that of the species *monades* we have sixteen ($\bar{\iota\varsigma}$) and of the species *dynamis* we have four ($\bar{\delta}$), while, at the same time, of the unknown itself, the *arithmos*, sixteen ($\bar{\iota\varsigma}$) are lacking in the collection.

I think it is useful to think here once again with the help of the insightful explanation of Jacob Klein (quoted in Section 3.3) about the nature of the *arithmoi* in Greek arithmetic: "However different they may be, [they] are taken as uniform when counted; they are, for example, either apples, or apples and pears which are counted as fruit, or apples, pears and plates which are counted as 'objects'." This way of seeing Diophantus' expressions and the various species appearing in them, also help us understand the role of the symbol $\pitchfork$. In modern terms, it appears to be a kind of "minus," but for Diophantus it is not intended as an operation of *subtracting* x^2 from 16. Rather, he takes the symbol to indicate, in abbreviated form, the square of the unknown as a term that is *wanting* rather than joined. It is like saying that we have a certain amount of money, say, "three dollars less a penny," or that the time is "ten minutes to six." The origin of the symbol $\pitchfork$, by the way, is not clear, but historians speculate that it may be an inverted allusion to λειψις ("leipsis"), which is the Greek word for omission or lack.

After this somewhat detailed explanation about Diophantus' symbols and their usage, it is a rewarding exercise to read again the core of the solution of II.8, now in his own terms. This reading helps us put many pieces together into a coherent picture. The text reads as follows:[4]

Let it be proposed to divide the *tetragônos* 16 into two *tetragônoi*.

And let us put the first *tetragônos* to be $\Delta^Y \bar{\alpha}$; then the other *tetragônos* will be $\overset{\circ}{M} \bar{\iota\varsigma} \pitchfork \Delta^Y \bar{\alpha}$. Therefore, $\overset{\circ}{M} \bar{\iota\varsigma} \pitchfork \Delta^Y \bar{\alpha}$ must be equal to a *tetragônos*. I form the *tetragônos* from any number of *arithmoi* lacking as many units as there are in the side of $\overset{\circ}{M} \bar{\iota\varsigma}$; be it $\varsigma \bar{\beta} \pitchfork \overset{\circ}{M} \bar{\delta}$. The *tetragônos* itself will be $\Delta^Y \bar{\delta} \overset{\circ}{M} \bar{\iota\varsigma}$. These are equal to $\overset{\circ}{M} \bar{\iota\varsigma} \pitchfork \Delta^Y \bar{\alpha}$.

[4] (Ver Eecke 1959, p. 54).

Now, this is precisely the point where the equation-like expression has been set up, by equating the two collections: $\Delta^Y \bar{\delta} \overset{\circ}{M} \bar{\iota} \bar{\varsigma}$ and $\overset{\circ}{M} \bar{\iota} \varsigma \wedge \Delta^Y \bar{\alpha}$ (or, in modern terms, $4x^2 - 16x + 16 = 16 - x^2$). (Occasionally Diophantus indicated the act of equating with the help of yet another symbol $\iota\sigma$, which is an abbreviation of the Greek word for "equals," $\iota\sigma o\varsigma$ - "*isos*.") In all of Diophantus' problems, this is the climax of the solution procedure. Only after the equation-like expression has been clearly set can the next step be undertaken, and, in the case of Problem II.8, it is formulated as follows:

Add to both sides the missing terms and take like from like. Then 5 *dynamis* equal 16 *arithmoi*, and the *arithmoi* becomes 16/5.

In other words, Diophantus simplifies the expression by narrowing down the multiple appearances of same species: positive amounts of a given species are gathered together, while amounts of the same species that are wanting are subtracted, the lesser from the larger. The aim is to reach a simpler expression in which either a one-term or a two-term expression is equated with another one-term expression (e.g., "5 *dynamis* equal 16 *arithmoi*"). At this point, the solution is readily obtained. I will just indicate here that these two ways of simplifying ("to add the wanting species on both sides" or "to subtract like from like on both sides") are similar in their essence to the techniques that we will later see as fundamental to the approach developed in Islamic mathematical culture under the names of *al-jabr* ("restoring") and *al-muqābala* ("compensating"). This is explained in greater detail in Chapter 5.

In the end, Diophantus also summarizes, and checks that the solution fulfills the condition stated in the initial formulation (this latter step is not taken in all of the problems):

The one will therefore be $\frac{256}{25}$, the other $\frac{144}{25}$ and their sum is $\frac{400}{25}$ or 16, and each is a *tetragônos*.

Problem II.8 is but one example of what Diophantus does in the *Arithmetica*. The book displays the solutions to hundreds of numerical problems of increasing difficulty and complexity. Essentially, once the equation is set, each problem is solved by an ad hoc method especially suited to it. So, an important feature of Diophantus' approach is that he never developed *general algorithms*, but always specific solutions to given problems. Moreover, he neither looked for *all* the possible solutions nor discussed the possibility of existence of more than one solution to a given problem.

Diophantus' techniques were innovative, highly sophisticated and original, and they allowed him to address and successfully solve fairly complex problems. Tracing down his direct influence on the development of algebraic thought of later traditions, particularly those of Islamic and early Renaissance mathematics, has proved to be a challenging task for historians who continue to debate the question to this day. When it comes to algebraists in the sixteenth and seventeenth centuries, however, particularly the groundbreaking works of Viète (discussed in Chapter 7), the influence of Diophantus already appears as both indisputable and crucial.

We can at this point summarize the main traits of Diophantus' views on numbers. In the introduction to his treatise, which is actually a letter to a certain friend Dionysius, Diophantus defined numbers without deviating in any way from the classical definition found in Euclid's *Elements*: numbers are formed by a certain collection of units. These numbers, moreover, can be classified according to properties specified in advance the way the Pythagoreans did. In Diophantus' problems, we search for numbers of certain classes, such as square or cubic numbers, which in addition satisfy some conditions. In the solution, Diophantus plays around with various "species" of numbers that are part of his arithmetic theory and that he teaches how to manipulate. Diophantus also teaches how to manipulate the reciprocals of the species, and his manipulations bring to the fore interesting ideas about fractional numbers.

4.5 Diophantus' reciprocals and fractions

When discussing the reciprocals of species, Diophantus explicitly stressed the analogy with the way in which the sequence of reciprocals of numbers naturally arose in Greek mathematics from the idea of the sequence of numbers as collections of units. So, much as "the third corresponds to three, and the fourth to four"—Diophantus wrote—so the *dynamoston* corresponds to the *dynamis* and the *kyboston* to the *kybos*. In Section 3.5, I have already discussed the idea of fractional numbers in Greek arithmetic, while emphasizing the need to avoid the possible confusion derived from considering reciprocals of numbers as cases of "unit fractions." It is clear that a similar, cautious attitude must dominate the way in which we see Diophantus' understanding of the reciprocal of species. Diophantus' use of fractional numbers together with the fractional species is similar in some respects to that appearing in earlier texts of Greek mathematics in expressions such as "of the *m* the *n*th." At the same time, however, he did display a very elaborate command of the operations with such fractional quantities, and in this sense he went much further than any of his Greek predecessors.

A straightforward example of Diophantus' use of fractions appears in the solution to II.8, discussed above, and similar ones appear all along the treatise. Such examples appear, moreover, not only as part of the *solutions*, but also as intermediate results within the process leading to them. Often Diophantus' fractions are written with the denominator over the numerator, but in places where the numbers involved are larger, he writes them simply as "781543 units in part of 9699920." Diophantus explicitly indicated that these fractions are *numbers* (*arithmoi*), and, like numbers, some of them could be square or cubic. He did not develop, however, a full theory of how exactly to operate with these fractions, but such operations do appear, in various ways and non-systematically discussed, throughout the text.

Diophantus was very careful in using these fractions. It is noteworthy, for instance, that fractions never appear in the equation-like expression where Diophantus put to use his innovative symbolic abbreviations. There we find that the collections of species gathered together or those found wanting are always standard (i.e., integer) collections. Consequently, fractions do not appear either in the simplified expressions. In the example of II.8, after adding "to both sides the missing terms and taking like from like," Diophantus obtained the simplified equation "5 *dynamis* equal 16 *arithmoi*." It was only after this step that fractions began to appear in the way to solving the problem. Something similar is found in most other examples developed in the treatise.

Another important point to notice is that specific fractions are not used in *enunciations* of problems, also when fractions are implicit in the very idea underlying the problem. Take, for example, the case of Problem V.9, which is formulated as follows:

To divide the unit into two parts, such that if a certain number is added to each part, then the results are square numbers.

Since we are asked to divide the unit, it is evident that the numbers sought after are fractions, and fractions will indeed appear in the solution. But if we look at the specific instance that Diophantus chooses at the beginning of his solution, then we will see that he adds *an integer and not a fraction*. This is typical of his way to solve problems of this kind. The fractions will appear only once we have started with the detailed solution. In this example, he follows these three steps:

1 Diophantus writes: "Let it be proposed to add 6 *units* to each part and thus to form squares." In other words, he divides the unit into two, and adds two *integers*.
2 Adding the two parts together with twice the 6 units, we then obtain 13 units. In this way, we remain within the realm of integers.
3 These 13 units are now seen as a sum of two square numbers, each of which is larger than 6. Obviously, these two square numbers cannot be integers.

The details of Diophantus' solution to this problem are illuminating as to where and how fractional numbers and reciprocals do or do not appear in his text. They may be of interest for some readers, and so I have presented them in Appendix 4.1.

The bottom line is that Diophantus' approach to handling numbers of various kinds is unsystematic and ad hoc. Like most of the mathematicians discussed in this book, Diophantus was not a philosopher interested in a *systematic discussion* of the nature of numbers. Rather, he was trying to *use* them in a clear and coherent but, above all, *mathematically fruitful* manner. The introduction to the treatise was a didactic overview intended for motivating his friend to read it, rather than an attempt to provide a formal foundation for arithmetic at large. Much as he presented a mixed collection of problems variously solved with ad hoc techniques of increasing sophistication and difficulty, so the kinds of numbers and species he used in each case and the ways to operate on them were introduced as the need arose.

Still, two clear limitations to the kinds of numbers used in the text are easily discernible. They concern negative and irrational numbers. When translated into modern algebraic symbolism, some of his procedures could be further elaborated to yield irrational solutions. When stated in Diophantus' own terms, however, there is no place where such numbers could really arise. As for negative numbers, nothing of the like appears either as possible *solutions* to the problems or in the route to them. One might perhaps be tempted to claim that Diophantus did use negative "coefficients" in the *formulation* of problems, when he introduced the symbol ⋏ as a way of separating them from the positive ones in his symbolic rendering of the word problems. But, as already stated, this is a misreading of Diophantus' own views.

It is precisely because of the sophistication of Diophantus' methods that we can realize the limitations still appearing in them in the journey to a more general and flexible conception of numbers. The decisive idea that symbols can be used not just for *denoting* numbers in an efficient manner, but also for *operating* on them freely *as if* they were actually numbers is still absent in his works, and it will continue to emerge only gradually, as we will see. Moreover, the fact that Diophantus consistently used one and the same symbol for representing the unknown quantity was of course a main innovation of his work, but it was not favorable to the introduction of the broader idea of letters standing for general coefficients or for more than one unknown number that could be represented by different symbols. Another measure of the limitations implied in his symbolic approach is manifest in his description of the operations with the various species of numbers and their reciprocals already mentioned above. It had to remain at the rhetorical level and could not provide a general, number-like rule holding for all multiplications of one species by another.

But the issue of the "dimensions" of the species is also interesting from yet another perspective. Diophantus was willing to consider, without restrictions, quantities that represent more than three dimensions, such as square-square or square-cube. It is important to remember that in Greek mathematics, dealing with continuous magnitudes is a part of geometry, in the most direct and intuitive meaning of the concept. For that reason, entities of more than three dimensions present an inherent conceptual difficulty, and, properly speaking, they have no natural place in Greek mathematical thought. For example, the multiplication of a square by a square can have no geometric counterpart and hence is devoid of meaning. In contrast, when working with numbers, that is, with discrete magnitudes, there seems to be no apparent limitation to repeated multiplication of quantities. It seems possible to multiply a number by a different number and obtain yet another number that can be further multiplied by another one, and so on indefinitely.

And yet, if we look attentively at what the Greeks actually did in connection with these operations with numbers, we come to realize that in many cases they appeared under the same kind of constraints that typically apply to geometric situations. Indeed, this general attitude manifested itself in Diophantus' case, when it came to choosing names for the different species that appear in his symbolic abbreviations. Not only did he use "square," for instance, to denote the multiplication of a number by itself, with the clear geometric connotation behind it, but he also stated that in such a case, "the number itself is called the side of a square." Diophantus' cube, which he denoted by K^Y, is obtained when a square is multiplied by its side. But this geometric analogy did not prevent him from going on and saying that the cube-cube, indicated as $K^Y K$, is obtained when "cubes are multiplied by themselves." Apparently, in the arithmetic realm, there should be no inherent limitations to this kind of higher powers of a number. Still, by affinity with the geometric realm, we do not find them very often in the Greek mathematical discourse in general. When we find them appearing in additional contexts, it is mostly in a hesitant manner, always with qualifications and restrictions.

4.6 More than three dimensions

Deviations from the limitations imposed by the three-dimensional framework appear in a different, and more straightforward manner in the work of Heron of Alexandria, who lived before Diophantus, in the first century CE. Heron is a unique figure in many respects in the history of Greek mathematics. He was involved in a considerable amount of practical calculations and in the design and construction of mechanical devices of various sorts. He developed methods for calculating with fractional numbers and used them in various contexts. One of his well-known geometric results comprises a very practical and original way for calculating the area of a triangle, based on the lengths of the sides alone, without involving the height. Indeed, if we write the sides of a triangle as a, b, c and if we set $s = \frac{1}{2}(a + b + c)$, Heron's value for the triangle's area T is given by

$$T = \sqrt{s\cdot(s-a)\cdot(s-b)\cdot(s-c)}.$$

The product inside the root, as can be seen here, is a product of *four* magnitudes, indeed of *four geometric* magnitudes. Of course, Heron did not write his formula in this abstract algebraic way, using letters as here, but he did give an example meant to illustrate a set of instructions to be followed in similar cases.

When he presented his method of calculation, he added no word of explanation, warning, or doubt concerning this rather unusual multiplication of four lines. Moreover, according to his example, where the sides are the lengths 7, 8 and 9, the number whose square root needs to be calculated is 720. This root is not an exact integer, and Heron showed how to calculate its estimated value, in a way that actually embodies a general algorithm for calculating a square root. It is customary to think that Heron's method, and the accurate geometric proof of its validity, were already known to Archimedes. Consequently, to this earlier mathematician may also be attributed deviations from the accepted canon. Perhaps he himself allowed products that represent some kind of four-dimensional entity. The truth is that Archimedes, like Heron, deviated from these norms in many places. In general, Archimedes tended to make detailed calculations more often and more intensely than most other Greek mathematicians, and he did not refrain from using special methods, even if this implied disregarding accepted norms. Indeed, in Section 3.6, I have already mentioned the important case of Archimedes' calculation of an estimated value for π. In the example of Heron, we see an ingenious method for a useful calculation, but then the apparent deviation is soon mitigated in the sense that the alien entity embodied in a product of the four magnitudes does not appear "autonomously" as it were, but rather as an intermediate step in a calculation, which is immediately followed by square-rooting yielding a standard, two-dimensional area.

There is evidence in some important Greek mathematical texts that deviations of this kind, involving magnitudes of more than three dimensions, caused unease among mathematicians. An enlightening example of this, with which I want to conclude this chapter, appears in the work of Pappus, specifically in a problem formulated in his *Collection*. The problem involves an instance of the more general idea of finding a "geometric locus", namely, a set of points defined by a specific property. Such problems were important in the Greek mathematical tradition, and it was common that the locus in question would be defined in terms of the distances of the points from other objects in the plane. An immediate example of such loci is the circle, which is defined as a line every point of which has the same distance to a given point, namely its center. Pappus' problem deals with the geometric locus of four straight lines, and it can be understood

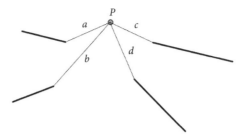

Figure 4.4 Pappus' problem on "the geometric locus of four straight lines." Given four straight lines and a point P, the segments a, b, c, d, connect the point P in given angles to the four straight lines given in place.

with reference to the diagram depicted in Figure 4.4 (This diagram does not appear in the original sources. I have added it here for ease of reference).

The problem is defined as the task of characterizing all points P that satisfy the following condition:

The rectangle constructed on the sides a, b and the rectangle constructed on the sides c, d should be in a predetermined fixed ratio. Symbolically expressed, $ab:cd :: m:n$.

Notice that it is possible to formulate the same problem for three magnitudes, a, b, c. In this case, the condition defining the locus becomes $ab:c^2 :: m:n$.

In the seventeenth century, Pappus' problem will attract Descartes' attention and his attempt to deal with it will become central to his formulation of analytic geometry. It turns out that the locus that solves this problem is always a conic section. But that will come later. Still staying with Pappus, the interesting point is that his problem can be easily generalized to the case of six straight lines. In this case, the points P defining the locus connect to two sets of *three* lines each. This is represented in Figure 4.5 (and of course, this diagram does not appear in the original text of Pappus). Rather than two rectangles constructed on two sides each, as above, we need to consider now two straight *prisms*, each of which is built on the three distances to the given straight lines. So, the problem is defined as the task of characterizing all points P that satisfy the following condition:

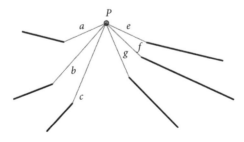

Figure 4.5 Pappus' problem on "the geometric locus of six straight lines." Given six straight lines and a point P, the segments a, b, c, d, e, f, connect the point P in given angles to the six straight lines given in place.

The prism constructed on the sides *a, b, c* and the prism constructed on the sides *e, f, g* should be in a predetermined fixed ratio. Symbolically expressed, *abc:def* :: *m:n*.

The actual solution of the problem came, as I have said, in the seventeenth century. Pappus was not able to solve even the simpler problem for four lines (which is quite hard, to tell the truth). But let us focus on some *conceptual* issues involved here. Both problems, with four or with six lines, do not give rise to any conceptual problem, because both the rectangle and the prism that define the locus are legitimate geometric figures. But Pappus' mathematical instincts led him one step further, and he suggested the possibility of exploring the situation involving *seven* or *eight* given straight lines. Despite its being just a natural continuation of the previous steps, this case gives rise to a new fundamental problem, since it demands reference to a *product of four segments*. But Pappus immediately warned that when asked to multiply a rectangle built on two segments of a straight line (two dimensions) by another rectangle built on two additional segments (also two dimensions), we are referring to something that is devoid of any geometric meaning. What do we do then? Pappus' suggestion was that instead of working with the ratio *abcd:efgh* (which is meaningless because it is a comparison between two products of *four* straight lines), we may introduce something that he called a "composite ratio." From the text, it is possible to guess that he had in mind something like

$$(a{:}e)\cdot(b{:}f)\cdot(c{:}g)\cdot(d{:}h).$$

Pappus did not explain what he meant by this kind of operation between ratios, yet, interestingly, while trying to avoid working with magnitudes that are a product of four segments, he was led to consider the ratios themselves as dimensionless magnitudes on which operations could be performed in a similar way as we do with numbers. The idea of ratios as numbers appeared here at least tentatively, but it was not further developed until much later. At any rate, Pappus clearly summarized a perception of number that was dominant in Greek mathematics and that was also passed over to Islamic mathematics and later on to that of the Renaissance in Europe. According to this perception, magnitudes that represent a geometric entity of more than three dimensions have no natural and legitimate place in mathematics. This was a major obstacle to be overcome on the way to the general idea of abstract numbers.

Appendix 4.1 Diophantus' solution of Problem V.9 in *Arithmetica*

Problem V.9 of Diophantus' *Arithmetica* asks us to separate the unit into two parts, such that if a certain number is added to each part, then the results are square numbers. In the instantiation stage, Diophantus proposes that 6 units be added to each part so that each is a square number (*tetragônos*). Thus, when adding the two parts together with twice the 6 units, we obtain 13 units. This number, 13, is then a sum of two square numbers, each of which is larger than 6, and whose difference is less than one unit. Let us see how Diophantus proceeds in order to solve the modified problem, namely, to find the two said square numbers whose sum is 13. My presentation here is quite detailed. It follows, more or less closely, the text as it appears in the 1921 French edition of Paul Ver Eecke (pp. 197–198), but sometimes I also give the abbreviated symbolism as appearing in the Greek–Latin edition of Paul Tannery, dating from 1893–1895 (pp. 334–336). I also give, in some places, a translation into modern algebraic symbolism.

Diophantus solves the problem by taking half of the units, $6\frac{1}{2}$, and looking for a fraction that when added to this yields a square number. Equivalently, we can multiply by 4, and look for a fraction such that when added to 26 yields a square number. Notice that in this step Diophantus is relying on a property of square numbers, of the kind that was well known in classical Greek mathematics, namely, that four times a square numbers is a (different) square number.

In the next step, Diophantus translates the problem into the search for an unknown *arithmos*, ς. The problem is thus restated using the reciprocal of the square of the unknown (*dynamoston*), Δ^{YX}. One dynamoston together with 26 units ($\overset{\circ}{M}\bar{\kappa}\bar{\varsigma}\Delta^{YX}\bar{\alpha}$ or, roughly, $26 + \frac{1}{x^2}$) should be a square number (*tetragônos*). From here, Diophantus deduces that 26 *dynamis* together with one *monades* ($\Delta^Y\bar{\kappa}\bar{\varsigma}\overset{\circ}{M}\bar{\alpha}$) is itself a square number.

Now, in modern algebraic terms, the passage from one expression to the next is always quite straightforward. Also, if $26 + \frac{1}{x^2} = a^2$ (with a integer), it is easy to see that if the unknown is also assumed to be an integer, then $26x^2 + 1 = x^2 a^2$ is also a square number. Diophantus does not give a detailed argument why $\Delta^Y\bar{\kappa}\bar{\varsigma}\overset{\circ}{M}\bar{\alpha}$ must itself be a square number, but we can more or less follow his train of thought along the lines of the operations with species as described in the introduction to the treatise.

In order to find the *arithmos*, ς, that satisfies the condition, Diophantus takes once again a particular instance. In this case, he chooses a specific kind of square number, namely, one whose side is of the form $5x + 1$. He concludes right away that in this case the value of the unknown is 10, and hence that the reciprocal of the square of the unknown is $\frac{1}{100}$ (ρ^Y). If we now add $\frac{1}{100}$ to 26, or if we add $\frac{1}{400}$ to $6\frac{1}{2}$ units, we then obtain the number $\frac{2601}{400}$, which is indeed a square number whose root is $\frac{51}{20}$.

Diophantus explains neither the choice of the square number whose side is $5x + 1$ nor the conclusion that it yields the value 10. But, in modern algebraic terms, we can come up with some ideas that speculatively point to the kind of considerations that Diophantus may have plausibly relied upon. Thus, in the first place, numbers of the form $mx + 1$ appear in many cases in the treatise in instantiations of problems. Of course, if we want to use one such number here, it is necessary to find a convenient value of m. Given that we are interested in the identity $26x^2 + 1 = (mx+1)^2$, or $26x^2 + 1 = m^2x^2 + 2x + 1$, we readily realize that $x = \frac{2m}{26-m^2}$. On the other hand, we have chosen $\frac{1}{x^2}$ to be a part of the unit, $\frac{1}{x^2} < 1$, and hence $26 - m^2 < 2m$, or $m^2 + 2m + 1 > 27, (m+1)^2 > 27$. It follows that the minimal value for m to yield a workable solution is $m = 5$. And, in this case, we obtain the following:

$$26x^2 + 1 = (5x + 1)^2 \quad \rightarrow \quad 26x^2 + 1 = 25x^2 + 10x + 1 \quad \rightarrow \quad x^2 = 10x.$$

Whether or not Diophantus worked out his way along these lines, the value of the *arithmos*, ς, has been found up to this point with the help of an equation-like expression and some quick manipulations. But this is just an intermediate result indicating that in order to separate the number 13 into two square numbers, the root of each of these numbers has to be close to $\frac{51}{20}$. Hence, Diophantus says, we need to consider the two numbers that when subtracted from 3 and added to 2, respectively, yield that number, i.e., $\frac{51}{20}$. These two numbers are $(3 - \frac{9}{20})$ and $(2 + \frac{11}{20})$, but Diophantus does not mention them. Rather, he goes on to complete the solution of the problem by formulating a new equation-like expression in which the unknown is again the *arithmos*, representing now a different unknown but still denoted by the same symbol, ς. He reasoned as follows:

Assume that the two square numbers we are searching for are (a) that of 11 *arithmoi* together with 2 units (ςιᾱ$\overset{\circ}{M}$β̄) and (b) that of 3 units wanting 9 *arithmoi* ($\overset{\circ}{M}$γ̄ ⋀ ςθ̄). The sum of their squares is 202 squares of the *arithmos* together with 13 units wanting 10 *arithmoi*, and this is equated with 13 units.

In Diophantus' symbolic language, this last "equation" is expressed as follows:

$$\Delta^Y \bar{\sigma}\bar{\beta}\overset{\circ}{M}\bar{\iota}\bar{\gamma} \;\wedge\; \varsigma\bar{\iota} \text{ equals } \overset{\circ}{M}\bar{\iota}\bar{\gamma}.$$

Without any further explanations or details, Diophantus concludes that the value of the *arithmos* is $\frac{5}{101}$. He goes on to state (again without any detailed calculation) that the roots of the fractions sought after are, respectively,

$$\frac{257}{101} \text{ and } \frac{258}{101} \quad \left[\text{i.e.,} \left(11 \times \frac{5}{101} + 2\right) \text{ and } \left(3 \times \frac{5}{101} - 9\right)\right].$$

Finally, the two parts into which the unit is separated are

$$\left(\frac{257}{101}\right)^2 - 6 = \frac{4843}{10201} \quad \text{and} \quad \left(\frac{258}{101}\right)^2 - 6 = \frac{2358}{10201}.$$

In this specific problem, Diophantus writes the fractions as a denominator over a numerator: $\frac{\alpha.\sigma\alpha}{\delta\omega\mu\gamma}$ and $\frac{\alpha.\sigma\alpha}{\epsilon\theta\nu\eta}$. He states that it is clear that each of the values when added to 6 are square numbers, but he does not check or mention that, indeed, $\frac{4843}{10201} + \frac{2358}{10201} = 1$.

What remains rather unclear is how Diophantus moved from the statement that the solutions is related to the numbers $(3 - \frac{9}{20})$ and $(2 + \frac{11}{20})$ to the choice of the squares of $(11x + 2)$ and $(3 - 9x)$ for establishing the equation-like expression. Again, using modern algebraic symbolism, we may speculate about some of the ideas underlying his reasoning. The starting point is that $2^2 + 3^2 = 13$ and that the numbers that we are seeking for are close to these values. The two specific numbers that Diophantus mentions, $(3 - \frac{9}{20})$ and $(2 + \frac{11}{20})$, cannot themselves be the sought-for numbers, since $(3 - \frac{9}{20})^2 + (2 + \frac{11}{20})^2 = 2 \times (\frac{51}{20})^2 > 13$. Hence, if we choose $(11x + 2)^2$ and $(3 - 9x)^2$ to be the sought-after squares, we will surely obtain a value for the unknown that is close to $\frac{1}{20}$. Finally, by setting $(11x + 2)^2 + (3 - 9x)^2 = 13$, we obtain, as in the text, $202x^2 - 10x + 13 = 13$, and hence $x = \frac{5}{101}$.

CHAPTER 5

Numbers in the Tradition of Medieval Islam

The mathematical culture of the lands of Islam spans a very long period of time that one can roughly date between the seventh and the late fifteenth centuries, and even beyond. Likewise, it covers an enormous geographical region. At times, it extended in the East as far as China, Indonesia and the Philippines, and in the West up to the Maghreb (Northwest Africa) and sub-Saharan Africa, and into Spain. It also comprised territories in what are today Turkey and France. Scholars working in these vast geographical areas were mostly Muslims, but at some periods there were also considerable numbers of Jews, Christians, Zoroastrians and members of other religious communities.

Very much as with ancient Greece, one should not imagine this culture as a monolithic and homogeneous body of knowledge. Within this broad cultural context and over such a long historical period, it goes without saying that mathematicians working at different places and at different times focused on diverging topics of interest and developed a wide variety of techniques, methodologies, symbolic languages and strands of thought. One must also be aware of the simultaneous existence of more learned and theoretically oriented traditions that were active alongside more popular or practically oriented ones. In addition, one must keep in mind that within this rich panorama of mathematical traditions, knowledge developed and was transmitted in written text, but also within oral environments.

Still, for short, I will refer to this great variety simply as "medieval Islamic mathematics" or "Islamicate mathematics" (a term commonly used nowadays to indicate regions in which Muslims are culturally dominant). In this chapter, I will try to present, in a somewhat schematic fashion, some of the main characteristic threads of the ways in which numbers were conceived and handled within this many-sided mathematical culture.

5.1 Islamicate science in historical perspective

Until quite recently, it was common in historiography to look at Islamic science in general, and mathematics in particular, as not much more than a pipeline for the transmission of Greek science from the ancient world to Renaissance Europe. The main contribution of Islamic culture, under this view, consisted in the translation of the great treasures of the Greek tradition, thus helping preserve and subsequently transmit them to Europe. And indeed, it is true that Islamicate scholars translated many such works and that, moreover, some of these are known to us nowadays only thanks to Arabic translations that remained where the Greek originals disappeared. But more recent historiography has modified this essentially limiting outlook on Islamicate science, and the latter is seen now within a broader, more complex and more interesting perspective. The many translations of Greek sources are only a part, albeit significant, of a rich and many-sided spectrum of intellectual activity that yielded original, meaningful contributions to the sciences and particularly to mathematics. As a matter of fact, historians continue to reinterpret the historical significance of the classic works of Islamicate mathematics, while at the same time uncovering and investigating new documents and manuscripts of the Islamicate mathematical culture, found in many places around the world, and containing previously unknown contributions to mathematics.

Islamicate mathematics developed at different research centers, often distant from each other. Ideas and techniques practiced in a Western corner of the Islamic world, say in the Maghreb, were frequently unknown in the East, and vice versa. As we will see, some of the important ideas developed in the medieval Islamicate world reached Latin Europe in the twelfth to the fifteenth centuries, and they played a crucial role in the mathematical revival that took place there. But, at the same time—and this has typically been less emphasized—a considerable corpus of original mathematical knowledge never came to the attention of Europeans involved in mathematical activities and thus fell into complete oblivion. This is particularly the case concerning the development of the concept of number, and more generally of algebraic and arithmetic knowledge. Much attention was actively devoted to this kind of knowledge in various regions of the Islamic world while individuals at various places in medieval Europe were only gradually starting to engage in more intense intellectual (and mathematical) activity.

As with Greek mathematics, the historian trying to reconstruct the basic ideas lying at the hearth of Islamicate mathematics must take into account some specificities of this culture, particularly concerning the ways in which texts were produced and transmitted. For one thing, this was, especially in its earlier stages, a predominantly oral culture revolving around the personal relationship between student and teacher. Recitation and memorization of texts were the main vehicle of this mathematical culture. In some cases, written books were ancillary to the latter, and were seen essentially as aids to memorization. Some texts, as a matter of fact, have come down to us in the form of poems or of other memory-friendly formats such as questions-and-answers. Moreover, some of the extant manuscripts were probably direct or indirect transcriptions of orally presented lessons, and they may date from much later than the time when the ideas presented in them developed and started to circulate more or less widely.

There are of course many issues related to changing terminologies and to nuances in the use of language that have attracted the attention of historians but that I will not be able to consider at all in our discussion here. At any rate, inasmuch as the texts known to us derive directly from oral traditions, or were transmitted orally after they had been written, we find that many are devoid of any kind of numerical notation. They were written in their entirety in words, including the names for the numbers. Actual calculations related to the texts were often worked out on a dust-board or on some other kind of ephemeral medium, so that we are not really aware of the notation that may have been used as part of them. There was also an interesting tradition of mental calculation and finger-reckoning known to have been in use in trade and commerce. In texts where numerals do appear, especially in the earlier periods, these typically are not as part of the running texts, but rather as added, illustrative figures. Remarkably, when texts of this kind were eventually translated into Latin, the translators continued to handle the numerals appearing in them in the same manner that they handled geometric diagrams. They often kept the same directionality of the numerals, writing them from right to left as in the original Arabic texts of later periods. In astronomical and astrological texts, on the other hand, Islamic authors did include the numerals as part of the running text.

Keeping in mind these general background comments, and in particular the overall heterogeneity that characterized Islamicate mathematics, we can proceed to discuss in greater detail some examples of how numbers appeared and were conceived in these cultures. With a view to the topics discussed in later chapters of this book, I pay here closer attention to uses of Hindu–Arabic numerals and the development of decimal positional techniques. Historians often refer to the reign of the Abbasid caliph Abū Ja'far al-Manṣūr (754–775) as a significant turning point in this thread of the story. In the early 770s, an Indian scholar traveling as part of an embassy that visited Baghdad is reported to have brought with him a Sanskrit astronomical text. This text aroused great interest among his hosts and was soon translated into Arabic. The decimal positional arithmetic used in these texts became central to education in the Islamic world, and part of a well-seated tradition that came to be known as Hisāb al-Hind ("Indian calculation").

Al-Manṣūr was an innovative kind of leader who turned Baghdad into a very active commercial and intellectual center, and who began an impressive tradition of support for science and scholarship. This support was further encouraged by al-Manṣūr's successors, and particularly by his grandson, the caliph Hārūn al-Rashīd, who reigned between 786 and 809. His name has become associated with the creation of the famous center of learning *Bayt al-Ḥikma* ("House of Wisdom"), although historians have more recently questioned his actual role in establishing this institution. The *Bayt al-Ḥikma* is reported to have housed a large collection of manuscripts assembled from all over the Middle East and beyond.

Translation activity of scientific texts into Arabic reached a climax, probably in connection with *Bayt al-Ḥikma*, during the reign of Hārūn's son, Abū Ja'far al-Ma'mūn (r. 813–833). Many important works were translated, mainly from Greek but also some from Sanskrit. Over the next two hundred years, these, as well as other scientific traditions, continued to be assimilated into Islamicate mathematical cultures. New ideas about the social role of the caliph and an increasing interest in the secular sciences on the side of the local elites prepared the ground for this unprecedented movement of translation. Scholars like Abu Yūsuf ibn 'Isḥāq al-Kindī (ca. 800–ca. 873) or Thābit ibn Qurra (826–901) can be counted as the most prominent intellectual leading forces of this movement. By the end of the ninth century, the works of Euclid, Archimedes and Apollonius had already been translated into Arabic, along with works from the Hindu traditions. At the same time, other traditional techniques that had hitherto been transmitted orally began to be gradually collected into written texts.

I do not mention these institutional landmarks purely as a passing remark within our story about numbers. Scientific ideas do not simply evolve out of thin air. In each historical context, there may be public or private institutions, and social arrangements or political conditions, that allow, encourage, or hinder the creation and propagation of science. Only by reference to them can we understand why scientific ideas—ideas about numbers included—emerged, flourished and spread in certain cultures, whereas in other contexts they disappeared and left no lasting impact.

The case of the adoption of the Hindu system of positional notation for numbers and its eventual transmission to Europe is a notable case in point. Systems of numeration based on positional techniques are known to have existed in Hindu cultures at least since the sixth century, and perhaps even earlier. In the year 662, the Syriac bishop Severus Sebokht reported about the "science of the Indians," which he described as being "more subtle and more ingenious" than that of the Babylonians and Greeks. He specifically referred to their valuable methods of calculation based on the use of nine signs (presumably he did not count the dot as a separate symbol denoting zero). Some of these ideas on numbers may have reached the Islamicate world following the early expansion of Islam, and may have been used in commercial and administrative contexts. But in order to turn an incidental collection of ideas into an accepted, broadly known, and consistently used system, it was not enough that positional techniques be suggested and even used in a more or less sporadic manner. Rather, some deeper institutional and cultural changes were needed for this. The developments in Baghdad around the *Bayt al-Ḥikma* may have been among the earliest significant crossroads on the complex way to the full adoption and elaboration of this notational system. Following the completion of the translations that al-Manṣūr had commanded, the system began to be taught and gradually adopted in various frameworks, and the undisputed institutional primacy of *Bayt al-Ḥikma* played no doubt a fundamental role in bestowing official prestige upon it.

5.2 Al-Khwārizmī and numerical problems with squares

One of the most prominent scholars active in the time of al-Ma'mūn, and to whose work and influence I want to devote here particular attention, was Muḥammad ibn Mūsā al-Khwārizmī (ca. 780–ca. 850). His activities covered a wide variety of scientific fields, including mathematics, astronomy and geography. His most famous text, written around 825, is *Al-kitāb al-muhtaṣar fī ḥisāb al-jabr wa'l-muqābala* ("The Compendious Book on the Calculation of *al-jabr* and *al-muqābala*"). Both the name of the author and the title of the book are well known as being the source of important terms in our mathematical lexicon, namely, "algorithm" and "algebra." But, of course, his influence is much more significant than this, and taking a closer look at his treatise is a most appropriate way to begin our exploration of concepts of number in Islamicate mathematical culture.

Al-Khwārizmī's treatise is devoted to a systematic presentation of certain methods for solving numerical problems that involve an unknown and its square. Ancient mathematical cultures, the early Islamicate traditions included, developed various kinds of techniques for solving problems of these kinds. The methods compiled from various sources and systematically presented in al-Khwārizmī's treatise focused, as stated in its title, on two special kinds of techniques known as *al-jabr* (often translated as "restoring") and *al-muqābala* ("compensating"). The sources of these techniques are still a matter of debate among historians. Solving practical problems of land measurement, inheritance, accounting and trade was surely a direct motivation. Eventually, the same techniques also began to be developed for their own sake as well as a powerful tool for the solution of theoretical problems in geometry. The mathematical tradition of problem-solving that developed around the use of techniques of *al-jabr* and *al-muqābala* is what is usually called "Arabic algebra," and I will follow this terminology here too.

Representative of the typical problem discussed in al-Khwārizmī's treatise is the following: "squares and ten roots equal thirty-nine numbers." Retrospectively interpreted in terms of modern algebraic symbolism, if we rely on the idea of a general quadratic equation, $ax^2 + bx + c = 0$, this problem can be seen as particular case of it, namely, the case $a = 1, b = 10, c = -39$, or, simply, $x^2 + 10x = 39$. But when we look more closely at the way in which al-Khwārizmī operated with numbers, and how he handled the various mathematical situations he addressed, we readily realize the differences between his techniques for solving these "problems with squares" and the more general idea of a quadratic equations that we learn to solve in high school.

From al-Khwārizmī's perspective, the above-stated problem is a representative example of one out of six specific kinds of problems discussed in the book, a kind that he dubbed "squares and roots equal numbers." The six different kinds of questions are the following:

squares equal roots	$(ax^2 = bx)$;
squares equal numbers	$(ax^2 = c)$;
roots equal numbers	$(bx = c)$;
squares and roots equal numbers	$(ax^2 + bx = c)$;
squares and numbers equal roots	$(ax^2 + c = bx)$;
roots and numbers equal squares	$(bx + c = ax^2)$.

I have added here symbolic expressions on the right-hand side, but, as you should be aware by now, these were never used by al-Khwārizmī himself. He formulated his problems in completely rhetorical terms, where even the numbers are written in words. Still, the main difference concerns not so much the way of formulating the problems but rather the way of *solving* them. Rather than by providing a symbolic expression, a formula for the solution, al-Khwārizmī presented, well . . . an algorithm.

Let us consider, for instance, the exemplary problem "squares and ten roots equal thirty nine numbers." Al-Khwārizmī's solution is the following:[1]

You halve the number of roots, which, in this problem, yields five; you multiply it by itself; the result is twenty-five; you add it to thirty-nine; the result is sixty-four; you take the root, that is eight, from which you subtract half of the number of roots, which is five. The remainder is three, that is the root of the *square* you want, and the *square* is nine.

Notice, first of all, the rhetorical formulation of the steps to be followed, where, as already mentioned, even numbers are denoted with words. Efficient systems of numeration were already known at the time, as I have stressed, so this rhetorical formulation provides interesting evidence about the oral roots of the practices within which the solution was conceived and possibly taught. Moreover, it is evident that the specific wording adopted in the text is meant to facilitate the reader memorizing it. And, more importantly, there is no manipulation of symbols whatsoever en route to the solution. A reader wanting to solve a new problem of the same kind ("squares and roots equal numbers") would realize that the solution is reached by following exactly the same steps, while taking different values according to the problem.

A direct, and most important consequence of this purely rhetorical approach is the inherent difficulty in realizing that the solution to the problem involves a direct relationship (a formula, in our current view) connecting the value of the result, 3, with the values that define the problem, 10 and 39. To solve other instances of the problem, one is expected to repeat each step in this algorithm using the new values, rather than just replacing these values in a general formula that would yield the solution. If, as an exercise, we reconsider this example but replace numbers with symbolism, $b = 10$, $c = 39$, and if we then repeat the steps indicated by al-Khwārizmī, using the letters b and c, we will end up with the following value for the unknown quantity:

$$\sqrt{\left(\frac{b}{2}\right)^2 + c} - \frac{b}{2}.$$

This of course corresponds to the result known to us if we were to write al-Khwārizmī's problem as we do nowadays, namely, $ax^2 + bx + c = 0$. The general solution of such a quadratic equation is, as we all know,

$$\frac{-b \pm \sqrt{b^2 - 4ac}}{2a}. \tag{5.1}$$

And it is indeed easy to see that the solution obtained with the method of al-Khwārizmī corresponds to one of the solutions obtained with this formula. Simply set $a = 1$, and write c in the formula with a negative sign, as $-c$ (since we move it from one side of the

---

[1] Quotations and figures in this section are all taken from the Rashed edition (2009), of al-Khwārizmī's text. This quotation is from p. 100.

equation to the other). As al-Khwārizmī considered only one solution, so we may also take only one (e.g., the one obtained by writing + instead of ± in equation (5.1)). We thus obtain precisely

$$\sqrt{\left(\frac{b}{2}\right)^2 + c} - \frac{b}{2}.$$

But we should not see in the lack of symbolism just a technical inconvenience that involves a more cumbersome writing of the same, general underlying idea of a quadratic equation. On the contrary, our very ability to think of and use a general, symbolic formula such as expression (5.1) embodies *per se* several significant ideas that developed historically. These ideas do not appear in al-Khwārizmī's mathematical practice the way they appeared in later texts, and this explains why he needed to speak about six different kinds of equations. For him, any two instances in his list (for example, "square and roots equal numbers" ($ax^2 + bx = c$) and "square and numbers equal roots" ($ax^2 + c = bx$)), were two truly different problems that could not be directly intertranslated via symbolic expressions.

In order to reach that point where the six problems were seen as just different manifestations of the more general and abstract idea of the quadratic equation $ax^2 + bx + c = 0$, a long and complex process was still needed. This process involved the consolidation of a flexible and unconstrained idea of number that does not distinguish between positive and negative numbers, or fractions and roots; a general idea in which quantities that represent length and areas are equally considered as just numbers, and for which discrete and continuous magnitudes are just different cases. It was also necessary to develop an unconstrained ability to denote numbers by letters and to operate freely with these letters, manipulating them in such a way as to solve the equation. These kinds of ideas were not yet a distinct component of al-Khwārizmī's conceptions. We will see them arising as we go through our story.

The influence of al-Khwārizmī's treatise over the next generations of mathematicians manifested itself at several levels, and the very separation into six cases was among the most prominent of them. References to "the six problems of al-Khwārizmī" continued to appear as a standard concept, in the same way that direct references to the theorems of Euclid's *Elements* appeared in all main mathematical texts in Islamicate and European mathematics well into the seventeenth century. Anyone versed in the mathematics of the time knew exactly which were the six cases and what was involved in solving them. The idea that these were six *different* types of problems was abandoned only gradually as the fully fledged idea of equation slowly emerged together with the new conceptions about numbers that were necessary to allow for it.

A main issue related with the centrality of the six cases is that the numbers defining each of the problems (the "coefficients") were always *positive* numbers. Had al-Khwārizmī possessed, for example, a clear idea of negative numbers as possible coefficients, then the number of exemplary cases would immediately be reduced. And of course this is also the case with zero, which was neither a possible coefficient nor a solution to a problem. Interestingly, in the positional decimal system of numeration that was already in widespread use at the time, zero does play an important role. A clear idea of zero as a placeholder for the positional system was well known (the symbol 0 in 103, for instance, indicates that the place for the tens is left empty). The idea of zero as a number in other respects, however, was still absent, so that zero was not yet considered to be a possible coefficient. Had it been accepted as such, then the three cases "squares equal roots" ($ax^2 = bx$), "squares equal numbers" ($ax^2 = c$), and "roots and numbers equal squares" ($bx + c = ax^2$) would all be one and the same case. And similar, though not identical, was the case for the solutions: solutions were always positive, and zero was not accepted as a possible solution. Fractions and even roots of natural numbers or combinations thereof appeared occasionally as solutions, though with some reticence and only in a few cases as coefficients.

5.3 Geometry and certainty

These characteristic features of al-Khwārizmī's use of numbers are related to a more basic, underlying attitude, which pervaded significant parts of Islamicate mathematics, and certainly of "Arabic algebra." This is the continued, fundamental readiness to rely on geometry as common ground and safe source of certainty in mathematics. The fact that al-Khwārizmī considered positive values as the typical, possible solutions to his problems is a significant manifestation of this view. But, above all, the underlying attitude toward geometry as the source of mathematical certainty is manifest in the way that he justified the validity of his methods by reliance on geometric properties of figures. The status of geometry as a safe source of mathematical certainty, and the doubts about the status of certainty in arithmetic and in algebra, is an important issue that will continue to accompany our story until the nineteenth century, and it deserves some further elaboration at this point.

In the framework of a fully fledged algebraic way of thinking, the accepted explanation for the use of the solving formula of the general quadratic equation derives from a formal manipulation of symbols according to certain rules that are stipulated in advance. The crucial step in the derivation of expression (5.1), for instance, is the step called "completion of the square." This involves the manipulation of the algebraic expressions that appear in the successive steps of handling the equation, so that in a final step, an expression appears that we can recognize as "the square of another algebraic expression." If one of the steps leads to the expression $x^2 - 6x + 9$, for example, then we can easily recognize that this is the square of $(x - 3)$. If we were able to derive such a square algebraic expression as part of a process of formal manipulation of symbols, then the next natural step would be simply to take its "square root." For the benefit of readers who have forgotten how this is done, I give in Appendix 5.1 a detailed explanation of how expression (5.1) is obtained, by formal algebraic completion of the square, as part of the formula for solving a quadratic expression. Now, also al-Khwārizmī "completed a square" whose side leads us to the value we are looking for. But in his case the said square is not just the "square" of a formal expression. Rather, this is literally a *geometric square*. Let us see a detailed example of how this idea appeared in his text.

Al-Khwārizmī's solutions to the six problems typically comprised two parts. The first, for which I just gave an example, consisted in providing an algorithm leading

D		
six and one quarter	H	six and one quarter
C	A the square B	K
six and one quarter	I	six and one quarter
		E

Figure 5.1 Al-Khwārizmī's first geometric justification of the solution to a problem with squares.

to the solution. In the second part, al-Khwārizmī reformulated *in geometric terms* the case that he was handling. This was meant as the necessary way to explain *why* it is that the procedures yield correct results, and at the same time to provide them with undisputed *legitimation*. In the case "squares and ten roots equal thirty-nine numbers," for example, al-Khwārizmī presented two different geometric reformulations. The first is depicted in Figure 5.1. The small square in the center of the diagram, AB, is the square whose side we are looking for. Since the problem states that we add ten roots to the square, al-Khwārizmī adjoined to each side of the square a rectangle. These four rectangles taken together add up to ten times the side of the square. In other words, we take in the figure the number of roots, ten, and divide it into four (that is, two and a half), and we build four rectangles (C, H, K and I) with one side being equal to the root and the other being equal to 2.5. This yields the larger square in the figure, DE, whose side is still unknown, but each corner of which we can easily calculate to be the product of 2.5 by itself. Hence, every small square in the corner is 6.25, and the four of them taken together yield 25. Now, the problem tells us that the inner square (x^2) and the four rectangles ($10x$) taken together add up to 39. Thus, adding to this the four little squares that together represent 25, we obtain 64 for the large square, as indicated in Figure 5.2.

Figure 5.2 The arithmetic behind al-Khwārizmī's geometric justification.

In other words, we can put together the components of the problem in a geometric setting where we "complete the square" representing 64, and then we know that the side of the larger square is 8. Subtracting from this side twice the side of the corner squares (2 × 2.5 = 5), we realize that the length of the side of the small square (which is what we are looking for in the problem) equals 3. Al-Khwārizmī explicitly stressed that "the fourth of any number multiplied by itself and then by four is equal to the product of half the number multiplied by itself," and hence the large square is built by adding to 39 the square of half the roots. Rendering in algebraic symbols what al-Khwārizmī says in words, we obtain the following:

$$4 \times \left(\frac{b}{4}\right)^2 = \left(\frac{b}{2}\right)^2.$$

A crucial step in the solution turns out to be the "geometric completion of the large square" to an area whose square can be calculated exactly, in this case 64. From here, we can also understand why all the numbers involved are positive and why, consequently, al-Khwārizmī considers the six cases as involving truly different situations. But al-Khwārizmī's second geometric explanation is even closer to his actual algorithm. It is represented in Figure 5.3.

Here we have a square with side AB, AB being the unknown number. To the square built on this unknown, we adjoin two rectangles, C and N. One side of each rectangle equals the unknown, AB. In order for this figure to represent what is formulated in the problem, we need to take the second side of each rectangle to be half the roots, that is 5. Hence the remaining square, in the bottom-right corner, is 25. So, from this easily manageable geometric situation we now understand why the algorithm begins by squaring "half of the roots", and why the entire procedure does indeed yield a correct result. Half of the roots have been used to "complete a square", DE. Since, according to the problem, the square of the unknown and 10 roots is 39, it follows that the entire

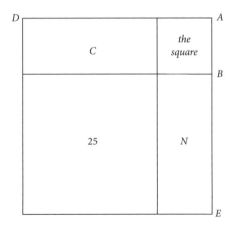

Figure 5.3 Al-Khwārizmī's second geometric justification of the solution to a problem with squares.

square is the sum of 39 and 25, namely 64. Similar reasoning to the above yields the result 3 for the unknown *AB*.

The two geometric situations used for explaining the method in this case are both very simple. In other cases, al-Khwārizmī had to rely on somewhat more complex geometric results. His successors often relied for this purpose on results taken from the classical texts of Greek mathematics, such as Euclid's *Elements* or Apollonius' treatise on conic sections (an intriguing historical issue is that al-Khwārizmī himself never directly referred to or otherwise mentioned the *Elements*, although it seems unlikely that he was not aware of some of the existing Arabic translations of Euclid's treatise). Some problems required that restricting conditions be added, so that in the course of solving the problem, roots of "negative quantities" could not arise. Also in such cases, al-Khwārizmī translated the restricting condition into a geometric constraint. In most cases, once al-Khwārizmī's had found one solution to a problem, the problem was considered to have been fully solved. But in one of the six cases, "squares and numbers equal roots," al-Khwārizmī did show that two solutions could possibly arise, and he also explained this situation in terms of the "discriminant of the equation." The problem, symbolically rendered, is $x^2 + b = cx$, and the solution is

$$x_{1,2} = \frac{b}{2} \pm \sqrt{\left(\frac{b}{2}\right)^2 - c}.$$

Al-Khwārizmī explained that if $\left(\frac{b}{2}\right)^2$ is less than c, then there is no solution to the problem. This makes sense when we consider that for him a magnitude that is less than zero can have no meaning as a number, and much less can it have a square root. The interesting point, however, is that in saying this, he also suggested, at the same time, that the problem had only one solution, but there were *two possible ways* to achieve it: "try its solution by addition, and if that does not serve, then subtraction certainly will." As far as we know, in the Hindu tradition, there are examples of problems for which more than one solution was actively sought, but it is not clear to what extent this tradition had a direct impact on al-Khwārizmī. Negative solutions were acceptable in Hindu mathematics, but not for al-Khwārizmī. Here we have again an idea that we take for granted, but whose full adoption was slow and gradual: that a quadratic equation has two solutions and that solving the equation means finding both of them.

5.4 Al-jabr wa'l-muqābala

The detailed solution of the six possible cases and their justification on the basis of geometric arguments provided the foundation of the entire approach to problem-solving in al-Khwārizmī's treatise. From this point on, any given problem should be solved as either being an instance of one of the six canonical cases or by reduction to one of them. But what does it really mean to reduce a given problem to a known one, and how this is done? It was here that the two basic kinds of operations of "Arabic algebra," namely, "restoring" (*al-jabr*) and (*al-muqābala*) "compensating," came to play their crucial role. Let us consider an example to see how exactly they worked:[2]

> You divide ten into two parts; you multiply each part by itself; you add <the products>, and then you add to them the difference between the two parts before they are multiplied, so that you get fifty-four dirhams.

In this problem, we are asked to divide the number 10 into two parts that satisfy a certain condition. This is of course quite similar to the kinds of problems solved by Diophantus. Diophantus was translated into Arabic somewhat later (see below) and al-Khwārizmī was not aware of his texts. But these problems were of a rather general kind, which in several mathematical cultures were known and solved in various ways. A main innovation involved in al-Khwārizmī's treatise was the systematic way in which he approached, analyzed and classified all possible cases of those problems involving squares of the unknown. It is also interesting to notice the use of the term "dirham" here to mean just "numbers." The issue of the terminology used in Arabic mathematical texts is very interesting in itself and revealing about underlying attitudes towards numbers, but it goes beyond the scope of this book, and I will not go into the details.[3]

Let us follow the steps of al-Khwārizmī's solution to this problem. This is where the focus of our interest lies now. I describe the steps rhetorically, as he did, and I add to them symbolic formulations that help us follow his way of reasoning. After we have gone through this, however, I will stress the places where the symbolic diverts from what al-Khwārizmī did rhetorically:

1. You multiply ten minus a thing by itself; the result is one hundred plus a square, minus twenty things. $[(10 - x)^2 = 100 + x^2 - 20x]$
2. You multiply the thing that remains from ten by itself; the result is a *square*. $[x^2]$
3. Then, you add all of that, and the result is one hundred plus two *squares* minus twenty things. $[100 + 2x^2 - 20x]$
4. But he said: you add to them the difference between the two parts before they are multiplied. You say: the difference between them is ten minus two things. $[(10 - x) - x]$
5. The sum of this is therefore one hundred and ten plus two *squares* minus twenty-two things, equal to fifty-four dirhams. $[110 + 2x^2 - 22x = 54]$

Al-Khwārizmī has arrived at this point at an "equation," which is a rhetorically formulated condition. He obviously displays a remarkable ability to conduct this kind of complex reasoning in a purely rhetorical manner. But now the question is how do we proceed from this expression to one that we can recognize as embodying one of the six canonical cases. He does it, using "restoring" and "compensating," as follows:

6. One hundred and ten dirhams plus two squares are equal to fifty-four dirhams plus twenty-two things. $[110 + 2x^2 = 54 + 22x]$

---

[2] (Rashed 2009, pp. 158 ff.).
[3] See (Oaks and Alkhateeb 2007).

7 Reduce the two *squares* to one single *square* by taking half of what you have: the result is fifty-five dirhams plus one *square* equal to twenty-seven dirhams plus eleven things.   [$55 + x^2 = 27 + 11x$]

8 Take twenty-seven away from fifty-five: the remainder is one *square* plus twenty-eight dirhams, equal to eleven things.   [$28 + x^2 = 11x$]

We have thus transformed the initial "equation" that represents the given problem into one of the canonical cases, namely "squares and numbers equal roots." And this case is one that we already know how to solve. By applying the required procedure, we will immediately reach the result 4, "which is one of the two parts." Notice, by the way, that al-Khwārizmī does not mention that the value 7 is also a solution. And indeed, it is a solution of the "equation," but certainly not of the problem, because then the sum of the two squares would be 58.

Now, the operation of "restoring" (*al-jabr*) is the one applied in passing from step (5) to (6). What is restored here are the one hundred and ten plus two *squares* that had been diminished by twenty-two things. The operation of "compensating" (*al-muqābala*), in turn, was applied in passing from step (7) to step (8). It handles terms of similar kind that appear on opposite sides of an equation. After compensating, we remain with their difference on the side of the larger one. In this case, 55 and 27 are compensated, and we remain with 28.

On the face of it, there is not much of a difference between these manipulations and what we do with our symbolic algebra. But this is true only to a limited extent. With symbolic algebraic methods, we perform the following transformation, which corresponds to what al-Khwārizmī did here:

$$(100 + 2x^2 - 20x) + [(10 - x) - x] = 54 \rightarrow 2x^2 + 110 = 54 + 22x.$$

But the same symbolic methods would equally allow for the following transformation:

$$(100 + 2x^2 - 20x) + [(10 - x) - x] = 54 \rightarrow 2x^2 - 22x + 56 = 0.$$

And this is something that al-Khwārizmī would *not* do. This would simply be foreign to the nature and aims of what he actually did. In the same vein, we can take an expression that we write symbolically as

$$(10 - x) \times (10 - x) = 100 + x^2 - 20x.$$

Al-Khwārizmī's kind of rhetoric "multiplication" could allow for this to be expressed as follows:

"Ten numbers with one root removed" multiplied by itself yields "hundred and a square with twenty things removed".

But then, from our perspective, we could equally go the opposite direction and assert that $100 + x^2 - 20x$ can be written as the product of two binomials $(10 - x) \times (10 - x)$. Again, we do not find anything of the kind in al-Khwārizmī's methods, and indeed it can be said that going this opposite way was devoid of meaning for him.

More generally speaking, in spite of al-Khwārizmī's remarkable ability to handle these complex situations so ingeniously in a purely rhetorical manner, there are important differences between what we find in his texts and what we do in our modern algebraic manipulations. And what is really important for us is that these differences are closely connected to al-Khwārizmī's understanding of what were numbers, and the way in which geometry provided the underlying justification to his way of handling them. When speaking, for instance, of "one hundred and ten numbers and two squares equal fifty-four numbers and twenty-two roots" (symbolically, $110 + 2x^2 = 54 + 22x$), al-Khwārizmī was comparing two *separate* "quantities." Each of the quantities compared might comprise more than one kind (squares, things, numbers, roots, dirhams), but these kinds were seen as being aggregated to one another while preserving their separate identities ("fifty-four numbers and twenty-two roots"). They were not operated upon, as we do nowadays, in order to produce a new, single entity, which is an abstract number representing their added result.

Similar considerations arise when we look at al-Khwārizmī's "equations." When equating two expression involving aggregations of quantities, he was not creating yet another, single, autonomous mathematical entity, like the one we nowadays call an "equation." In our equations, we move elements freely from one side to the other according to rules of manipulation, and with no constraints beyond what the rules stipulate. All the time, we consider this to be just the same equation, albeit differently written. We can reduce an equation equally to either $28 + x^2 = 11x$ or to $x^2 - 11x + 28 = 0$. Indeed, we will typically prefer the latter form over the former. Not so al-Khwārizmī, for whom the latter would simply make no sense. Not only because of the appearance of a negative coefficient in the canonical form, but also, and especially so, because instead of comparing two quantities of the same type with each other, he would be equating some abstract quantities to zero. This is not what his understanding of numbers and magnitudes would lead him to do.

The geometric kind of justification provided for each procedure, as exemplified above, makes it clear that the magnitudes that al-Khwārizmī was comparing and manipulating represented squares and rectangles, and were not abstract numbers. Dimensional homogeneity was an implicit, underlying principle in his approach. His comparison of two quantities and the way to reduce the "equation" is more conveniently understood in analogy to searching for equilibrium in a balance, where two separate entities are placed on each side of the lever. One can visualize his manipulations in terms of "cut-and-paste" procedures being applied to a geometric figure. This is why his reduced equations contained one term for each power and no subtracted terms, and always compared a magnitude (or two magnitudes taken together) with some other magnitude. While arithmetic in content, then, there was always a geometric (one is tempted to say, even physical) substrate underlying or accompanying al-Khwārizmī's equations and the two basic kinds of operations of his "Arabic algebra."

5.5 Al-Khwārizmī, numbers and fractions

Al-Khwārizmī's treatise on *al-jabr* and *al-muqābala* was very influential for generations to come, in both the Islamicate and European mathematical cultures. But no less influential on Islamicate conceptions of numbers was another of his treatises, *Kitāb al-jam' wal-tafrīq bi-hisāb al-Hind* ("Book of Addition and Subtraction According to the Hindu Calculation"). This is the oldest Arabic work known to have systematically presented the Hindu decimal positional system and its associated arithmetic, and it became a main vehicle for its early propagation within the Islamic world. No extant original Arabic manuscript of this treatise is known. All we have are Latin versions dating from about the twelfth century and based on an earlier translation now lost.

In this treatise, Al-Khwārizmī was not handling algebraic methods but rather focusing on arithmetic and numbers as such. He showed how to write numbers by means of the nine digits and a small circle to indicate the zero and how to perform all the basic arithmetic operations with integers as well as with fractional numbers. The idea of the unit and the possibility of adding it repeatedly to itself was for him, just as in Pythagorean mathematics, the basis of all of arithmetic. It is instructive to quote directly from the original text his very clear statement in this regard:[4]

Because one is the root of all number and is outside number. It is the root of numbers because every number is found by it. But it is outside number because it is found by itself, i.e., without any other number. . . . Number is therefore nothing else but a collection of ones, and as we said, you cannot say two or three unless one precedes; we have not spoken about a word, so to speak, but about an object. For two or three cannot exist, if one is removed. But one can exist without second or third.

Al-Khwārizmī obviously did not intend this definition to apply to fractions or irrational lengths. In this sense, in this treatise at least, fractions or irrational lengths would not be formally considered "numbers." At the practical level, however, al-Khwārizmī presented a more or less elaborated arithmetic of fractions, whose basic ideas derived from various sources that he was trying to combine. Al-Khwārizmī's technical explanations about the operations with fractions are clear and quite straightforward. Conceptually, however, there are many difficulties that are worthy of attention. This tension between technical proficiency and conceptual lack of clarity makes the *Kitāb* a text of particular historical interest.

The starting point of al-Khwārizmī's use of fractions is in the treatment of reciprocals of the natural numbers: the half, the third, the tenth, the thirteenth, etc. We have already seen this basic idea arising in the context of Greek arithmetic. It probably existed in practical Islamic arithmetic from very early on. Into this basic idea, al-Khwārizmī incorporated a second one taken "from the Hindus," and embodying a sexagesimal approach to fractions. The underlying context of this idea is obviously astronomical. The Hindus, al-Khwārizmī wrote, divide the unit, called "degree," into sixty parts, each of which is called a "minute." These are further divided into sixty seconds, these into sixty thirds, these into sixty fourths and so on.

Multiplying these sexagesimal fractions is not like multiplying an integer times an integer, which yields another integer. One needs to multiply the values of each of the kinds involved, but also to "collect their places." For example, in order to multiply "7 seconds *times* 3 minutes," we need to rewrite the quantities as same kinds. This is "7 seconds *times* 3×60 seconds," or "1260 seconds." But this is "483 minutes," or

---

[4] Quoted from (Katz (ed.) 2007. p. 525).

"8 minutes *and* 3 seconds." Techniques of this kind were known in ancient cultures such as the Babylonian, but the point is that, at least as manifest in this treatise, we can see that "Hindu calculation" was not limited, in al-Khwārizmī's views, to the decimal positional system. Within the sexagesimal contexts, the arithmetical procedures for dealing with fractions appear as well seated and clearly understood. Al-Khwārizmī relied on these ideas in order to go on and handle fractional numbers in the decimal context as well. Curiously, he did so in a somewhat roundabout way by first translating the decimal into sexagesimal terms, as in the following example:

Multiplying one and a half by itself is equivalent to multiplying 1 degree and 30 minutes by itself, and, since one degree is 60 minutes, this is equivalent to multiplying 90 minutes by itself. This is 8100 seconds, which is the equivalent of 135 minutes, or of two degrees and 15 minutes. This, in turn, is the equivalent of 2 and one quarter.

If calculated directly in decimal terms, we would obtain the same result, of course, namely, $1.5 \times 1.5 = 2.25$. So, the reason for using this roundabout approach is not that it yields more accurate results. Similarly roundabout is al-Khwārizmī's method of dividing decimal fractions. His explanation of addition and subtraction, on the other hand, is quite sketchy and at times unclear.

Al-Khwārizmī also explained a procedure for extracting roots. His method applies equally to integers, to sexagesimal fractions, and to reciprocals and common fractions. He also introduced some ingenious techniques for making all calculations more accurate, and here again the sexagesimal representation plays an important role. A simple example is the calculation of the square root of 2. Al-Khwārizmī calculated it as the square root of 7200 seconds (i.e., $\sqrt{2} = \sqrt{2 \times 60 \times 60}$), which yields 84 minutes plus a remainder (because $84 \times 84 = 7056$). This, in turn, is equivalent to 1 integer and 24 minutes plus a remainder.

Yet more efficient methods where the decimal and the sexagesimal are mixed are based on adding zeroes and working out the results with the help of various types of sexagesimal fractions. To see how this approach works, consider the example of the square root of 2. Al-Khwārizmī did it as follows (I write the procedure here in modern symbols in order to make it easier for the reader to follow the steps):

$$\sqrt{2} = \sqrt{\frac{2.000.000}{10^6}} = \frac{1}{1000} \times \sqrt{2.000.000} = \frac{1}{1000} \times [1414 + \text{remainder}]$$

$$\approx \frac{1}{1000} \times 1414 = 1 \text{ (unit)} + \frac{414}{1000} = 1 + \left(\frac{414 \times 60}{1000}\right)' = 1 + \frac{24840}{1000}$$

$$= 1 + 24' + \left(\frac{400 \times 60}{1000}\right)'' = 1 + 24' + \left(\frac{50400}{1000}\right)''$$

$$= 1 + 24' + 50'' + \left(\frac{400 \times 60}{1000}\right)''' = 1 + 24' + 50'' + 24'''.$$

The text continually establishes analogies between reciprocals, common fractions and sexagesimal fractions, and suggests that all of these may be treated similarly. The analogies are stated sometimes implicitly and sometimes explicitly. Sometimes they help clarify the underlying concepts, and sometimes they obscure them. The text is uneven in other respects as well, partly because the necessary terminology had not yet been clearly established. Some sections are long and excessively detailed, others short and confused.

All of this can be seen as direct evidence of how the arithmetic of fractions was practiced in the early stages of Islamicate mathematics and how fractions were conceptualized. Ingenious ideas and innovative techniques taken from a variety of sources were tried out in original but hesitating ways, while their full integration into the broader body of existing arithmetic knowledge was yet to be formulated in a more coherent way that could also be combined with the classical conception of number as a collection of units.

5.6 Abū Kāmil's numbers at the crossroads of two traditions

Another interesting example of the conceptual tensions manifest in the Islamicate use of numbers appears in the work of Abū Kāmil ibn Aslam (ca. 850–ca. 930). He was one of the earliest commentators of al-Khwārizmī, and in his influential treatise on algebra, *Kitāb fī al-jabr wa al-muqābala*, he elaborated in greater detail the methods and results that his predecessor had introduced. But in order to do so, he took a step that was both original and bold, when he followed the style and the systematic kind of presentation typical of the arithmetical books of the *Elements*. The combination of these two sources of influence, al-Khwārizmī and Euclid, created an interesting situation in the way that numbers were treated and conceived.

I want to illustrate the tensions arising in Abū Kāmil's work by considering an example where he explains an arithmetic rule in two different ways: first as a specific instance of calculating with two numbers ($\sqrt{9} \cdot \sqrt{4} = \sqrt{9 \cdot 4} = 6$) and then as a general statement about numbers ($\sqrt{a} \cdot \sqrt{b} = \sqrt{a \cdot b}$). The specific case he proved along the lines of al-Khwārizmī's reliance on geometric constructions (which in turn rely on *geometric* ideas taken from Books II and VI of the *Elements*), and the general arithmetic statement he proved along the lines of the *arithmetical* books of the *Elements* (Books VII–IX). The interesting point, however, is that in the former numbers are seen as continuous magnitudes, whereas in the latter they are seen as collections of units. And the attempt to combine these two views together did not always work very smoothly. Let us see why.

The purely geometric argument that Abū Kāmil followed in order to show that $\sqrt{9} \cdot \sqrt{4} = \sqrt{9 \cdot 4} = 6$ is this: construct a square *AGFH*, as in Figure 5.4, where *AB* and *BG* represent, respectively, $\sqrt{9}$ and $\sqrt{4}$. From the construction, it follows that the square *ABEK* is nine whereas *EMFZ* is four. Now, reasoning in terms of proportions on the basis of the same construction, Abū Kāmil obtained the following:

		ME : *EK*	::	*ZE* : *EB*	and
		ME : *EK*	::	*EMFZ* : *ZHKE*.	
But		*ZE* : *EB*	::	*ZHKE* : *EKAB*;	
Hence		*EMFZ* : *ZHKE*	::	*ZHKE* : *EKAB*.	

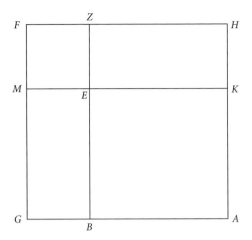

Figure 5.4 Abū Kāmil's geometric proof that $\sqrt{9} \cdot \sqrt{4} = \sqrt{9 \cdot 4} = 6$.

Considered from our modern point of view, the last part of the argument is simple and straightforward. The square *EMFZ* is 4 and the square *EKAB* is 9. Hence, if we read proportions as equality of fractions, what Abū Kāmil has shown is that 4/*ZHKE* = *ZHKE*/9, or, in other words, that the rectangle *ZHKE* represents the geometric mean of 4 and 9, which is $\sqrt{9 \cdot 4}$ or simply 6, as we wanted to show. But this is precisely what Abū Kāmil does not have the tools to do. Ratios are not for him fractions, and proportions are not equalities of fractions. The proportions were not even written in the simple symbolic language that I have used here, but in a completely rhetorical manner: "The ratio of surface *EMFZ* and surface *ZHKE* is the same as the ratio of surface *ZHKE* and surface *EKAB*." So, the only tools he had for handling the proportions and deducing the result in a rigorous way (and this is what he wanted to do) were those afforded by the theories of proportion appearing in Euclid's *Elements*. Now, in principle, there are two propositions in the *Elements* that could be used to handle this case, namely, the following:

VI.17: If three straight lines be proportional, the rectangle contained by the extremes is equal to the square on the mean; and, if the rectangle contained by the extremes be equal to the square on the mean, the three straight lines are proportional.

VII.19: If four numbers be proportional, the number produced from the first and fourth will be equal to the number produced from the second and third; and, if the number produced from the first and fourth be equal to that produced from the second and third, the four numbers will be proportional.

In modern symbols, they can be formulated as follows:

VI.17: $A{:}E :: E{:}D \Leftrightarrow AD = E^2$.

VII.19: $A{:}B :: C{:}D \Leftrightarrow AD = BC$ (and hence $A{:}E :: E{:}D \Leftrightarrow AD = E^2$).

Although both symbolic expressions would seem to provide the step that was necessary for Abū Kāmil to complete his proof, both propositions (as originally conceived) actually present difficulties, and neither of them can be directly applied to the situation here. First, VI.17 is a proposition about *lines* that are proportional, whereas Abū Kāmil's proportions involve *surfaces*. Clearly, there is no direct translation in this case to what the proposition stipulates, "the rectangle contained by the extremes is equal to the square on the mean," since "the extremes" of the proportion are two surfaces, and it does not even make sense to speak about their multiplication. And as for VII.19, this is a proposition about *numbers*, seen as collections of units, and this is not exactly the case in Abū Kāmil's proof. It is true that "surface *EMFZ*," for example, is taken to be 4, but this is because it is the square built on a side that was taken to be $\sqrt{4}$ and not because it was defined as a collection of four units.

Abū Kāmil was aware of these difficulties, and of course he was also aware that the specific case he was considering was somewhat artificial, given that he could have simply said that $3 \times 2 = 6$ and that's all. But his point was to show that he was able to go beyond both al-Khwārizmī and Euclid and to provide a sound justification for operational rules in arithmetic. In the end, however, he was not truly careful in handling all these different entities, and he ended up mixing together various meanings of the idea of number and magnitude. Retrospectively, we see that his text actually highlights the difficulty created by the fact that integers were handled as discrete quantities, while their roots could only be handled as continuous magnitudes. So, this is how Abū Kāmil rounded up his argument:[5]

> The product of the number that corresponds to surface *ZEMF* by the number that corresponds to surface *EKAB* is equal to the number that corresponds to surface *ZHKE* by itself. And Euclid showed this in Book VI of his work. He said that for three proportional numbers the product of the first number by the third number is equal to the product of the second number by itself. So we multiply what is in surface *ZEMF* in units, which is four, by what is in surface *EKAB* in units, which is nine, so it yields thirty-six. So the product of what is in surface *ZHKE* in units by itself is thirty-six. So surface *ZHKE* is the root of thirty-six, which is six, which comes from the multiplication of the root of nine by the root of four, since *KE* is the root of nine and *EM* is the root of four.

Abū Kāmil could not have failed to know, of course, that what Euclid proved in Book VI related to "lines" and not to "numbers" as he wrote here, or even to "continuous magnitudes" in general.

But now Abū Kāmil also wanted to prove the general rule, $\sqrt{a} \cdot \sqrt{b} = \sqrt{a \cdot b}$, and here he changed his approach so that the statement will refer to numbers seen as discrete quantities. The proof now was fully in the style of the arithmetical books of Euclid, even though the rule was meant to apply to the general case that considered roots of any integer (and hence irrational numbers as well). In the diagram accompanying the proof, which is depicted in Figure 5.5, the numbers are represented by lines, but this is just schematic and not intended as the basis for any geometric construction. The multiplication of two numbers yields another number (i.e., another line) and not a surface.

In the diagram, *B* is a number, *A* is another number, *G* represents the root of *B*, and *D* represents the root of *A*. The product of *B* by *A* is represented by *Z*, whose

---
[5] Quoted (with slight modifications) from (Oaks 2011b, pp. 239–240).

Figure 5.5 Abū Kāmil's arithmetic proof that $\sqrt{a} \cdot \sqrt{b} = \sqrt{a \cdot b}$.

root is H. Finally, the product of G by D is F, and the aim is to prove that F is equal to H. The arithmetic context would seem to suggest that all of these are treated as discrete numbers, while in fact not all of them are such. The general rule is formulated rhetorically, of course, and not symbolically. It reads as follows:[6]

One multiplies a number by a number, then one takes the root of what results. So it yields the same as the product of the root of one of the numbers by the root of the other number.

The argument of the proof is in essence the same as in the geometric proof, but it is based on VII.19 rather than on VI.17. Curiously, however, while Abū Kāmil explicitly mentioned Euclid in the geometric proof, he did not mention him here, nor in any other of his proofs of this arithmetic kind. Perhaps this has to do with his awareness that the arithmetic for which he wanted to provide Euclidean type of foundations was broader than that of the integers and that it included fractions and irrational roots as well.

In his book, Abū Kāmil applied al-Khwārizmī's techniques to problems that included various kinds of "numbers," including roots of non-square numbers as well as roots within roots. He used his rules to calculate relatively complex cases such as

$$2\sqrt{10} \cdot \frac{1}{2}\sqrt{5} = \sqrt{50}.$$

But, more importantly, he was able to apply his arithmetic rules to the various kind of quantities appearing in the algebraic problems of al-Khwārizmī. Thus, for example, by multiplying "the root of half a thing by the root of a third of a thing, yields the root of a sixth of a square." In symbols,

$$\sqrt{\frac{1}{2}x} \cdot \sqrt{\frac{1}{3}x} = \sqrt{\frac{1}{6}x^2}.$$

A quantity such as $\sqrt{1/6 x^2}$ would make sense in the numerical world of Abū Kāmil and most Islamicate mathematicians, meaning that, of the magnitude of type "squares," we have a certain amount (1/6 in this case), and then we take the square root of that. In

---

[6] Quoted from (Oaks 2011b, pp. 241–242).

contrast, the number $\sqrt{1/6}\,x$ would not make sense, because it is not clear what would it mean to have of the magnitude of type "the unknown" an amount of $\sqrt{1/6}$.

5.7 Numbers, fractions and symbolic methods

I return now to the important topic of fractions and in particular to the idea of decimal fraction. Glimpses of this idea appear in various ways in Islamicate mathematics. This is true for the very idea of decimal fraction, as well as for the possible notation used for indicating them. A treatise written about a hundred years after al-Khwārizmī's text on Hindu arithmetic provides an interesting example of both aspects. This is the *Kitāb al-Fuṣūl fī al Ḥisāb al-Hindī* ("Book of Sections on Hindu Arithmetic") by Abū al-Ḥasan Aḥmad ibn Ibrāhīm al-Uqlīdisī (ca. 920–ca. 980), who was active in the area of Damascus. In al-Uqlīdisī's view, Hindu methods of calculation were more powerful and easier to verify than the finger reckoning techniques then still ubiquitous in Islamicate culture. He considered the latter not only less sophisticated from a mathematical point of view, but also open to cheating. At the same time, he saw in the continued use of dust-boards a main obstacle in the reception of the new techniques of calculation. Erasing texts written on these boards was easy, and hence it facilitated working out problems with traditional methods that involved overwriting digits. Al-Uqlīdisī was very explicit in his dislike, because he associated the method with the work of street astrologers. When intermediate stages of the calculations are erased—he said— the only way to check the procedure is by reworking it in its entirety. In his treatise, he taught the use of pen and paper as the correct way to adopt and promote the Hindu system of numeration and calculation.

The increased insistence on developing a standard, convenient symbolic language within Islamicate mathematics is also related to this gradual shift in the material basis of the entire culture, from dust-board to pen-and-paper practices. In the former, only the result is required, and the intermediate steps do not really matter. In the latter, the focus on the importance of the various stages of the procedure, as well as the result, call for a more systematic approach to writing them. This is interestingly visible in al-Uqlīdisī's treatment of fractions, which occupies considerable parts of his treatise. A simple example that illustrates how this is manifest in the text is that of "multiplying a number with fractions by fractions." The example that he considered is "769 and two-thirds, and two thirds of one-ninth" times "a half, and a fourth, and a fifth." This is "drawn" in the text as follows:[7]

$$\begin{array}{ccc} 769 & \therefore & 0 \\ 2\ 2 & \therefore & 1\ 1\ 1 \\ 27\ 3 & \therefore & 2\ 4\ 5 \end{array}$$

Following the techniques explained earlier in the treatise, "the two-thirds of one-ninth" is transformed into $\frac{60}{81}$, and the same goes for "a half, and a fourth, and a

---

[7] (Saidan 1978, pp. 72–73).

fifth," which transforms into $\frac{57}{60}$. Notice that the transformation of fractions is based not on finding the *least* common denominator, as we do nowadays, but just *a common denominator*. This result is represented as follows:

$$\begin{array}{cc} 769 & 0 \\ 60 & 57 \\ 81 & 60 \end{array}$$

The multiplication can now be performed just by transforming 769 and $\frac{60}{81}$ into $\frac{62349}{81}$ and then multiplying by $\frac{57}{60}$. This yields $\frac{3553893}{4860}$, which is $731\frac{1233}{4860}$.

From our perspective, the outcome of the multiplication in this example could be further reduced to $731\frac{137}{540}$. But this is not a kind of reduction that al-Uqlīdisī would consider to be necessary or interesting in all cases. So, again, while his level of technical competence in handling operations with these fractions in a wide variety of situations was very high, his conceptualization of fractions was far from detailed. His text does not present a fully elaborated theory of fractional numbers under which the arithmetic of the integers is also subsumed as a particular case. What I want to show here are two interesting examples where al-Uqlīdisī introduced a notation very similar, in retrospect, to our decimal fractions and yet conceived the situation quite differently.

One of these two cases appears in a chapter dealing with "halving and doubling" a fraction. Doubling and halving were considered, here as well as in many other Arabic treatises, to be two basic arithmetic operations that require a separate treatment alongside root extraction and the four elementary operations. For example, in order to halve 99999, al-Uqlīdisī does not just divide by 2, but rather applies the special operation "halving," whose steps are as follows:[8]

One-half the eight is 4; we insert it in place of the last nine. We join nine to one and get 19. We take half of the 18, which is 9. The nine remains in its place. The remaining one is joined to the nine. We go on, the nine remains in its place, until we reach the first nine; we put a half below it. We get $\begin{smallmatrix} 49999 \\ 1 \\ 2 \end{smallmatrix}$.

Now, a subtle but significant twist comes when al-Uqlīdisī explains how to work out the halving of a fraction. Earlier in the treatise, he had shown how the operation of halving works in the context of sexagesimal quantities, that is, magnitudes involving the degree, the minutes, the seconds, etc. Like al-Khwārizmī previously in his treatise on Hindu calculation, al-Uqlīdisī also discussed the cases of sexagesimal and of decimal quantities separately, sometimes (but not always) transferring the techniques from one context to the other. Clearly, for him, these are not just two different ways to represent a more general, underlying idea of number, but actually various separate kinds of entities: the "degrees," the "numbers," and the "fractions." The halving of the degree requires special attention when "in the place of integers, we assume any number of degrees, provided it is odd." The "secret of the work" lies in halving the one. This is done by taking the one degree to be 60 minutes, whose half is then 30, and the same

[8] (Saidan 1978, p. 108).

is true for each of the successive subdivisions. If we take, for example, 17 degrees, the process of halving successively several times will yield

$$\frac{08}{30}, \frac{04}{15}, \frac{02}{07}, \frac{01}{03}, \frac{00}{31}\\\frac{}{30}\frac{}{45}\frac{52}{30}.$$

And then, in the case of halving common fractions, al-Uqlīdisī suggested to imitate this approach, together with a new and important notational convention:[9]

If we halve an odd number we set the half as 5 before it, *the unit place being marked by a sign´ above it, to denote the place.*

This is of course a basic principle that we typically associate with the idea of decimal fractions: write the integer part as we usually write integers, then write a point or a comma (al-Uqlīdisī wrote ´) to separate the integer from the fractional, and then write the fractional part. If we look at al-Uqlīdisī's example for this part, he halves 19 five successive times. In modern notation, this would yield

9.5, 4.75, 2.375, 1.1875, 0.059375.

Incidentally, in the extant manuscript of the treatise, the only term where we actually find a special mark separating the integers from the fractions is 2´375. But al-Uqlīdisī was very explicit in his explanation of the use of this sign, and hence it seems reasonable to assume that it was the copyist who, inadvertently, dropped the sign from the other places where it originally appeared. Nevertheless, it is no less clear than separating the integer and the fractional part with a sign, though ingenious, appears in the treatise in a rather fortuitous manner. Here, as in other cases discussed in this book, one should not jump too soon to conclusions about the way in which an idea is used only because it is similar to what we do nowadays. Al-Uqlīdisī's ingenious notation is not integrated into a full arithmetic of fractions, and there is no obvious way to operate with numbers written in this way. Moreover, even within the treatise itself, the idea is used in a very limited context and it is not nearly fully exploited in many other places where it would be relevant.

Arabic texts of later generations began to handle ideas related with fractions in a more systematic manner that allowed for a natural and more complete integration into the broader fabric of arithmetic. Notation became more standard, and calculation procedures were simplified and made more efficient. Mathematical texts of the twelfth-century Maghreb presented a well-developed notation for fractions, using a horizontal line to separate numerator from denominator. Ibn Yaḥyā al-Maghribī al-Samaw'al (ca. 1125–1174), for example, calculated—in a way that resembles a decimal expression—the square root of 10. He expressed the result in an entirely rhetorical formulation, without using any kind of digits: "three, and one part of ten, and six parts of hundred, and two parts of thousand, and two parts of ten thousands, and six parts of

---

[9] (Saidan 1978, p. 110).

one hundred thousands, and seven parts of a million" (that is, in modern decimal expression, 3.162277). A more complete version of a system of decimal fractions appears much later in the work of the Persian Ghiyāth al-Dīn Jamshīd ibn Mas'ūd al-Kāshī (1380–1429).

Al-Kāshī's work was known in Venice in 1562, but similar ideas started to appear in Europe much earlier than that, albeit in a very sporadic, non-systematic fashion. As we will see in Chapter 7, the full use of decimal fractions did not become standard and mainstream in Europe before the beginning of the seventeenth century, particularly following the work of Simon Stevin. The discussion of his work will allow a better understanding of what I mean here by suggesting that, in spite of all the ingenious ideas elaborated by Islamicate mathematicians in this regard, the full idea of decimal fractions, in the most important senses of this word, was still absent from this culture.

The development of symbolic languages concerned not only the way in which fractions could be written. While the earlier texts of Islamicate mathematics were characterized by the purely rhetorical formulation of the methods described in them, elaborate symbolic methods began to appear in later stages, especially in the Maghreb and in al-Andalus on the Iberian Peninsula, between the late twelfth and the fifteenth centuries. These methods, however, developed more from textbook traditions in arithmetic than from the mainstream tradition of algebraic problems associated with al-Khwārizmī and Abū Kāmil. Abū Bakr al-Karajī (ca. 953–ca. 1029), to mention another prominent name, while providing geometric justifications for solution procedures along the lines of al-Khwārizmī, also added some new kinds of justifications involving incipient techniques of symbolic manipulation based on general rules for operating with higher powers of the unknown. Sharaf al-Dīn al-Ṭūsī (ca. 1135–1213) solved problems involving cubes of the unknown, assisting himself with geometric diagrams, but at the same time he also relied on some existing methods for obtaining approximate values of the unknown.

Among the arithmetic treatises from which these symbolic traditions developed, one of the most prominent and influential was the *Talkhīṣ 'amal al-ḥisāb* ("Summary of Arithmetical Operations") written by Ibn al Bannā' al Marrākushī (1256–1321), who was active in the area of Morocco. Without itself introducing any symbolic notation, the *Talkhīṣ* described procedures for performing arithmetic operations on integers, fractions and square roots, as well as procedures for problem-solving such as the rule of three, double false position, and rules for the solution of equations of the first and second degree. Early commentators of the *Talkhīṣ* introduced their symbolic notation in a rather hesitant and non-systematic way, and essentially as additions to the rhetorical presentation. Gradually, the symbolism became more pervasive and flexible, allowing operations with relatively complex numerical expressions such as (written in current symbolism):

$$\sqrt{\sqrt{1+\frac{1}{2}}+\sqrt{\frac{1}{2}}}+\sqrt{\sqrt{1+\frac{1}{2}}-\sqrt{\frac{1}{2}}}.$$

The increasing ingenuity displayed by these notational methods are of great historical interest. Still, it seems fair to state that they remained mostly as means to express symbolically the basic ideas and approaches previously underlying the Islamicate conceptions of numbers. They did little to modify or expand these conceptions. Most Arabic treatises on arithmetic written after the fourteenth century were mainly commentaries and epitomes of earlier works, and brought about little conceptual innovation. At any rate, they possibly had little, if any, impact on the renewed mathematical activity in Europe, which is the main focus of the forthcoming chapters of my account.

While the decimal positional system continued its development and became increasingly dominant in Islamicate mathematics, other systems of numeration did not completely disappear from this culture. Traditional systems of finger-reckoning continued to be used in trade and commerce, and dust-boards and other devices such as wooden slats remained popular in schools. On the other hand, in the astronomical context that flourished under the influence of works like Ptolemy's *Almagest*, the sexagesimal system of numeration dominated the scene. Also, while the algebraic–algorithmic approach to problem-solving typical of Al-Khwārizmī became the emblematic representative of this entire culture, there were many other approaches that were cultivated.

5.8 Al-Khayyām and numerical problems with cubes

We gain additional insights into the algebraic methods of Islamicate mathematics and their connection with the understanding of number when we examine the work of the Persian 'Umar al-Khayyām (1048–ca. 1131). Al-Khayyām was a multi-faceted scholar, a celebrated poet and a scientist with significant contributions to various fields. He composed astronomical tables and promoted an important calendar reform. He is also known for his attempted proof of Euclid's parallel postulate. In 1077, al-Khayyām published his *Risāla fī'l-barāhīn 'alā masā'il al-jabr wa'l-muqābala* ("Treatise on Demonstration of Problems of Algebra"), where he did for canonical cases of problems involving *cubes* of the unknown what al-Khwārizmī had done more than two centuries earlier for similar questions with *squares*. This was, however, a treatise addressed to readers with some previous knowledge of algebra (algebra in the sense of al-Khwārizmī, that is, not symbolic algebra). It contained no detailed discussion of how to set up equations and how to reduce them to the canonical cases using the techniques of *al-jabr* and *a'l-muqābala*, and it only discussed the canonical cases themselves.

Questions involving cubes—and even higher powers—of an unknown quantity, or the reciprocals of these powers, had come to the attention of Islamicate mathematicians especially in the wake of Qusṭā ibn Lūqā's Arabic translation of Diophantus' *Arithmetica*, toward the end of the ninth century. In his treatise on algebra, al-Khayyām presented 25 different cases of questions involving cubes, squares, roots and numbers, and showed how to solve each of them. As in al-Khwārizmī's earlier treatise, al-Khayyām solved an exemplary instance of each case. Al-Khayyām separated all cases into three groups, according to the number of "terms" appearing in the question. In modern algebraic terms, the cases can be described as follows:

(a) Simple questions:

$$x = a \qquad x^2 = ax \qquad x^2 = a$$
$$x^3 = c \qquad x^3 = ax^2 \qquad x^3 = bx$$

(b) Questions involving three terms:

$$x^2 + bx = c \qquad x^2 + c = bx \qquad x^2 = bx + c$$
$$x^3 + ax^2 = bx \qquad x^3 + bx = ax^2 \qquad x^3 = ax^2 + bx$$
$$x^3 + bx = c \qquad x^3 + c = bx \qquad x^3 = bx + c$$
$$x^3 + ax^2 = c \qquad x^3 + c = ax^2 \qquad x^3 = ax^2 + c$$

(c) Questions involving four terms:

$$x^3 + bx + c = ax^2 \qquad x^3 + ax^2 + c = bx \qquad x^3 + ax^2 + bx = c$$
$$x^3 = ax^2 + bx + c \qquad x^3 + ax^2 = bx + c \qquad x^3 + bx = ax^2 + c$$
$$x^3 + c = ax^2 + bx$$

As with al-Khwārizmī, the detailed classification into cases and the separate treatment of each case are closely connected with al-Khayyām's idea of number. Only positive numbers appear as coefficients and as solutions. If the possibility of using negative numbers as coefficients existed, then, obviously, not three groups but just three kinds of equations would appear here. And, if also zero could appear as a coefficient, then, of course, he would only consider a single equation, namely, the general cubic equation as we currently conceive of it:

$$ax^3 + bx^2 + cx + d = 0.$$

For his problems with cubes, al-Khayyām did not provide *formulas* that yield solutions in each case. But neither did he describe a procedure, or a rule, of the kind previously found in al-Khwārizmī's treatise. Al-Khayyām's solutions were *geometric*: he showed how, by intersecting two conic sections specifically constructed according to the problem, one can find a *line segment* that is the solution to the problem. In what sense is a segment a solution to an arithmetical problem? This is indeed an interesting issue. If we take the number obtained as a result of following the steps prescribed in al-Khwārizmī's procedures, we can directly check that it satisfies the conditions stated in the formulation of the problem. This cannot be done with al-Khayyām's solutions. In Appendix 5.2, I give the details of one of his solutions. It requires some effort on the part of the reader, but the effort will be gratified by its beauty and ingenuity. Here, I want to stress that the strictly geometric context of al-Khayyām's solution had obvious consequences for the way that numbers were presented and manipulated. Al-Khayyām took pains to carefully preserve dimensional homogeneity in all constructions appearing in his proofs. In the question of the type "cube and root equal number" ($x^3 + bx = c$), for example, it is clear from the text that the coefficient b represents for him an area (which when multiplied by x, which represents a length, yields a solid), whereas c represents a cube. Otherwise, we would find ourselves adding and comparing magnitudes, x^3, bx and c, that are not of the same type. This would make no sense from the point of view of the classical Greek tradition, and in this particular sense al-Khayyām closely followed this tradition.

Al-Khayyām was aware of the inherent conceptual tension created by treating in a strictly geometric manner problems that can also be conceived arithmetically. In the introduction to the treatise, he defined its subject matter as the treatment of unknowns that might be either numbers or measurable magnitudes. The problems to be solved involve relations between unknown and known quantities, and the unknowns are to be determined either numerically or geometrically. Now, while we look at the quantities from an arithmetical point of view, we can speak about the thing (or "root"), its square, and its cube, but also, by continued multiplication, about its square-square, its cube-cube and so on indefinitely. Moreover, this sequence of successive powers of a number was well known to Euclid (as a geometric progression of numbers), and he studied many of its properties in Book IX of the *Elements*. We also saw how Diophantus handled systematically the higher powers of the unknown (which for him was a number) as well as their reciprocals.

The meaning of this, I stress once again, is that while we are in the realm of the numbers, considering higher powers creates no real difficulty. But since al-Khayyām was also considering continuous quantities in his equations, it was necessary to warn about the nature of the entities involved. Since there are only three dimensions, the square-square cannot be a measurable magnitude, and much less can the higher powers. So, while working within the numerical realm, one enjoys, in principle, a greater freedom when it comes to considering higher powers of the unknown. In the realm of the continuous magnitudes, on the other hand, one has clear limitations.

But then, when it comes to solving the problems with the help of methods such as those developed by al-Khwārizmī, it is exactly the other way round: al-Khayyām was considering only numbers meant as collections of units, and hence only positive integers. Unlike continuous magnitudes, "numbers" (i.e., integer numbers) will not always fit the procedures of solution. For example, when considering the solution of the problem "A square and ten roots equal thirty-nine numbers" ($x^2 + 10x = 39$), al-Khayyām indicated that in terms of arithmetic solutions two conditions need to be met if we want the known procedure to work: (a) the number of the roots (here 10) must be an even number, so that its half is also a number; (b) the sum of the square of half the roots plus the number (i.e., $5^2 + 39$) must be a square number, so that its root is again an integer. If, alternatively, we approached the same problem geometrically, we would not reach the case of impossible solutions. Al-Khayyām expressed his disappointment that neither he nor any of his algebraic predecessors was able to find, for numbers, the kind of purely geometric solution that he presented in his treatise for the cubic equations. Hopefully, he said, someone else would succeed at that in the future.

There is another, shorter text written by al-Khayyām previous to his treatise on algebra, which provides additional insights on his hesitations about the right way to conceive the relations between numbers and continuous magnitudes. This treatise addressed a very specific geometric problem: to divide the quadrant of a circle into two parts that satisfy a certain predefined condition. The problem was important for dealing with issues arising in the astronomical calculations of Ptolemy's *Almagest*.

It belonged to a somewhat different tradition within Islamicate mathematics, where algebraic techniques were applied to solving problems not in arithmetic, but in theoretical geometry. I want to explain in a very sketchy way how he solved this problem, just in order to see what he had to say here about numbers.

In order to solve the problem of the quadrant, al-Khayyām proved that it may be reduced to a simpler one, namely that of constructing a specific right triangle that satisfies certain properties (Figure 5.6).

The task here is to find the length of the segment DB, given some specific conditions satisfied by the sides of the triangle. The first stage in solving this problem is to set up an "equation." Al-Khayyām begins by associating here the length 10 with AD, while taking DB to be the unknown, or the "thing." In order to find the value of the unknown, he used some geometric properties of the figure. On the one hand, using the Pythagorean theorem, we see that the square on AB is 100 plus the square of the unknown. On the other hand, the specific conditions stipulated for the triangle, according to the problem, translate into the following: the same square on AB is found to be equal to "one hundred in numbers and two squares and one tenth of one tenth of a square-square minus twenty of the unknown minus a fifth of the cube." This allows setting up the required "equation." Symbolically, this can be written as follows:

$$100 + 3x^2 + \frac{1}{10}\frac{1}{10}x^4 - 20x - \frac{1}{5}x^3 = 100 + x^2.$$

Already at this stage we see the conceptual problem that will arise, because of the appearance of a "square-square" (x^4). But let us go one step further and complete the explanation before returning to this point. In the second stage, al-Khayyām used the basic techniques of *al-jabr* and *al-muqābala*, in order to reduce this equation to a simpler one: $x^3 + 200x = 20x^2 + 2000$. In the third and last stage of the solution, al-Khayyām returned to the geometric context by restating the equation in purely geometric terms:[10]

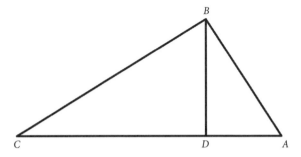

Figure 5.6 From al-Khayyām's treatise on the quadrant of the circle. The angle at B is right, and BD is drawn perpendicular to AC.

[10] (Rashed and Vahabadzeh 1999, p. 256).

To find a cube, which together with two hundred of its sides, is equal to twenty squares of its side with two thousand in number.

Relying on a result of Euclid's *Elements*, Proposition II.14, and with some further manipulations, al-Khayyām finally reaches the result: *DB* is 10. This is the solution to the problem of the quadrant, and at the technical level it is perfectly acceptable. But let us go back now to the conceptual problem, namely, that in the equation that was derived in stage one, al-Khayyām had to include a square-square. Unlike the numerical situation appearing in the treatise on algebra, in which the use of higher powers of the unknown does not give rise to any conceptual problem, here al-Khayyām is operating in a purely *geometric* context. How can he make sense of this situation, where more than three dimensions are involved? Remember that this is a situation already encountered in late Greek mathematics, and it will continue to accompany us in the next chapters as well, because it is at the heart of the question of what is a number and what is the right way to define the relationships between discrete and continuous quantities.

Al-Khayyām was aware that there is a real problem here. He tried to provide a satisfactory explanation, by defining a number as being abstracted by the mind from material things. The thing, its square, and its cube can all be associated with geometric entities of the corresponding number of dimensions up to three. But what about the higher dimensions? What he wrote here in this regard is very interesting:[11]

I say: that which the algebraists call square-square is something imagined in the continuous magnitudes that does not exist in any way in the individuals. . . . The square-square, which algebraists consider to be the product of a square by itself, has no meaning whatsoever in the continuous magnitudes, because, if the square is a surface, how can it be multiplied by itself? . . . A body can have no more than three dimensions. All what is said in algebra is said about these four kinds [number, line, square, cube]. And those who believe that algebra is an artifice devoted to the determination of unknown numbers believe the impossible. . . . No doubt, *al-jabr* and *al-muqābala* are geometrical things that have been demonstrated in Book II of the *Elements*, Propositions 5 and 6.[12]

There is a problem, then, with entities such as square-square, but at the same time, it was known that the techniques of al-Khwārizmī allowed solving a problem such as "square-square and three squares equals twenty-eight numbers" (i.e., $x^4 + 3x^2 = 28$). All that was needed was to write "square" instead of "square-square" and "root" instead of "square" (i.e., $y^2 + 3y = 28$, or what we call nowadays "change of variable"). This is easily solved, and the unknown is found to be 2, its square is 4, and the square-square is 16. But al-Khayyām warned that this solution should not lead to confusion:

This person is convinced that he found the square-square with the help of algebra. But this opinion is feeble. He has not really been dealing with "square-square," but rather just with "square." This is a kind of mystery that will lead you to unveil many others.

---

[11] (Rashed and Vahabadzeh 1999, pp. 248–250).

[12] In Appendix 8.2, I discuss Euclid's Proposition II.5 and its relations to geometry and algebra. The details of the discussion there clarify why al-Khayyām refers here specifically to it (as well as to II.6, which is quite similar to II.5).

Since in this treatise on the quadrant, al-Khayyām presented not only the standard procedures for solving the canonical cases (as he did in his treatise on algebra), but also the initial stages of setting up and then reducing the equation, we gain here additional insights into his views of magnitudes and numbers. All the entities involved in the first stages (the numbers, the thing, the square, and the cube) are used, equally and indifferently, to measure lines, surfaces, or bodies. Moreover, when al-Khayyām sets up the equation, we notice that a geometric square, namely the square on AB, appears as an aggregation of five different powers: "one hundred in *numbers* and two *squares* and one tenth of one tenth of a *square-square* minus twenty of the *unknown* minus a fifth of the *cube*." In other words, in these two stages, al-Khayyām considered all the magnitudes involved to be *abstract, homogeneous, dimensionless* quantities. And then, only in the final step was the purely geometric context of the problem restored and both the dimensionality of the magnitudes and the requirement of homogeneity regained their importance.

Al-Khayyām claimed in this treatise on the quadrant that the unknowns are abstractions of the *continuous* magnitudes. Hence, algebra can only be used to solve problems in geometry. Later on, however, in his treatise on algebra, he wrote that algebra can also be used to solve problems in arithmetic. We saw above how he handled such arithmetical problems with the help of continuous magnitudes. For problems involving the square of the unknown he gave two solutions: a numerical rule and a geometric construction. But besides the specific context of the problem and its final solution, al-Khayyām played around interestingly with the idea that in the specifically algebraic part of the procedure, all the quantities involved were neither numbers nor concrete geometric magnitudes of specific types, but rather "dimensionless" and hence abstract magnitudes.

These views are quite remarkable when seen retrospectively from the perspective of later developments. In the following chapters, we will be seeing further developments toward a more general and abstract notion of number, disconnected from either a specific geometric context (where the dimension of the continuous entities is considered of fundamental importance) or a view of them as collections of discrete units. In this sense, al-Khayyām's views go precisely in the direction that will eventually became central and dominant. However, it seems that his incipient insights were somewhat vague and had little resonance among Islamicate practitioners of algebra. Consequently, they were also unknown to European mathematicians in the middle ages and the Renaissance. The requirement of dimensional homogeneity continued to be central to the mainstream view on what numbers are and how equations have to be handled.

5.9 Gersonides and problems with numbers

As already said, some Jewish and Christian scholars participated in Islamicate mathematical culture. As with Arabic mathematical texts, historians have recently devoted increased attention to Hebrew medieval texts, including some that were never previously published or known. These texts contain many original ideas, including those related to numbers and arithmetical practices. Although mainly rooted in the mainstream traditions of contemporary Islamic mathematics, Jewish scholars who were active between the eleventh and fifteenth centuries drew their ideas from a variety of sources. In general, they did not feel constrained by, or exclusively committed to, any one of their sources. This allowed them to develop original approaches to various issues, and to devote attention to topics that were of special interest to them, such as Kabbalistic debates that elicit combinatorial questions. Many, but hardly all, of the Hebrew mathematical texts were translations—sometimes with comments and more or less substantive additions—of Arabic ones that were then in extended use. The story of these medieval Jewish scholars working in mathematical topics is a fascinating and little-known one.

Two important names to be mentioned in the Hebrew mathematical traditions are those of Abraham Bar-Ḥiyya ha-Nasi (1070–1136) and Abraham Ibn-Ezra (1092–1167). Both men worked in the Iberian peninsula and published influential and learned treatises in a variety of topics, mostly connected with Biblical exegesis. In 1116, Bar-Hiyya published his *Ḥibbūr ha-meshīḥah we-ha-tishboret* ("Treatise on Measurement and Calculation"), a book that also appeared in a Latin version in 1145 as *Liber Embadorum*. This Latin text introduced for the first time in Europe the techniques of Arabic algebra for solving quadratic equations, antedating in this way even the first Latin translation of al-Khwārizmī's *Algebra* by Gerard de Cremona (ca. 1114–ca. 1187). The *Liber Embadorum* was widely read, and it is known to have directly influenced Leonardo Fibonacci. Bar Ḥiyya's methods were even more geometrically oriented than al-Khwārizmī's. His unknown was always a geometric object. Ibn-Ezra's *Sefer ha-Ehad* ("Book of the One") and *Sefer ha-Mispar* ("Book of Numbers") were among the first to transmit to Europe (though without attracting much attention) the basics of the positional–decimal system and the Hindu ciphers.

Other interesting scholars were also active in later times, such as Yitzhak Ben Shlomo al-Aḥdab (ca. 1350–ca. 1429). Following a request from the Jewish community in Syracuse (Sicily), he translated Ibn al Bannā's *Talkhīṣ* into Hebrew. He added commentaries and explanations, which included original ideas concerning the system of numeration, solutions to problems with numbers, and solutions to arithmetical questions related to day-to-day life. He adapted some of the symbolic notation developed by the successors of Ibn al Bannā' while using Hebrew letters instead of the Arabic ones.

But if we have to devote some more detailed attention here to one medieval Jewish scholar of particular interest, then the choice will of necessity fall on Levy Ben Gerson, Gersonides (1288–1344), who was active in the area of Provence. Gersonides published many books in various fields of interest, scientific as well as Biblical exegesis. The most mathematically interesting of them, published in 1321, is *Ma'aseh Hoshev* (roughly, "The Work of the Calculator"). Gersonides was well acquainted with the main mathematical sources from both the Greek and Islamicate traditions, but his books also contain many original ideas that, like other works of the Islamicate tradition, probably remained unknown in Europe.

Ma'aseh Hoshev deals with arithmetical topics. It relies explicitly on the arithmetic books of the *Elements*, but goes well beyond the contents of this classical treatise. Gersonides, for example, begins by proving some elementary properties of arithmetical operations, such as the associative property of the product, $a \cdot (b \cdot c) = (a \cdot b) \cdot c$. This is a far from trivial point, since, as already mentioned, even in Euclid there is nothing similar to an axiomatic presentation of arithmetic. Thus, Gersonides is among the few mathematicians who thought that elementary properties of this kind have to be discussed in a general way, to begin with. On the other hand, he was far from systematic or exhaustive in his treatment, and it seems that he only proved properties, such as associativity, when this was immediately needed for proving more complex propositions in the text. The entire question of the axiomatic presentation of arithmetic and of what needs to be assumed of the natural numbers in order to develop arithmetic as a whole would remain essentially unattended until the late nineteenth century.

Like al-Khwārizmī and Abū Kāmil, Gersonides based some of his proofs of arithmetical propositions on geometric results taken from Book II of the *Elements*. But, at the same time, he also introduced new types of proof, including proofs that come truly close to the modern idea of mathematical induction. In this, he may have been influenced by earlier mathematicians such as al-Karajī of al-Samaw'al, but this cannot be ascertained with certainty.

What is of particular importance for our story here is the way in which numbers appear in Gersonides' book, and the interesting results he was able to achieve even in the absence of a general, flexible system of notation. In retrospect, a modern reader faced with his text may only wonder how far Gersonides could have taken these ideas had he been able to rely on such a symbolic language. Two results that clearly exemplify the tension between the originality of his ideas and the limitations imposed on them by the lack of an adequate symbolic language are Propositions 26 and 27 of Book I of *Ma'aseh Hoshev*, which can be formulated as follows:[13]

I.26: If we add all consecutive numbers from one to any given number and the given number is even, then the addition equals the product of half the number of numbers that are added up times the number that follows the given even number.

I.27: If we add all consecutive numbers from one to any given number, and the given number is odd, then the addition equals the product of the number at half way times the last number that is added.

In modern terms, we may write these two results as follows:

- If n is an even number, then $1 + 2 + 3 + \ldots + n = \frac{n}{2} \cdot (n + 1)$.
- If n is an odd number, then $1 + 2 + 3 + \ldots + n = \frac{n+1}{2} \cdot n$.

But notice that if we have already decided to translate the two results into modern algebraic language, then there is no apparent reason why we should not simply write them as a single arithmetic proposition, namely,

$$1 + 2 + 3 + \ldots + n = \frac{n \cdot (n+1)}{2}.$$

Why then did Gersonides present this single result as two separate ones? Well, this example of Gersonides is a wonderful illustration of the point I have been stressing all along, namely, that the mathematical language used in a specific historical context

---

[13] Quoted from (Lange 1909, pp. 15–16). My translation from the Hebrew text.

is far from being neutral and that it has a meaningful impact on the ideas that are being handled. For Gersonides, the two cases were really different, and there was no way he could realize that the two situations (which he indeed described differently in his purely rhetorical rendering) were one and the same as they are for us. And, more interestingly, if we look at the *proofs* he gave for the two propositions, then their difference becomes even more perspicuous. The proofs are, of course, totally rhetorical, and no symbolic manipulation appears in them. As with Euclid's proof of the infinity of prime numbers, which we discussed in Chapter 3, the first difficulty here is to express the idea of "any given number" or, as we would say nowadays, "an arbitrary number of elements that are added." Gersonides denoted his numbers by Hebrew letters, adding a dot on top of each. Wherever he needed "an arbitrary even number," he always took six: "Let the successive numbers be אבגדהו, and the number right thereafter ז." Of course he intended his argument not to be limited to six numbers, but rather to be universally valid.

The proof of Proposition 26 is based on the idea of forming pairs of numbers with equal sums: if we add 1 and 6, we obtain 7 (which is the number that follows the last given number), and, since 2 is obtained from 1 by adding 1 whereas 5 is obtained from 6 by subtracting 1, then it follows that adding 2 and 5 is the same as adding 1 and 6 and hence 2 + 5 equals 7. And a similar argument is valid, of course, in the case of 3 and 4. And how many such pairs are there here? Precisely half of the total of numbers. Hence, the total sum equals half the total number of numbers times the number that follows after the number of numbers (7 in this example), as stated in the proposition. Gersonides made sure to indicate that his argument is valid without limitation for any even number of elements added up. But, obviously, this argument is not valid if one is summing an odd number of elements—hence the need for Proposition 27. This time, Gersonides took the numbers 1 to 7, which he indicated as אבגדהוז, and he argued as follows:[14]

The difference from 3 to 4 equals the excess of 5 over 4. Hence 3 and 5 add up to twice 4 (which is the number in the middle of the given sequence). Likewise, the difference from 2 to 4 equals the excess of 6 over 4. Hence 2 and 6 add up to twice 4, and the same argument is valid for the remaining pair, 1 and 7. Hence the overall sum is obtained by adding to the middle number the number of such pairs times twice the middle number. And of such pairs we have as many as the middle number minus one. From here we conclude, as stated in Proposition I.27, that if we add all consecutive numbers from one to any given number and the given number is odd, then the addition equals the product of the number at half way times the last number that is added.

All of this is more transparent and easier to follow when written in the symbolic language that Gersonides could not take advantage of:

Each of the mentioned pairs add up to half the middle number, or up to $2 \cdot \frac{(n+1)}{2}$; there are $\frac{(n-1)}{2}$ such pairs, hence the overall sum is $\frac{(n-1)}{2} \cdot 2 \cdot \frac{(n+1)}{2} + \frac{(n+1)}{2}$, or $\frac{(n+1)}{2} \cdot n$, as stated.

Again, neither the statement nor the proof make sense if n is even.

---

[14] (Lange 1909, p. 16).

An additional, interesting detail that one finds in these proofs is that when Gersonides presented the successive numbers, א ב גונה, he explicitly stressed that א is the unit and "nonetheless, we will call it here number." Why so? Remember that Euclid defined number as a "collection of units", and this definition involves an ambiguity concerning the status of the unit itself. There is no doubt that Gersonides' comment here reflects the continued impact of this ambiguity. When one is interested in properties of, say, primality and divisibility of numbers, it may be more convenient to not consider the unit to be a number. But then, in the context of problems such as Gersonides was dealing with, if we sum a series of successive numbers, it is more convenient, and indeed necessary, to consider the unit as a number like any other natural number. And indeed, this is what Gersonides did here as well as in other cases.

Appendix 5.1 The quadratic equation. Derivation of the algebraic formula

Most readers of this book may certainly remember that the formula

$$\frac{-b \pm \sqrt{b^2 - 4ac}}{2a}$$

provides a general algebraic solution to the general quadratic equation $ax^2 + bx + c = 0$ (with $a \neq 0$). Many, I guess, may have forgotten how the formula is derived, and hence a reminder is in place here. It takes a few simple steps of symbolic manipulation, as follows:

move c to the right-hand side of the equation: $ax^2 + bx = -c$;
multiply both sides by $4a$: $4a^2x^2 + 4abx = -4ac$;
add b^2 to both sides: $4a^2x^2 + 4abx + b^2 = b^2 - 4ac$;
the expression on the left is a "perfect square",
and hence the equation can be re-written as $(2ax + b)^2 = b^2 - 4ac$.

This last step is, of course, the crucial one in the derivation, because we have obtained on the left-hand side of the equation the expression $(2ax + b)^2$, which contains the unknown x and at the same time is a perfect square, whereas on the right-hand side we only have the coefficients, and the unknown does not appear at all. This is why one says that the derivation of the formula is based on "completing squares." If we now take square roots on both sides of the last expressions, we obtain the following:

$$2ax + b = \pm\sqrt{b^2 - 4ac},$$

and from here it is trivial to isolate the unknown and obtain our formula:

$$\frac{-b \pm \sqrt{b^2 - 4ac}}{2a}.$$

From the point of view of our historical account, it is important to stress that the crucial step of completing squares is performed differently at various historical stages. But before the introduction of a full algebraic language affording the kind of flexible symbolic manipulation displayed in this short derivation, completion of the square was always performed in the geometric fashion described in Section 5.3 in relation to the example of al-Khwārizmī.

Appendix 5.2 The cubic equation. Khayyam's geometric solution

Al-Khayyām in the eleventh century found a geometric solution to a problem of the type "cubes and squares and roots equal numbers," or, in modern algebraic notation, $x^3 + ax^2 + b^2x = b^2c$. The letters a, b, c in this equation represent what for al-Khayyām were three given straight-line segments, and also x is a segment, namely, the unknown one we are requested to find. The algebraic formulation that I adopt here is far removed, of course, from the geometric context in which the problem was originally conceived, but I have made sure to preserve in this formulation the dimensional homogeneity that is typical of the original context. All the terms in the equation represent three-dimensional magnitudes.

In presenting al-Khayyām's proof in this appendix, I will keep this kind of compromise between the modern algebraic and the original geometric, to make it easier for the modern reader to follow and at the same time not to lose all of the original flavor. At the end, I will add a fragment translated from the original text. The geometric proof is based on a very good command of the properties of conic sections, which al-Khayyām obtained from Apollonius' books and which he studied carefully. Euclid's *Elements* presented only knowledge corresponding to "straightedge and compass" techniques, and al-Khayyām was certainly aware that such techniques were not appropriate for solving the problem in question. Al-Khayyām's solution is based on the diagram in Figure 5.7.

The diagram represents the following construction. Let BD, GB and BH be given, and place them as in the diagram (*GB* and *BD* on the same straight line; *BH* orthogonal). With *GD* as diameter, draw circle *GZD*. Draw *HK* parallel to *GD*, and *GK* on the tangent to *GZD* at *G*. Extend *HB* in the direction of *A*, and draw a hyperbola through *G* such that *KH* and *HA* are its asymptotes. *Z* is the intersection of the hyperbola with the circle *GZD*, and *ZA* is parallel to *GD*, while *TZ* is parallel to *HA* and intersects *GD* at *L*. Notice that to the original diagram I have added here an indication for the segments that are represented in the equations by magnitudes a, b, c, as follows: $a = BD$, $c = BG$, $b = BH$ (here *BH* is orthogonal to *GD*).

Notice that *KH* and *AH* are by no means axes of coordinates in the sense known to us in analytic geometry, even if they look so. They are, instead, straight lines that are defined in purely geometric terms, on the basis of Apollonius' research on conic sections. Given a hyperbola (in our case *GZ*), Apollonius proved the existence of two straight lines (here *KH* and *AH*), with the following property: for any given point in the hyperbola, the rectangle constructed with the point as vertex and the two lines as sides has a fixed area (in our case, from point *G* we draw the rectangle on *BH*, *BG*,

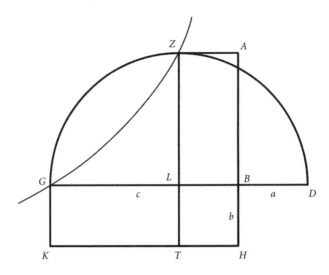

Figure 5.7 Khayyam's geometric solution of a cubic equation.

and from Z we draw the rectangle on ZT. ZA, having equal area). Apollonius also proved (in *Conics* II.4) that given any point and two straight lines in position, it is always possible to draw a hyperbola through the point with the lines as asymptotes. Moreover, he showed that if the hyperbola intersects a circle at G, then it also cuts it at a different point, say Z. Clearly, al-Khayyām's hyperbola ZG in the diagram is defined on the basis of this property. The segment ZA (or equivalently BL), stated al-Khayyām finally, is the solution to the problem (or the unknown x in our equation). This he proved as follows.

First, following Apollonius' result, we have $BH \cdot BG = ZT \cdot ZA$. Using the lengths a, b, c, we can also express this as $BH \cdot BG = ZT \cdot ZA = bc = ZA \cdot (ZL + b)$. Hence,

$$ZL \cdot ZA = b \cdot (c - ZA). \tag{5.2}$$

Second, from a very basic property of the circle, we have $DL \cdot LG = ZL^2$. Hence,

$$ZL^2 = (a + ZA) \cdot (c - ZA). \tag{5.3}$$

From the identities (5.2) and (5.3), we can immediately derive the following two proportions:

$$ZL : (c - ZA) = b : ZA, \tag{5.4}$$

$$ZL^2 : (c - ZA)^2 = (a + ZA) : (c - ZA). \tag{5.5}$$

But from Eudoxus' theory, al-Khayyām knew well that if four magnitudes are in proportion, then so are their squares, and hence, from (5.4), we obtain the following:

$$ZL^2 : (c - ZA)^2 = b^2 : ZA^2. \tag{5.6}$$

Then, by combining the proportions (5.5) and (5.6), we obtain

$$b^2 : ZA^2 = (a + ZA) : (c - ZA) \tag{5.7}$$

And finally, transposing denominators and multiplying by numerators in (5.7), we obtain

$$ZA^3 + aZA^2 + b^2 ZA = b^2 c. \tag{5.8}$$

This is perhaps not transparent yet, but notice that (5.8) is precisely the statement that ZA satisfies the equation of the problem: $x^3 + ax^2 + b^2 x = b^2 c$! Thus, al-Khayyām's construction indeed yields the value of the requested unknown! But what kind of solution have we actually achieved? Proposition II.4 in Apollonius' *Conics* does not show a constructive procedure that actually yields the hyperbola, and certainly not a straightedge and compass construction of it. The proposition simply asserts the existence of a certain hyperbola "in position." The problem can only said to have been solved in a "theoretical" way, in the sense that we are aware of the existence of a certain line segment that satisfies the property stipulated in the problem and that, moreover, this segment is located at the intersection of a certain circle and a certain hyperbola. Clearly, this is not a "practical" solution, in the sense that if we are given three lengths a, b, c, we will know how to build the segment of length x that fits the equation.

But, as already mentioned, even in my description of the solution, I have used many elements of modern algebraic manipulation. I would therefore like to cite a fragment of the original text, so that the reader can face with no intermediaries al-Khayyām's style of proof:[15]

We draw BH to represent the side of the square equal to the given sum of the edges, and construct a solid whose base is the square of BH, and that equals the given number. Let its height BG be perpendicular to BH. We draw BD equal to the given sum of the squares and along BG produced, and draw on DG as diameter a semicircle DZG, and complete the area BK, and draw through the point G a hyperbola with the lines BH and HK as asymptotes. It will intersect the circle at the point G because it intersects the line tangential to it [the circle], i.e., GK. It must therefore intersect it [the circle] at another point. Let it intersect it [the circle] at Z whose position would then be known, because the positions of the circle and the conic are known. From Z we draw perpendiculars ZT and ZA to HK and HA. Therefore the area ZH equals the area BK. Now make HL common. There remains [after subtraction of HL] the area ZB equal to the area LK. Thus the proportion of ZL to LG equals the proportion of HB to BL, because HB equals TL; and their squares are also proportional. But the proportion of the square of ZL to the square of LG is equal to the proportion of DL to LG, because of the

---

[15] Quoted from (Fauvel and Gray (eds.) 1998, pp. 233–234).

circle. Therefore the proportion of the square of *HB* to the square of *BL* would be equal to the proportion of *DL* to *LG*. Therefore the solid whose base is the square of *HB* and whose height is *LG* would equal the solid whose base is the square of *BL* and whose height is *DL*. But this latter solid is equal to the cube of *BL* plus the solid whose base is the square of *BL* and whose height is *BD*, which is equal to the given sum of the squares. Now we make common [we add] the solid whose base is the square of *HB* and whose height is *BL*, which is equal to the sum of the roots. Therefore the solid whose base is the square of *HB* and whose height is *BG*, which we drew equal to the given number, is equal to the solid cube of *BL* plus [a sum] equal to the given sum of its edges plus [a sum] equal to the given sum of its squares; and that is what we wished to demonstrate.

CHAPTER 6

Numbers in Europe from the Twelfth to the Sixteenth Centuries

The mathematical activities of Gersonides and of his fellow Jewish scholars, discussed in Chapter 5, are directly connected with the Islamicate tradition in terms of their content, style and sources. From the point of view of their geographical context, however, these activities took place in the framework of late medieval Europe. In the centuries following the fall of the Western Roman Empire in 476, scholarly activity in general, and mathematics in particular, was drastically reduced in Europe. A first important turning point came in the twelfth century, when translations of scientific texts, especially those connected with the Islamicate tradition, started to appear in various European centers.

A particularly active school of translation developed in Toledo, Spain, after the Catholic reconquest of 1085. A large number of Arabic scientific manuscripts were found by that time in Toledo, where also a large and prosperous Jewish community existed. Many among the local Jews were proficient in Arabic, and in other languages such as Spanish, Hebrew or Latin, and so they became significant agents of transcultural exchange. Often, scientific texts were translated in two stages: first from Arabic to Spanish, and at a later stage from Spanish to Latin.

By the mid thirteenth century a great number of Arabic scientific texts had been translated into Latin, partly translations of Greek works and partly original works of the Islamic mathematical tradition. With the help of these translations, the works of many prominent mathematicians—Euclid, Archimedes, Apollonius, Ptolemy, al-Khwārizmī, Abū Kāmil, Bar-Ḥiyya and others—started to be studied by European scholars.

These texts had a great impact on the world of learning at large, not just in its content but also in its institutions. At this time, existing cathedral schools developed into universities at Bologna, Paris and Oxford, and new universities were created at Cambridge and Padua. These, as well as other budding centers of intellectual activity, provided the crucial institutional support for the Latin scholastic tradition of the Late Middle Ages. Students intent on entering the higher faculties of law, medicine, or theology had first to earn a degree of Master of Arts, by studying the seven liberal arts, which comprised the *quadrivium* (arithmetic, geometry, astronomy and musical theory) and, of greater importance at the time, the *trivium* (grammar, logic and rhetoric). Aristotelian philosophy was also a core topic of the curriculum.

Euclid's *Elements* enjoyed a prominent position in the intellectual world of the Latin West, and a great number of versions of it circulated in Europe at the time. Among the most influential translators of the treatise were Adelard of Bath (ca. 1080– ca. 1150), Robert of Chester (fl. c. 1150), and, somewhat later, Campanus de Novara (1220–1296). Campanus' text was the one that in 1482 appeared as the first printed version of the *Elements*. It became the standard source of reference for the Latin world until the sixteenth century, when new versions were printed, based on direct translations from the Greek.

Under the influence of some general trends, typical of the scholarly activity of time, the most widely circulated versions of the *Elements* differed in interesting ways from the Euclidean original. Prominent among these was a marked didactical spirit that directly influenced the style of presentation. In most medieval versions of the *Elements*, we find a careful labeling of the various sections of the proofs, direct references to earlier propositions on which a proof is based, and hints about other propositions that are directly related to the one considered. Nothing of this can be found in the original, very austere Euclidean version. This increased attention to the explicit analysis of the logical structure of the text gave rise to original discussions about the nature of incommensurable magnitudes and the double treatment of proportions in the *Elements*: once in the realm of magnitudes in general (Book V) and once in the realm of numbers (Book VII). The rigid separation between these two realms in the original treatise was highly unsatisfactory for Latin medieval mathematicians. They looked for new approaches in which a unified treatment could be reached, and the search after such a unified treatment would be critical for the continued development of ideas about numbers.

Another, related topic that attracted the attention of translators and commentators was the lack of an axiomatic foundation for the arithmetic books of the *Elements*. Attempts at conceptual clarifications of the foundations of arithmetic appear in several texts, or in commentaries on individual results in the *Elements*. The most elaborate of these appears in a text by Jordanus Nemorarius, who lived in the early thirteenth century, but about whose life very little is known with certainty. His treatise, *De elementis arismetice artis*, continued to appear as a printed textbook up until the early sixteenth century. This text contains a systematic attempt to present the body of arithmetic as known at the time—including the basic properties of the arithmetic operations with numbers—based on a short list of axioms, such as Euclid had provided for geometry many centuries before. This part of the book, however, was perhaps its weakest, and it did little to further the discussion on the nature of numbers. It seems to have aroused little interest among the very few readers of Jordanus' book.

The turn of the fifteenth century saw the decay in the vitality that characterized the early stages of Scholasticism over the twelfth and thirteenth centuries. It also saw the rise of Renaissance humanism as a main trend of intellectual activity in Western Europe. This trend promoted the rediscovery and re-evaluation of the main cultural values of classical civilization seen as worthy of pursuit and study for their own sake, rather than for their relevance to Church doctrine. Renaissance humanists developed an active interest in manuscripts recording the intellectual and scientific treasures of the classical world, some of them long held in monasteries and forgotten for centuries. In the wake of the fall of Constantinople in 1453, many new manuscripts reached Italy, along with a considerable influx of expatriate Byzantine scholars that helped translating them. The invention of the printing press, of course, was another seminal turning point that strongly contributed to the dissemination of the classical texts cherished by the humanists, alongside their cultural values.

As an outcome of the activities of the Renaissance humanists, and side by side with the many versions of the *Elements*, and the other mathematical treatises that had initially been translated from Arabic, larger numbers of Greek mathematical manuscripts now started to be translated in Europe directly into Latin and also distributed in print. The first printed version of the *Elements*, for instance, appeared in 1482 in Venice. The first published Latin version, which was a direct translation from a Greek source, appeared in 1505, also in Venice.

The works of Diophantus were not known in Europe before 1463. When the Renaissance humanists started to read Diophantus' *Arithmetica*, they assumed that this was the basis out of which Arabic algebra had developed. Many of them even considered that this was the main, if not only, source of the algebraic-like ideas that were increasingly developed in Europe beginning in the twelfth century. In this way, a rather pervasive myth evolved as part of the mathematical culture of the Renaissance. It postulated a continued European mathematical tradition with its beginning in ancient Greece and with a somewhat lengthy, but not very significant, hiatus of some centuries during which the Arab mathematicians merely translated, "preserved," and "transmitted" this purely European heritage.

More generally speaking, the picture of the mathematical body of knowledge of earlier cultures, as conceived in Europe at this time, was far from historically accurate. For one thing, some of the earlier texts became available only partially and gradually. For another thing, the very act of translation was challenging not only from the point of view of language, but also mathematically. Debates about the actual meaning of the terms appearing in Arabic mathematical treatises continue to occupy historians to this very day. Early European translators obviously had to invent new terms as part of their work, or to rely on existing ones that did not always really fit. But they also incorporated, as a matter of course and without much fuss, new mathematical ideas, new terminologies and new notational approaches that were never part of the original texts. Underlying their difficult task was an inherent tension between two basic forces that struck different kinds of balances in each case: on the one hand, the drive to remain faithful to the classical sources, and on the other hand, the desire to produce more streamlined mathematical texts that would be easier to comprehend by their intended readerships. This underlying tension was often a fruitful source of mathematical innovation, even if the translators themselves were not fully aware of it.

6.1 Fibonacci and Hindu–Arabic numbers in Europe

The mathematical knowledge embodied in the many classical texts translated into Latin was initially available to rather reduced audiences. These were the audiences connected with the big cathedrals and universities during the scholastic period, and with the secular intellectual elites of Western Europe during the humanist period. These limited audiences, however, grew and diversified gradually and in parallel with the great intellectual trends of the period, particularly as books devoted to the more practically oriented traditions of calculation started to appear. Some of these texts were initially written in Latin, but texts in the vernacular languages began to circulate in Europe as early as the thirteenth century. Of course, this also added to the breadth and variety of their readership. Such books differed in content and style from the contemporary circulating translations of the classical texts, and they also implicitly conveyed a different view of numbers. For one thing, they incorporated arithmetic knowledge derived from the practical traditions of calculation related to trade, especially in the Mediterranean context. One of the earliest, and most well known of these books was the *Liber Abbaci* (1202), written (still in Latin) by Leonardo Pisano, Fibonacci (ca. 1170–ca. 1240).

Fibonacci was well acquainted with Greek, Byzantine and Arabic sources. He may have acquired this knowledge in his many trips to the lands of Islam. He played a key role in compiling and helping disseminate in Western Europe innovative mathematical techniques and concepts, particularly the use of the Hindu–Arabic system of numeration and methods for solving equations. A reader of the book would find there a systematic presentation of this system, including the nine ciphers and zero, and an explanation of how to perform the four arithmetic operations with them. Fibonacci also introduced in the book the techniques of al-Khwārizmī for solving problems that involve squares of the unknown. Likewise, he also taught how to solve various kinds of arithmetic problems, of a more practical character, intended for use in commerce: gain-and-loss calculations, currency exchange and similar problems. Such problems were often formulated so as to exemplify certain general techniques.

Here is one example: A pit is 50 feet deep. If a lion climbs up 1/7 of a foot each day and drops 1/9 of a foot in the evening, how long will it take him to get out of the pit? This problem was solved with the help of the technique known as "false position," a method with roots going back at least to the Egyptian traditions. It starts by assuming a certain value, which is then adjusted to reach the right solution. In this case, Fibonacci started by assuming 63 days to be the answer, given that both 7 and 9 divide 63. In 63 days, the lion would climb up 9 feet and fall down 7. From here, he deduced that for climbing 50 feet, the lion would need 1575 days. His answer, by the way, happens to be wrong, given that after 1571 days, there only remains 8/63 of a foot to reach the top, so that 1572 will be enough. Another illustrative example is solved by following an approach more akin to the methods of Arabic algebra: to calculate the amount of money that remains in the hands of two persons after they have exchanged two sums that are in a given ratio. In this context, Leonardo taught calculation techniques with fractions, and not only with integers.

The kinds of techniques that Leonardo taught the Europeans in the *Liber Abbaci* ("Book of Calculations") were all part of the practice of Islamicate mathematics before the tenth century. He wrote other books that reached more reduced audiences and that dealt with more complex problems, both in geometry and in arithmetic. Such books provide additional evidence of the historical complexity of the processes that led to the consolidation of the modern understanding of number and of algebraic equation. In a book of 1225 entitled *Flos*, Fibonacci dealt with problems involving cubes of the unknown, similar to those we saw in the case of al-Khayyām. Possibly, Fibonacci learnt of these problems in Abū Kāmil's books. But, unlike al-Khayyām or Abū Kāmil, Fibonacci also taught how to obtain approximate numerical solutions, rather than only purely geometric, theoretical ones. Faced with the exemplary problem "ten roots, and two squares, and one cube equal twenty numbers" (symbolically, $10x + 2x^2 + x^3 = 20$), Fibonacci stated that "it is not possible to solve it," by which he meant that there is no solution that is an integer or a fraction, or a square root of a fraction (or, to state it in Pappus' language, the problem cannot be solved with plane methods).

Al-Khayyām, as we saw, solved such problems with the help of conic sections (or, in Pappus' language, solid methods). Al-Khayyām had also explained that this is the only possible way to solve such problems. Leonardo suggested a solution that was essentially arithmetic, as befitted his overall approach. The details of his solution are not particularly important here, but what is really important is that he chose *not* to present his solution with the help of the decimal system that he had himself taught in his previous books. Rather, Fibonacci expressed an approximate numerical solution in a sexagesimal expression (base sixty), as follows: 1.22.7.42.33.4.40. This was Fibonacci's way to express the value $1 + \frac{22}{60} + \frac{7}{60^2} + \frac{42}{60^3} + \frac{33}{60^4} + \ldots$ Incidentally, this also happens to be a rather exact approximation, since, translated into decimal values, it yields 1.3688081079, which is a correct solution, up to the ninth decimal place. So, here we have this mathematician who is well aware of the technical details of the decimal system and of its advantages, but at the same time he does not think that all problems must be solved and expressed in terms of one and the same system. For a calculation arising in a more geometric context, perhaps given the traditional connections with trigonometry and astronomy, he preferred to use the sexagesimal system over the decimal one. The decimal system he preferred to keep for the arithmetic context of his earlier book.

6.2 Abbacus and coss traditions in Europe

New developments related to the use of numbers began to appear at different places in Europe—above all in Italy—in relation to new needs created by the accelerated renewal of commercial activity and urban development. In the second half of the thirteenth century, there appeared a new profession, *maestri d'abbacus*, masters of calculation. They were involved in the writing of textbooks and in the teaching of the new generations of merchants, so as to provide them with arithmetical tools necessary for their trade. This gave rise to a new type of mathematical culture, fully devoted to the solution of problems with numbers. Institutionally as well as in terms of language, it developed mainly outside the scholarly world. Its texts were written in vernacular Italian. The decimal system and the kind of techniques introduced in Fibonacci's book, following the Islamicate traditions, were central to this culture. The more problems they solved and the more books there appeared as part of the tradition, the more the decimal system became entrenched as the preferred one for dealing with those problems. Eventually, the decimal–positional system was sweepingly adopted, and other systems that were still in practice at the time (such as the cumbersome Roman system) soon fell into oblivion. In addition, new techniques for solving problems, alongside new ways to conceive of numbers, accompanied those processes.

The abacus tradition exhibits several threads of mathematical thinking that, as they further developed, became central to the new views on numbers that came to dominate mathematics in the seventeenth century. Among them, we can mention the following:

- the development of enhanced methods for problem-solving;
- the increased use of abbreviations in mathematical texts, which eventually turned into symbols that could be formally manipulated;
- the increasing legitimation of the use of broader classes of numbers (e.g., negative and irrational);
- the emergence of an autonomous realm of arithmetical knowledge with little connection with geometry.

These threads developed in a slow, non-linear and rather erratic manner, and with relative independence from each other. Also, they appeared differently combined in the works of various mathematicians. I would like to consider now some examples that illustrate this point.

Techniques for problem solving Such techniques were presented in a very large number of abbacus texts that spanned a period of about two hundred years. These texts typically contained a long list of up to two or three hundred problems. In the early stages, the problems were formulated in rhetorical terms, with no symbols or even abbreviations. Some typical problems are similar to those solved by Diophantus where we are asked, for example, to find two numbers given their product and the sum of their squares.

In the solution, the abbacus masters assumed the existence of an unknown quantity, the thing, or "*cosa*," and in general the adequate choice of this unknown was a key step leading to a correct solution. In the example above, the *cosa* would stand for the first of the two numbers. In most cases, only one unknown was defined in each problem, but in a few instances, two unknowns could also appear. Using the *cosa*, its square, and additional "numbers," the problem was translated into a sort of "equation," purely rhetorical and not involving symbols. For example, in the problem just mentioned, the first number is "the thing (*cosa*) minus the root of some quantity," whereas the second equals "the thing plus the root of some quantity."

Starting from this rhetorical equation, the relations among the terms were simplified with operational rules similar to *al-jabr* and *al-muqābala*, learnt from the Islamic tradition. Sometimes, Italian names were adopted for these rules, such as *ristorare* (restoring) and *ragguagliare* (confronting). Other authors preferred to adapt the Arabic words to their vernacular, as is the case in a manuscript of 1365 entitled *Trattato dell'alcibra amuchabile*. The simplified "equation" was always one of a series of standard forms with known solutions, along the lines taught by al-Khwārizmī and his successors. Rules for solution were not based on an ability for symbolic manipulation, but rather on a repeated application of accepted rules, typically cast in rhetorical formulations that were easy to memorize (sometimes they were even phrased in rhyme, as we will soon see in a very remarkable case). In some of the texts, the very repetition of the rule was taken as the justification of its validity, and certain pre-fixed, approving sentences were added at the end of the proposed solution, such as "*esta bene, ed è provata, e così fa le simigliante*" ("it is correct, it is proven, and so should be done in similar cases").

Increased use of abbreviations and symbolic notation Initially, the main sources for the problems treated by abbacus masters were in day-to-day life and in commerce. With time, however, under the influence of the Islamicate traditions, as well as of the intrinsic dynamics of the ideas, the abbacus masters also began to deal increasingly with problems of little practical significance. They also handled problems that were no more than arithmetical riddles.

Treatises began to include more theoretical problems involving cubes of an unknown, and sometimes even higher powers. Then, following a textual practice that originated with medieval manuscripts in general (i.e., not just mathematical), and that was also followed in early Renaissance texts, the use of standard abbreviations became increasingly pervasive in abbacus masters' treatises as a convenient way to handle these higher powers. Thus, we find abbreviations for denoting the unknown (*cosa*: *c*), or its square (*censo*: *ce*), or its cube (*cubo*: *cu*), or sometimes the root of the unknown (*Radice*: *R*). Later on, this train of thought was gradually extended to higher powers as well: *ce ce* for the fourth power (*censo di censo*), *ce cu* for the fifth (*censo di cubo*), or *cu cu* for the sixth (*cubo di cubo*). There is, however, an interesting gap between the gradually increasing use of symbols as *abbreviations* and their fully fledged adoption as standardly accepted *operational devices*. Moreover, this gap comes hand in hand with a parallel gap between the rather straightforward extension of the notational techniques to higher powers of an unknown and a possible systematic conceptual clarification of the mathematical entities involved.

Noteworthy examples of this situation are found in the *Deutsche Coss* tradition that developed in the fifteenth and sixteenth centuries, when the Italian abbacus treatises were transmitted to the German-speaking territories. The local traditions of problem-solving that subsequently developed there were generally designated with the German neologism *Coß*, derived from the Italian *cosa*. A prominent instance of this tradition is found in the treatise *Die Coss* published in 1525 by Christoff Rudolff (1499–1545).

Rudolff used in his text not only notational conventions for higher powers of an unknown but also other arithmetical signs. Some of these were his own original innovations, but some others had already appeared in earlier treatises, such as the horizontal line for fractions. The + sign for addition, for example, which he uses consistently alongside – for subtraction, appears in a manuscript of the treatise *Algorismus proportionum* written by the French scholastic philosopher Nicole Oresme (ca. 1323–1382) around 1360. The sign appears as an abbreviation for the Latin word *et* ("and") and it is not clear whether it was introduced by Oresme himself or by the copyist. It was already common practice at the time to abbreviate the word *et* with an ampersand, and the use of + in this contexts is simply a further step in simplifying the labor of the copyist. In other words, the use of signs like + and − involves in its initial stages much less a mathematical idea than a lexical practice.

Rudolff denoted the square root by $\sqrt{\ }$. This was possibly a stylized abbreviation, based on a script such as r, for *radix*, "root." With the help of dots, he could even handle more complex, binomial expressions involving roots. For instance, he wrote $\sqrt{.}12 + \sqrt{.}140$, where we now write $\sqrt{12} + \sqrt{140}$. Moreover, he ingeniously extended the same notational idea so as to be able to denote higher roots: $\sqrt{\sqrt{\ }}$ for cubic roots and $\sqrt{\sqrt{\ }}$ for the fourth root (the latter two-stroke symbol possibly being an abbreviation of $\sqrt{\sqrt{\ }}$). While indeed ingenious, this way to extending the notation of the square roots to higher roots has obvious limitations, which can only be overcome once we realize that one and the same sequence of natural numbers can be used to indicate the different powers (or roots, in this case).

This realization, however, was not easy to come by, and we find this difficulty even more poignantly in Rudolff's symbols for higher powers. These symbols embody another interesting attempt to extend existing ideas to cover broader cases, which immediately faced the same kinds of limitations as Rudolff's own notation for higher roots. They are summarized in Figure 6.1.

The systematic use of these abbreviations for higher powers obviously implied the introduction of new kinds of entities into the current mathematical discourse. Nevertheless, Rudolff's explanations indicate the extent to which some of his basic views about numbers remained unmodified by the introduction of the new techniques he developed to handle them. "*Dragma* or *numerus*," he wrote, "is taken here as 1. It is no

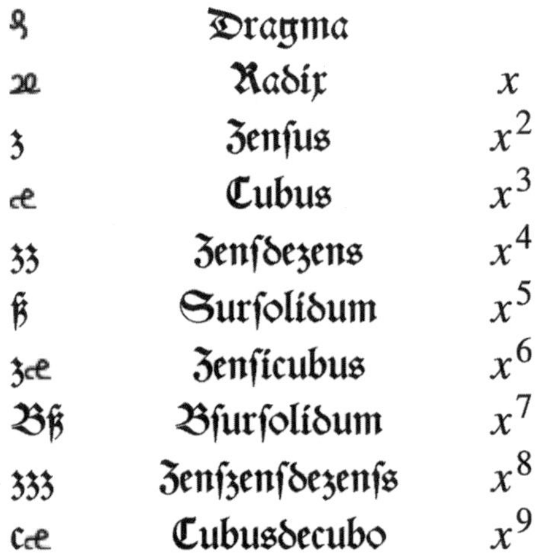

	Dragma	
ꝰ	Radix	x
ʒ	Zensus	x^2
ce	Cubus	x^3
ʒʒ	Zensdezens	x^4
ß	Surfolidum	x^5
ʒce	Zensicubus	x^6
Bß	Bsurfolidum	x^7
ʒʒʒ	Zenszensdezensſ	x^8
cce	Cubusdecubo	x^9

Figure 6.1 Symbols for the unknown, square, cube and higher powers, as they appear in the book *Die Coss* (1525) by Christoff Rudolff. Added, in the right-most column, are translations into modern algebraic language. As reproduced in (Katz and Parshall 2014, p. 207).

number, but assigns other numbers their kind. *Radix* is the side or root of a square. *Zensus*, the third in order, is always a square; it arises from the multiplication of the *radix* into itself. When *radix* means 2, for example, then 4 is the *zensus*." In other words, Rudolff continued to conceive numbers in terms that did not differ essentially from those of his predecessors, going all the way back to Euclid. As a matter of fact, in his explanations, he referred the reader to Book IX of the *Elements*. His quite systematic use of symbols did not require, or envisage, or promote a fully abstract notion of number that could apply—once and for all and in the same manner—to coefficients, to unknowns, and to designating the degree of the powers or roots of a number. Although his practice involved a rather abstract approach to the use of numbers, he continued to conceptualize the various powers or roots as different kinds of mathematical entities.

This gap between practice and conceptualization is interestingly manifest in Rudolff's treatment of arithmetic and geometric progressions, a topic that occupies a central place in the first part of the treatise. In the first place, Rudolff associated each of the successive members of a geometric progression with the numbers of their ordinal location. He even took the rather bold step of identifying the "zeroth" place as the one corresponding to the first term in the progression. He explicitly wrote, for instance,

```
 0    1    2    3    4    5    6    7     8     9     10
 1.   2.   4.   8.   16.  32.  64.  128.  256.  512.  1024.
```

Then, later in the text, he identified the ordinals of the subsequent terms of the progression with the symbols of higher powers, as indicated in Figure 6.2. Subsequently, he also introduced a table that associated the ascending sequence of higher powers written in his symbolic notation with the ascending sequence of the natural numbers:

```
 0    1    2    3    4    5    6     7     8     9     10
 𝟋    ϰ    ӡ    ce   ʒʒ   ß    ʒce   ʒß    ʒʒʒ   cce   ʒß
```

He used this table to justify rules for multiplying the various powers with each other, while implicitly connecting this operation with the parallel addition of the powers. Thus, he said, multiplying ϰ times ϰ yields ӡ. Likewise, ϰ times ce yields ʒʒ, and so on.

And yet, in spite of this ability to manipulate the higher powers, the progressions and the symbols, Rudolff did not take what for us would look to be a natural step, namely, to simply denote each power, somehow, by the natural number that

𝟋	ϰ	ӡ	ce	ʒʒ	ß	ʒce
1	2	4	8	16	32	64
1	3	9	27	81	243	729
1	4	16	64	256	1024	4096

Figure 6.2 Members of three geometric progressions ordered according to Rudolff's symbols.

represents either its place in the sequence of powers or the number of times that the basis has been multiplied by itself. We do this by taking the said natural number as a superscript, as in 2^3, but this, of course, is just a contingent fact. What is really important in our notation is that it reflects an awareness of a possible universal use of one and the same abstract idea of number to represent so many apparently different situations. The fact that Rudolff, for all of his calculational and notational abilities, did not take this apparently simple step (nor did any of his immediate successors) is very revealing about the real difficulty inherent to it.

By referring together and as part of one and the same idea to the various powers of the unknown, including those over three, Rudolff's notation had a favorable effect on the trend toward admitting the legitimate use of quantities lacking a direct geometric interpretation. At the same time, however, given the ad hoc character of the terminology and of the symbolism, where each of the powers was represented by a different letter, his case shows the natural difficulty involved in making explicit the already-mentioned link between the ascending sequence of the powers and that of the natural numbers.

While abbacists and cossists also explored more purely theoretical aspects of arithmetical knowledge, such as this link, their main motivation continued to be on the development of calculational techniques with a view to practical applications, particularly those of commercial accounting. They clearly understood the need to develop techniques for checking results of lengthy calculations. A well-known example is the so-called technique of casting out nines, which was known in Islamicate mathematics and which continued to be taught in elementary schools until just a few decades ago. These techniques are widely used in Rudolff's book, as well as in many of its contemporaries.

An interesting example of acknowledging the link between the sequence of powers and that of the natural numbers appears in the work of the Frenchman, Nicolas Chuquet (ca. 1445–ca. 1500). Chuquet developed numerical methods for solving problems and for calculating by approximation square and cubic roots. Lacking the ability to write decimal fractions, he did not always come up with actual values and simply left them indicated (but not completed) as calculation with common fractions. For this reason, he was led to develop a notation that would help him deal with this situation. For $\sqrt{14 + \sqrt{180}}$, for example, he wrote $R^2 14\bar{p}R^2 180$. He also developed an original way to write expressions containing an unknown. The expression $.3.^2\bar{p}.12.egaux.9^1$, for instance, would represent for him what is for us the equation $3x^2 + 12 = 9x$. Such an abbreviated way of writing was not amenable to formal manipulations of symbols leading to finding solutions to a problem with unknowns, but they allowed thinking about negative exponents. Indeed, these appear when one writes an expression such as $.72.^2\bar{p}.8.^3egaux.9^{2m}$, representing what is for us the equation $72x + 8x^3 = 9x^{-2}$. Chuqet, moreover, explained how to operate with such expressions. In order to multiply, for example, $.8.^3by.7.^1m$ (i.e., $8x^3 \times 7x^{-1}$), one has first to multiply 8 times 7, thus obtaining 56, and then to add up the exponents −1 and 3, thus obtaining 2. The outcome, then, is $.56.^2$. Chuquet's ideas are known to us via a text of 1484, *Triparty en la Science de Nombres*, which remained unpublished until the nineteenth century. Some of its contents were reproduced (without any attribution) in a textbook of 1520, *l'Arismetique* by Estienne de La Roche (1470–1530), and it is possible, though not certain, that Rudolff took some inspiration from them when writing his book *Die Coss*.

As a general statement about the introduction of accepted algebraic symbols, it can be said, to summarize, that in most cases this involved very gradual and hesitant processes that culminated relatively late in time. The symbol "=" for equality, to take an important example, was introduced by Robert Recorde (1510–1558) in his 1557 *The Whetstone of Witte*. This book also made consistent use of the signs + and – and had a decisive influence on their definitive adoption within the British mathematical context. Still, other signs such as < and > for inequalities appeared only some decades later in the work of Thomas Harriot (1560–1621).

Even in cases like this, it is important to remember that the fact that someone used signs that eventually became standard does not mean that their acceptance was smooth and universal. Often, there were various alternatives simultaneously in use in works of others. Abbreviations involving the letters *p* and *m* (for (*più*) and for (*meno*)) continued to be used in important algebraic treatises throughout the sixteenth century. In some texts, moreover, full rhetorical expressions continued to be used even though convenient symbols were already in wide use. The mere availability of symbols, therefore, was not always understood as a necessary indication that it would be convenient to put them to immediate use.

Gradual legitimation of broader classes of numbers The continued promotion of the use of symbols had additional consequences for the understanding of numbers. We find an interesting example of this in the work of the cossist Michael Stifel (1487–1567), who published a later edition of Rudolff's book, as well as a rather influential book of his own, *Arithmetica Integra* (1544). Stifel was among the last important practitioners of the cossic tradition, and a rather unique figure within it. He introduced many notions, methods and concepts, but did not use them in a systematic and consistent manner. His books never reached wide audiences and were only indirectly influential. Also, from the institutional point of view, Stifel's status was quite unique. He was ordained in 1511 in the Augustinian monastery of Esslingen, and later became a supporter of Luther. His early interest in numerology led him, only as late as 1535, to study mathematics seriously in Wittenberg and eventually became professor there. This is an important point, because Stifel, unlike other masters of the abbacus and coss traditions, introduced these kinds of traditions, which had arisen in trade circles, into the university curriculum.

Stifel was the first to formulate a single rule for solving all cases of quadratic equations seen as a single, general idea: "halve the number of roots, square it, add or subtract the given number, to the resulting number take its square root, and to this add or subtract half the coefficient" (and you can now translate this into symbols and check that this is exactly the general formula known to us). In solving equations with the help of this procedure, Stifel also gave clearly formulated, general rules for working with negative numbers and with common fractions. The latter were written in the already-standard way as $\frac{p}{q}$, and he also explored the idea of fractional and negative powers. So, his text displayed an increasing awareness of the prospect of considering them as numbers in every respect.

In some texts of the abbacus tradition, we also find extended and rather systematic treatments of irrational numbers, and, more specifically, of square and cubic roots of integers. Various Italian terms were used to indicate such numbers (*indiscreta*, *sorda*) and different authors had different attitudes towards them. Indeed, sometimes within one and the same text, one sees different attitudes to this issue. The treatment of irrational numbers typically appeared in relation to binomials such as $2 + \sqrt{5}$ or $3 - \sqrt{2}$. Such binomials and the need to learn how to handle them became strongly visible in the many arithmetized versions of Book X of the *Elements* that circulated at the time.

Book X deals with the classification of incommensurable magnitudes, and it is no doubt the most demanding one in the Euclidean corpus. Rendered in arithmetical terms, however, its propositions become much easier to understand, and the mathematical language recently developed by the cossists proved particularly well-suited to this end. Moreover, a main purpose of treatises in the abbacus tradition was to display the skills of the master, and hence irrationals were used excessively in some texts, even in places where examples with integers would have sufficed to illustrate the power of a method that was being discussed.

The increasing ability to operate with these binomials became an important factor in the development of techniques of symbolic calculations, since it drew attention to the existence of rules for sign multiplication. Moreover, operations with binomials carried with them an implicit willingness to accept negative numbers as a matter of course. Since the time of Fibonacci, negative numbers had appeared in solutions to problems and they had been treated with mixed reactions. Fibonacci himself, for instance, came up with various alternative interpretations of such solutions. If it was a problem related with money-lending, negative numbers could appear as debts. But in a problem related to pricing, no clear, accepted interpretation was always at hand. Often, Fibonacci spoke of negative solutions as "inconvenient" (*inconveniens*).

Formal rules for operations with signs The increasing appearance of subtractions in the framework of calculating with binomials gradually led to the systematic formulation of rules for sign multiplication. Remarkably, debates about these rules was purely pragmatic and did not necessarily involve a parallel debate about the nature of negative numbers or about the legitimation of their use. In a representative manuscript of 1380, attributed to Maestro Dardi of Pisa, for example, we find a detailed explanation of why the product of two subtractions yields a result of addition. His explanation was repeated in many texts, but it was likewise criticized in others. It is instructive to look at its details, since they are revealing about the ideas of number underlying it.

If we multiply 8 by 8, explained Dardi, we obtain 64. But this is also a multiplication of two subtractive binomials which, in modern terms, can be written as the product $10 - 2 \times 10 - 2 = (10 \times 10) - (2 \times 2 \times 10) + (2 \times 2)$. But multiplying these two binomials involved the need to refer to the multiplication of the two subtractions. Dardi proceeded roughly as follows: Multiply 10 by 10 and subtract, twice, 10 times 2. This yields 60. We are left only with the multiplication of "subtracting 2" by itself. The desired result of 64 implies, therefore, that the said multiplication (i.e., multiplication of "subtracting 2" by itself) yields an "addition" of 4.

Dardi drew from this example a general conclusion, namely, that multiplication of "a subtraction times a subtraction" is *always* an addition. I have written here "subtraction" rather than "negative number," since this was the way an abbacist like Dardi would regard what was at stake here, namely, an operation with numbers, rather than just a number of a special kind (i.e., a negative one). The idea of negative numbers, then, though gradually being admitted into the legitimate discourse of arithmetic, was still a source of debate and disagreement, and it would remain so for several centuries to come. This was not an obstacle, however, for discussing rules of operations with signs.

Another interesting example of this same trend appears in a famous book of 1494, *Summa de arithmetica geometria proportioni et proportionalita*, by Luca Pacioli (1445–1517). Pacioli was a close friend of Leonardo da Vinci (1452–1519), mathematically active in Italian universities and courts. In the *Summa*, Pacioli put together in a rather exhaustive and systematic way many of the main insights and techniques developed by abbacus masters previous to him. He also presented advanced accounting techniques, including double-entry bookkeeping, that were in use among traders in Venice. Given the wide audiences that his book reached, he became known in history as "the father of accounting."

Pacioli played an important role in disseminating the use of certain accepted signs for arithmetic operations, such as $\bar{p}$ (*più*) for addition, and $\bar{m}$ (*meno*) for subtraction. Still, at this point, we are talking about abbreviations as they were used at the time in many kinds of texts, rather than about abstract mathematical symbols that can be freely operated upon according to rules defined in advance. But Pacioli's systematic presentation did include the definition of certain rules for the multiplication and division with signs, as follows:[1]

A partire.   più per.   mē.   neven.   men.

A partire.   mē per.   mē.   neven.   più.

The meaning of this was "Divide plus by minus and you obtain minus. Divide minus by minus and you obtain plus." While there is no systematic and necessary use of abstract signs for the arithmetic operations here, Pacioli's texts exemplify the way in which formal rules were gradually established. The same Pacioli, however, was reticent to use negative magnitudes even in partial results of a given problem with binomials.

The rise of a realm of pure arithmetical knowledge   When irrational numbers appeared as solutions to problems, they typically appeared as part of binomial expressions, such as in the case

One of the numbers requested is 5 plus the root of $6 + \frac{10}{169}$, and the second one is 5 minus that root.

Cossists considered such statements to be "exact" solutions of the problem and typically did not attempt to calculate approximate values of the roots appearing in the expression, unless the context of the problem was clearly geometrical. In their

---

[1] Quoted in (Heeffer 2008, p. 14).

arithmetical problems, however, they typically treated problems of day-to-day life, such as those related to currency exchange, or to distances between towns in relation to the path to be followed by a postman. And yet, irrational solutions were accepted as legitimate for these problems as a matter of course and without much comment or explanation.

Additional texts in this tradition feature attempts to handle in a more systematic way the use of irrational numbers as well as discussions about their nature. Stifel, for instance, introduced the term "irrational," and he explained that "irrational numbers" appear in geometry and can be used there to solve problems in an exact manner, much the same as rational numbers are used for this purpose. This fact, he stressed, compels us to accept their "true existence." But, on the other hand, the fact that we do not know how to explain exactly the ratio between an irrational and a rational number raises doubts as to what is to be understood by such an "existence." Very much as an infinite number cannot legitimately be considered a number, he said, so irrational numbers cannot really be considered numbers, because they "lie hidden in a cloud of infinity." In spite of this explicit declaration, he continued to work out many properties of these irrational numbers and of the operations with them. He asserted, among other things, that between two integer numbers there is always an infinite number of irrational numbers as well as of rational numbers.

The case of Stifel exemplifies an interesting kind of attitude that we will see repeating itself from the Renaissance on, up to the nineteenth century: mathematicians call into question, or at least debate, the very existence of numbers of a certain kind, and try to understand the nature of such numbers by way of a reasonable (even if not perfect) physical analogy, but at the same time they continue to experiment and to attempt to come to terms with properties of these kinds of numbers, even under the assumption that they "do not exist" or that they are just "fictions."

We will see further examples of this, but here I want to stress that Stifel's attempts to clarify the status of the irrationals were not particularly representative of the attitude of abbacus masters or cossists, who simply used irrationals whenever they came up as solutions to problems they dealt with. At the same time, the abbacus masters consistently saw a much more serious problem in the use of negative numbers. At any rate, the multifarious threads of development of the concept of number in the period of time just considered converged into a handful of important books that were to leave their imprint on the forthcoming generations. In the remaining part of this chapter, I turn to consider two of the books that more emblematically represent the evolving views on numbers by the end of the Renaissance.

6.3 Cardano's *Great Art of Algebra*

One of the most fascinating figures in late Renaissance mathematics was Girolamo Cardano (1501–1576). A reputed physician and avid gambler who worked in the courtly circles of Milan, Cardano also published extensively on subjects such as algebra, probability, theology, astronomy, astrology and philosophy. His most well-known and influential mathematical treatise, *Ars Magna, sive de Regulis Algebraicis* ("The Great Art, or The Rules of Algebra"), appeared in 1545. It contained a systematic and rather comprehensive presentation of solution procedures for problems involving cubes and fourth powers of an unknown.

Cardano's ideas also appeared in other treatises that were well read at the time. In all of his texts, we find interesting attempts (both implicit and explicit) to clarify, and sometimes to modify and innovate, current ideas about the various kinds of numbers. Cardano was never weary of trying ever new and bold ideas even if they deviated from, or even contradicted, what he had written in previous texts. For this reason, it is very difficult to come up with a comprehensive synthesis of his arithmetical views. But for this same reason it is so interesting to review some of his ideas and the ways in which he explored the boundaries of the existing conceptions about numbers. We can get a basic grasp of this by briefly discussing some main traits of his algebraic work.

Cardano's solutions to problems involving cubes and fourth powers of the unknown were not all of his own creation. Other names in the story are Scipione del Ferro (1465–1526), Niccolò Fontana "Tartaglia" (1500–1557), and Ludovico Ferrari (1522–1565). At play behind the scenes was a story of cooperation, betrayal and other all-too-human feats. Del Ferro, working at the University of Bologna, was probably the first to come up with a solution to a problem involving cubes of the unknown. In this context, we are talking about solutions that lead to a numerical value by way of arithmetic operations alone (including root extraction), as opposed to the kind of geometric solution found in the work of al-Khayyām. Del Ferro's solution was for a problem of type "unknowns and cubes equal to numbers" (in modern notation, $x^3 + x = q$). He did not publish his method, and kept it as a secret, which he confided in his deathbed only to his student Antonio María Fior. Tartaglia, in turn, came up with a different method of solution for a different type of problem, and this is what gave rise to the story that ensued.

Tartaglia was a rather intriguing personality in his own right. He worked as an engineer in the design of fortifications, as a land surveyor, and as a bookkeeper. He was the first to develop a mathematical theory of ballistics. As a young child, his father was murdered by robbers. Then, in 1512, when French troops invaded his hometown of Brescia, he was injured in the face. He never recovered the full ability of speech, from whence came the nickname "Tartaglia" (stammerer). In 1543, he published an Italian translation of Euclid's *Elements*, which was actually the first to appear in any modern European language. He built for himself a reputation as a promising mathematician, but eventually, following his dispute with Cardano on cubic equations, he would die penniless.

In possession of del Ferro's method, Fior leveled a mathematical challenge against Tartaglia in 1535. Each of them prepared a list of 30 problems involving cubes of the unknown, which they sent to their rival. Tartaglia's method could solve one type of cubic equation, namely, "squares and cubes equal to numbers" (in modern notation, $x^3 + ax^2 = b$). But in a remarkable effort during the day of the confrontation, he found out how to solve the type of problem that Fior posed to him, while the latter was utterly unable to solve anything other than what he had directly learnt from del Ferro. The defeat was clear, but Tartaglia did not take his prize, since for him the honor of winning was enough.

Word of Tartaglia's abilities reached Cardano, who had also been working hard to find a solution to problems with cubes. He repeatedly tried to convince Tartaglia to reveal the secret of his method. He promised to include it in a book he was about to publish, but Tartaglia declined and stated his intention to publish the procedure in a book of his own. Cardano insisted and adduced all kinds of promises related to possible favors from the court of Milan. The temptation was too strong for Tartaglia, and he finally revealed the secret on March 25, 1539, with the oath of Cardano to never publish the findings.

Cardano began to make considerable autonomous progress in the general problem of the cubic equation, and even with the much more difficult quartic. In this, he was assisted by his servant and pupil Ferrari, who gradually became a real master of this art. Still, much of this progress depended quite directly on Tartaglia's methods. Now, in 1543, Cardano and Ferrari traveled to Bologna in order to inspect del Ferro's papers. They discovered a solution to one of the cases of the cubic in del Ferro's own handwriting. Based on this discovery, they adduced that they were no longer constrained by Cardano's oath to Tartaglia, and hence they were no longer prohibited from publishing their results.

Cardano published his *Ars magna* in 1545, containing the solutions to both the cubic and quartic equations. He duly credited del Ferro, Tartaglia and Ferrari with their respective discoveries and told his version of the story of the oath and the discovery of del Ferro's documents. Tartaglia, naturally, was exasperated on discovering that Cardano had disregarded the oath. He published a book the following year where he presented his side of the story. He accused Cardano of acting in bad faith, and openly insulted him.

Ferrari supported Cardano, of course, and challenged Tartaglia to a public debate. Tartaglia refused to debate Ferrari, as he was yet rather unknown, and insisted on challenging Cardano himself. The debate was not held until 1548, when Tartaglia was offered an important position in mathematics on the condition that he publicly debated Ferrari. The controversy and the challenge had aroused a strong popular interest by this time. The open debate took place on August 10, in Milan, in front of a large crowd. Ferrari showed his primacy very clearly, and Tartaglia decided to withdraw before ending the contest. In this way, Ferrari's reputation was greatly established as one of Italy's top mathematicians, while Tartaglia gradually fell into disgrace and oblivion.

When Tartaglia had disclosed the steps of his method to Cardano, he did so in rhyme. This was partly to keep an atmosphere of secrecy around his ideas, but it was also a way to follow the typical abbacus tradition already discussed above. It is worth quoting here the Italian original, taken from Tartaglia's 1546 book *Quesiti et Inventioni Diverse*,[2] which conveys the distinct flavor of this special approach to handling mathematical problems. The case treated here is "the cube and some things equal to a number" (in modern terms, $x^3 + px = q$). The solution is based on introducing auxiliary unknowns and then formulating an auxiliary equation that happens to be a quadratic one. The auxiliary equation can be solved using methods such as those developed by al-Khwārizmī, and the solution found in this way is used to solve the original problem. On the right-hand side of the text quoted here, I have added some modern symbolic expressions that help follow the logic of the algorithm:[3]

---

[2] The full text can be downloaded from <http://it.wikisource.org/wiki/Quesiti_et_inventioni_diverse>. See p. 241. This is from the 1554 second edition.

[3] Quoted in (Gavagna 2014, pp.168–169).

Quando chel cubo con le cose appresso	$x^3 + px$
Se agguaglia à qualche numero discreto	$= q$
Trovan dui altri differenti in esso.	$u - v = q$
Dapoi terrai, questo per consueto	
Che'l lor produtto sempre sia eguale	$uv = \left(\frac{p}{3}\right)^3$
Al terzo cubo delle cose neto,	
El residuo poi suo generale	
Delli lor lati cubi ben sottratti	$u^{\frac{1}{3}} - v^{\frac{1}{3}} = x$
Varra la tua cosa principale.	

Although in prose rather than in verse, Cardano also presented his solutions in purely rhetorical formulations, as follows:[4]

Cube the third part of the number of "things," to which you add the square of half the number of the equation, and take the root of the whole, that is square root, which you will use, in the one case adding the half of the number which you just multiplied by itself, in the other case subtracting the same half, and you will have a "binomial" and "apotome" respectively; then, subtract the cube root of the apotome from the cube root of the binomial, and the remainder of this is the value of the "thing."

With some patience, you can perform the algorithm by yourself. It yields a result that we can translate into modern algebraic symbolism, which is easier for us to read. What we get is this:

$$x = \sqrt[3]{\sqrt{\left(\frac{p}{3}\right)^3 + \left(\frac{q}{2}\right)^2} + \frac{q}{2}} - \sqrt[3]{\sqrt{\left(\frac{p}{3}\right)^3 + \left(\frac{q}{2}\right)^2} - \frac{q}{2}}.$$

The algorithm leads, indeed, to a correct solution, but, while in our algebraic formulation the specific contribution of each of the coefficients to the final result is evident (and indeed this is the meaning of having a formula for the solution), in Cardano's rhetorical rendering, this is far from being the case. Moreover, if we assign numerical values to the general formulation, as Cardano did, and refer to the concrete example $x^3 + 6x = 20$, we obtain the following result:

$$x = \sqrt[3]{\sqrt{108} + 10} - \sqrt[3]{\sqrt{108} - 10}.$$

In this numerical result, we cannot see the direct connection between coefficients and solution. Still, Cardano did realize that there exists some connection of that kind. He knew, for example, that if he had three different solutions for a given cubic equation, then their sum equals the "number of squares" mentioned in the formulation of the problem (i.e., the coefficient of x^2). Nevertheless, since he presented the solution in a completely rhetorical fashion and with little or no use of operational symbols, he was far from realizing all the relevant insights that could be achieved in this regard. He did

---

[4] I cite from the English translation in (Smith (ed.) 1929, p. 206).

not discuss in any systematic way this important topic, and in those places where he mentioned it, he did not formulate clear mathematical statements that deserved being proved. We will have to wait to the work of Descartes in the seventeenth century (and Chapter 8 in this book) for a much more systematic approach to this crucial issue.

Even more interesting is the fact that Cardano did not always bother calculating the exact value of the unknown. In the above example, he did not round up the procedure by stating that the solution is 2. At a different place in the text, he mentioned 2 as the value of the solution, but then he did not show the way in which he calculated the cubic and square roots leading to this result. Neither did he show to what extent this calculated result was close to the exact solution 2. He just hinted at a different text where he had provided this explanation. Still, in working out his numerical examples, Cardano introduced and used an incipient symbolism, as indicated in Figure 6.3. Notice that the symbols used here by Cardano are just abbreviations similar to those of the abbacus masters and cossists. They are not abstract symbols that could be operated upon. So, concerning style and content, one can draw a straight line connecting the work of al-Khwārizmī with that of Cardano, and passing through some of the significant developments in mathematics of early Renaissance Europe. Like his European predecessors, Cardano presented some concrete (albeit sometimes artificially framed) problems, with unknowns and numbers. He then showed how to solve these problems by presenting one specific, exemplary solution of each possible case.

Cardano's book followed the systematic kind of presentation found with al-Khwārizmī: where the latter had presented all *six* possible cases of quadratic equations, Cardano taught now how to solve *thirteen* possible cases of cubic equations: "the cube and some things equal a number" ($x^3 + px = q$), "the cube equal to some things and a number" ($x^3 = px + q$), "the cube equals some squares and a number" ($x^3 = px^2 + q$), etc. And also like al-Khwārizmī, side by side with the detailed, exemplary procedure to be followed in each of these cases, Cardano provided *geometric* arguments intended as justifications for the validity of the procedures. Such arguments were based on well-known results, mostly taken from the *Elements*, or being generalizations thereof. So, Cardano essentially abided by al-Khwārizmī's fundamental reliance on geometry as the source of legitimation for arithmetical procedures, but the increasing mathematical complexity of the problems addressed stressed some unsolved tensions inherent in this attitude. These tensions play a crucial role in our story, since the need to come to terms with them gradually led to significant conceptual breakthroughs.

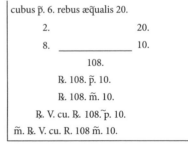

Figure 6.3 Cardano's shorthand for the solution of the problem "the cube and six things equal to twenty numbers." Reproduced from (Gavagna 2012, p.7).

The simultaneous consideration of the problems at both the geometric and the arithmetic level provides the most immediate source of tension. This becomes particularly evident when Cardano deals, in the *Ars Magna*, with twenty possible cases of problems involving square-squares of the unknown. At the technical level, when he described the procedures for solving these problems and provided geometric justifications for them, he was much less detailed than he was for the thirteen cases involving cubes of the unknown. And also at the conceptual level, he found some difficulties. In the introduction to the book, he emphasized that while the cube refers to a physical body, "it would be very foolish for us" to go beyond that, since "Nature does not permit it." But what is then the meaning of problems involving square-squares if it is foolish for us to pursue them? Well, in spite of the declared difficulty, such problems *are* included in the book—he said—"either by necessity or out of curiosity," but without going beyond "barely setting them out."

These same "necessity" and "curiosity" seem to be at the basis of Cardano's handling of other kinds of mathematical entities of unclear status, particularly negative numbers and their roots. Cardano stated very clearly that negative numbers should be avoided: "Subtraction is made only of the smaller from the bigger. In fact, it is entirely impossible to subtract a bigger number from a smaller one." And indeed, negative numbers are essentially absent from his treatment either in the enunciation of the problems or in their solutions. This was also, as we saw, the basic attitude underlying al-Khwārizmī's treatment of six separate cases of quadratic equations. Cardano certainly aimed at attaining a similar canonical status with his own presentation of problems with cubes and square-squares, and, as with al-Khwārizmī, the various cases he considered derived from the use of only positive numbers as possible coefficients.

If negative numbers and zero could be conceived as possible coefficients, then clearly he would have only one general equation of each degree: $ax^3 + bx^2 + cx + d = 0$ for the cubic equation and $ax^4 + bx^3 + cx^2 + dx + e = 0$ for the fourth degree. In this important sense, the general, abstract idea of an algebraic equation, which arises hand in hand with the modern understanding of number, is still absent from Cardano's highly accomplished treatment. The "coefficients" in Cardano's problems are always positive and mostly integer numbers. In some cases, they are also rational, and in a handful, they are even irrational.

But when it came to the *solutions* of the equations (as opposed to their coefficients), the need to consider negative numbers, and indeed even roots of such numbers, imposed itself, as it were, on Cardano. He found himself caught in a tension between his more stringent, declared concepts of number and a much more flexible, actual use of them. The same curiosity adduced in the case of square-squares seems to be strongly at play here. He contrasted the two types of solutions, positive (which he strongly preferred) and negative (which he nevertheless explored), calling them, respectively, "true" (*vera*) and "fictitious" (*ficta*). This was more than just a meaningless way to assign neutral names. Rather, it reflected a true belief in a different status generally accorded to the two kinds. As we will see in the forthcoming chapters, the ability to gradually reach a truly general and flexible idea of numbers will go hand in hand with an abandonment of value-loaded terminology and attitudes toward one or other kinds of numbers.

Let us see an example of how the need to consider the negative numbers arises in Cardano's work. If we take a quadratic equation of the kind "the square equal some things and a number" ($x^2 = px + q$), then the known solution to this problem can be expressed, in modern language, as follows:

$$x = \frac{p}{2} + \sqrt{\left(\frac{p}{2}\right)^2 + q}.$$

Cardano suggested that a problem of this kind can admit also a second, "false" solution. He did not just say that this false solution is obtained as

$$x = \frac{p}{2} - \sqrt{\left(\frac{p}{2}\right)^2 + q},$$

but rather suggested that it can be obtained by changing the sign of the true (*verae*) solution of a different, but closely related problem (namely $x^2 + px = q$). In modern terms, Cardano's solution to this latter problem is expressed as

$$x = \sqrt{\left(\frac{p}{2}\right)^2 + q} - \frac{p}{2},$$

which clearly, by taking its negative, yields the second, "false" solution to the original problem. Cardano lacked the adequate symbolic language and coherent concepts to realize the exact reason for the validity of this procedure. Moreover, negative solutions did not fit the support that Cardano sought for his procedures within the geometric books of the *Elements*. Still, he did check several numerical examples that confirmed the validity of the procedures. Although there is no evident geometric counterpart to these "false" solutions, Cardano did not fully reject their status as solutions of sorts to the problem. After all, it is not difficult to think of the negative numbers as such (*minus puro*), for example, in terms of debts in a commercial context. Moreover, Cardano's algorithm did lead to correct solutions of cubic equations where it was applied.

A different kind of "false" solutions were those involving square roots of negative numbers, which Cardano called *minus sophistico*. Cardano's example arose within a problem that did not seem to differ from standard abbacus arithmetic: "Divide 10 into two parts, such that their product is 30 or 40." If one solves the problem by following the known algorithm relevant to the case "some squares and a number equal some things" ($x^2 + q = px$), then one obtains the solutions $5 + \sqrt{-15}$ and $5 - \sqrt{-15}$. Following the well-known rule of the abbacus masters for calculating with binomials, and "putting aside the mental tortures involved" in doing so in the case of these two expressions, it is easy to check—as Cardano emphasized—that they do satisfy the conditions of the problem; indeed, their sum is 10 and their product is 40. But the geometric interpretation of the solution led to subtracting a rectangle of sides 4 and 10 from a square of side 5 (i.e., 25 – 40). The idea of a *negative area* arising here was certainly inadmissible. Moreover, the minus sign within the discriminant could not be bypassed with the help of some kind of associated problem, as he had done in the previous example.

How can one deal with a case that is both impossible and correct? Cardano declared this solution to be truly "sophistic" and "as subtle as it is useless," but he nevertheless considered such solutions of great mathematical interest. He understood that he must accept the legitimacy of square roots of negative numbers as solutions, even though he had no idea of how to interpret them. Moreover, the fact that their square is a negative number seemed to challenge the accepted ideas about the laws of multiplications with signs, and for this reason he thought that they are not really either positive or negative quantities, but rather a mathematical entity of some third kind.

Against the background of these kind of difficulties, Cardano raised the idea that what needs to be revised are actually, of all things, the abbacist rules for multiplying binomials. He discussed this matter in another well-known book of 1570, entitled *De Regula Aliza*. Cardano's argument referred back to the example discussed much earlier by Maestro Dardi, and which I already mentioned above. He approached the issue from a purely geometric perspective, with the help of a diagram as in Figure 6.4. Assume in this diagram the length of the side *ac* to be 10 and the area of the square *aefc* to be 100. Assume, further, that the side *bc* as well as the side *ga* are of length 2. The area of the square *egd* is 64, since its side is 8. On the other hand, 64 can be obtained by subtracting from *aefc* two rectangles, *bf* and *cg*, each of area 20, and restoring the square *cd* that was subtracted twice (i.e., 100 − 20 − 20 + 4). Now, the number 4, says Cardano, is not obtained from the multiplication of two negative numbers, −2 × −2, as the abbacist rule would have it. It followed from the geometrically based necessity to restore an area that was subtracted twice. In Cardano's view, therefore, the arithmetic situation involved here is not 64 = 100 + (10 × −2) + (10 × −2) + (−2 × −2) = 100 − (10 × 2) − (10 × 2) + (2 × 2), but rather 100 − (2 × 20) + (2 × 2) = 64. What appears to be a product of two negative numbers yielding a positive one, Cardano insisted, is just that, an appearance. The realms of positive and of negative quantities are separate, and cannot be mixed with each other. So, when positive numbers are multiplied, the

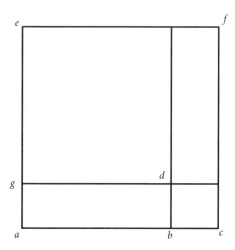

Figure 6.4 Cardano calls into question the abbacus masters' rule for sign multiplication.

result is a positive number, and when negative numbers are multiplied, the result is a negative number. Moreover, he said, when a positive number is multiplied by something that is outside its domain, for example a negative number, then the result must also lie outside its domain and is hence a negative number.

Cardano, then, had an interestingly ambivalent attitude towards negative numbers. He admitted negative results in specific situations, and in general he knew very well how to *operate* with negative numbers, using techniques developed by the abbacus masters. But he could not explain what is the *meaning* of a negative solution, and he was not consistent either in the way he denoted such numbers (or the positive ones, for that matter), nor in how he wrote them. In one and the same sentence, he could write both $3.\bar{m}$ and $\bar{m}.3.$, to express the same mathematical idea. Cardano spoke about differences in the behavior of even and odd powers. In the case of even powers, he emphasized, the negative and the numbers yield the same result: "both 3. and $\bar{m}.3.$ make 9., since minus multiplied by minus yields plus." But in the case of odd powers, he added, each number keeps "its own nature": what we call "debitum" cannot be produced as a power of "true number." As a consequence, if an even power equals some number, its roots have two values, "$\bar{m} \& \bar{p}$".

The kind of conceptual embarrassment appearing in the work of such a sophisticate mathematician as Cardano indicates how difficult it was to depart from the classical concept of number, seen as a collection of units, in spite of the continued progress in the technical expertise in matters arithmetical, such as the ability to solve difficult problems with unknowns in various powers. Cardano did not suggest the opposite and totally symmetrical argument whereby a negative number multiplied by something outside its realm (i.e., a positive number) yields a result that is outside its realm, namely, a positive number. Of course, such a symmetry is part of our current understanding of number. There was, for Cardano, a "realm" of numbers on the one hand, and on the other hand other things that were simply "out of that realm." What is really important in this story, however, is that neither conceptual, theoretical discussions about the nature of numbers nor some innovative philosophical arguments were the main motive forces leading to legitimizing the status of the negative or imaginary numbers, but rather *the actual practice of mathematics*, which, eventually, made it impossible to continue ignoring them. We have already seen some examples of this situation in which actual practice compels the adoption of new views, and we will see more in what follows.

6.4 Bombelli and the roots of negative numbers

Cardano's *Ars Magna* included material that went way beyond what the Islamic tradition had bequeathed to European mathematics several centuries earlier. Because of the broad readership it reached, the book became a milestone in the development of algebra and of the modern understanding of number. Cardano, whom no one could blame for excessive modesty, closed his book with a clear-cut statement of unbridled self-esteem: "Written in five years, may it last as many thousands." As a matter of fact, however, for all the mathematical innovations it did contain, the book was no easy reading, and it was lacking in terms of a clear organization of the material. So, it remained for a different book, published about twenty-seven years later, to succeed in spreading the word and in bringing the new algebraic gospel of Cardano to a broader audience of mathematical readers. This was the *Algebra* of Rafael Bombelli (1526–1573), published in 1572.

Born in Bologna, Bombelli had no formal university education. Rather, he received practical training in engineering and became involved in various works of marsh reclamation and lake draining, eventually gaining great expertise and professional reputation. In the decade of the 1550s, while working at the Val de Chiana area, in Tuscany, various temporary suspensions of the project allowed him free time, which he decided to use for studying algebra, and in particular Cardano's *Ars Magna*. He openly expressed his admiration for the book, but also his dissatisfaction with its lack of clarity. Between 1557 and 1560, he wrote his own book, *Algebra*, in Italian. Bombelli's text can be seen as both a culmination of the lineage of the abbacus tradition and a completely new starting point that incorporated the significant addition of the results presented by Cardano on third- and fourth-degree equations. It provided a rather coherent practical and conceptual synthesis of the latter that helped make sense of it and allowed its further dissemination.

Bombelli was also intimately acquainted with the ideas of Diophantus, whose book was then becoming increasingly popular among Italian humanists. Thus, he presented the new algebraic techniques learnt from Cardano in the form of abstract problems very similar in spirit and formulation to those of the Greek mathematician. In presenting a systematic and detailed account of Cardano's results, Bombelli engaged in an interesting dialogue with the ideas discussed in both the *Ars Magna* and the *De Regula Aliza* (and Cardano also had the opportunity to react to some of the ideas expressed in Bombelli's *Algebra*). Generally speaking, Bombelli's attitudes toward the negative numbers were similar to those of Cardano, but an interesting example of disagreement arises with Bombelli's exploration of the rules of sign multiplication. The latter followed quite closely that of Cardano, but reached the opposite conclusion and hence legitimized the accepted rules.

Concerning square roots of negative numbers, Bombelli declared that[5]

> It was a wild thought in the judgment of many; and I too for a long time was of the same opinion. The whole matter seemed to rest on sophistry rather than on truth. Yet I sought so long, until I actually proved this to be the case.

What Bombelli did in his book was to extend to the case of square roots of negative numbers (with the help of some new terminology) the abbacist rules for the arithmetic of binomials. The binomial that we write as $2 + 3\sqrt{-1}$, for example, he wrote as 2 *p di m* 3, which is an abbreviation of 2 *più di meno* 3. What we write nowadays as $2 - 3\sqrt{-1}$, he wrote as 2 *m di m* 3 (or 2 *meno di meno* 3). Using this nomenclature, he wrote down correctly all the rules of sign multiplication with numbers of this kind. Two examples (with modern symbols to the right) are the following:

più di meno via più di meno fa meno $[\sqrt{-1} \times \sqrt{-1} = -1]$,
meno di meno via più di meno fa più $[\sqrt{-1} \times \sqrt{-1} = 1]$.

---

[5] Quoted in (Kleiner 2007, p. 8).

Bombelli's willingness to find a consistent approach to handling square roots of negative numbers is directly connected to one of the cases discussed in his book, where he applied Cardano's methods to solving the problem "the cube equal to fifteen things and four numbers" ($x^3 = 15x + 4$). Bombelli knew that 4 is a solution to this problem, but Cardano's method yields the following solution:

$$x = \sqrt[3]{2 + \sqrt{-121}} + \sqrt[3]{2 - \sqrt{-121}}.$$

In other words, on the way to a fully legitimate solution, 4, one is compelled to deal with the root of a negative number, $\sqrt{-121}$, whose very legitimacy is questioned. In other cases where such roots appeared, Bombelli could say, like his predecessors, that the problem in question "has no solution" or that it is "subtle" or "sophistic". But in this case, there is by all means a legitimate and clear solution to the problem, and the mathematical practice simply leaves no room for hesitation: the strange entities that we come across on the way to the legitimate solution have themselves to be legitimized.

As with other examples we have been discussing in this chapter, Bombelli's shorthand for *più di meno* and *meno di meno* were not abstract symbols on which one can operate in any flexible and efficient way. It still took time before such symbolism developed in parallel to the completed idea of a complex number. But Bombelli did develop in his book some useful kinds of symbolisms for use in other contexts. To take one example, what we nowadays write as $\sqrt[3]{2 + \sqrt{-121}}$, he wrote as follows: R.c.⌊2p.R.q.21⌋. Another kind of interesting symbolic abbreviation found in his book is "*Agguaglisi* 1³ à.6¹ p.40.," meaning "Equate x^3 to $6x + 40$." The letter "*p*." stands here, of course, for *più*.

The adoption of this notation represents a very important step toward a more fully developed idea of abstract number, since it expresses the idea that also the *exponents* of the different powers of the unknown (cubes, squares, etc.) are themselves numerical expressions and can hence be operated upon according to rules, independently of any geometric representation or physical analogy. Compare it, for example, with the expression used by the cossists as described in Figure 6.1, where every power has its own symbolic expression.

Once we understand clearly how to *associate* the powers of the unknown with numerical expressions, the way is immediately open to the possibility of *operating* in an abstract way on the powers themselves. What we find in Bombelli's book is a nice illustration of the way in which, gradually but not without obstacles, it became clearer and clearer that coefficients, solutions to problems (as well as partial steps in the solution), powers, binomial expressions, lengths, areas and volumes, and several other ideas appearing in separate mathematical contexts are all expressions of a single, more general and thoroughgoing idea, namely, the abstract idea of number.

Bombelli's *Algebra* represents a high point in Renaissance mathematics in Europe, as it constitutes an elegant culmination and synthesis of the various traditions and threads that preceded it. It comprised five books, of which the first three were published the year of his death, 1572. The last two remained unknown until 1923, when Ettore Bortolotti (1866–1947) uncovered them in a library in Bologna and published them a few years later. They explore in original ways the relationships between algebra and geometry. The parts that were published back in the sixteenth century reached wide audiences. They were instrumental in allowing the passage to a new mathematical era in which strongly consolidated algebraic ideas and a broader concept of number unknown in previous times came to play a central role in the discipline. It is a well-established fact that the *Algebra* made a lasting impact on readers like Leibniz, as well as on Stevin, about whom I say more in the next chapter.

6.5 Euclid's *Elements* in the Renaissance

I would like to conclude this chapter with a particularly illuminating perspective on the contemporary conceptions of number that arises from looking at various editions of Euclid's *Elements* published in the fifteenth and sixteenth centuries in Europe. The many editions that appeared in print in Europe after 1450 often present original syntheses among different mathematical traditions, both scholarly (which were studied, taught, and disseminated in Latin, mainly among humanists and the Church institutions) and practical (which were developed and taught in vernacular languages). Those who prepared the editions typically added a preface as well as comments and explanations, but they did not work as historians or philologists preparing critical editions of classical texts, the way scholars would do nowadays. Rather, they simply saw themselves as making available a textbook for study at schools, and hence it was common that they would modify the original text to make it more easily understandable. For this reason, the editions of the *Elements* are a very useful source for the historian trying to assess changes in mathematical ideas throughout time. These editions show some significant changes in the understanding of number, and at the same time they show how pervasive the Greek basic ideas continued to be at the time. They continued to afford a necessary point of reference in the mathematical discourse of the time, particularly concerning the concept of number.

One important element that gradually entered the editions of the *Elements*, as well as many other contemporary mathematical books, was the implicit assumption that any magnitude can be measured. Accordingly, the issue of incommensurability, which had been the main motivation for Eudoxus' formulation of a new theory of proportions to replace the old Pythagorean one, lost its importance. Some of the new editions of the *Elements* did not even mention this issue. As a consequence, there was a gradual identification of "ratio" with the numerical value of the fraction defined by the two magnitudes compared. As always, this was a hesitant and slow process, and yet it is clearly visible by the end of the sixteenth century and it gradually acquires momentum. On some occasions, a ratio between an integer and an irrational (for e.g., a root of a non-square integer) was written as a fraction whose value is an approximation of the irrational ratio in question.

One remarkable example that can be mentioned here is that of the 1574 edition prepared by the German Jesuit Christopher Clavius (1538–1612). Clavius was the most prominent figure in the educational system of the Jesuit Order, and he greatly influenced the study programs they implemented. Mathematics was always at the focus of interest of these programs. Clavius' mathematical texts reached an enormous audience in Europe and, because of the missionary activity of the order, in many other places around the world as well. Clavius' edition of the *Elements*, then, was among the most well known in his time. In the arithmetical books (VII–IX), he allowed himself a high degree of liberty in formulating the definitions and propositions, for the purpose of incorporating more recent knowledge into the Euclidean treatment of the properties of numbers.

Where Euclid defined the multiplication of a first number by a second one ("when that which is multiplied is added to itself as many times as there are units in the second"), for example, Clavius added, without any cautionary note, a new definition, this time of division of two numbers (an operation that appears nowhere in the *Elements*). In doing so, he extended, by simple addition of one sentence, the kind of operations considered at the time to be part of Euclid's arithmetic, and the kind of numbers to which they apply. In a separate appendix added to the main text, he also explained in detail how to perform these operations and devoted lengthy pages to teaching in detail the arithmetic of common fractions. Also, this is foreign to the spirit of Euclid's original text, where we never find actual operations with numbers. Clavius' addition was in line with the practical and computational stress of the abbacist and cossic traditions that certainly influenced him.

Clavius' edition of the *Elements* is a remarkable example of how the new trends that developed in the practical mathematics of late Renaissance in Europe entered the scholarly tradition. Previously, the practitioners of this tradition saw themselves as drawing directly, and in many cases exclusively, from the classical texts. But as the *Elements* continued to provide the main source of legitimization for mathematical knowledge in the scholarly world, editions like that of Clavius were a main vehicle for the adoption of new ideas in future works. The many readers who learnt their *Elements* from these new editions were already capturing an image of geometry and of its connection with algebra that differed in many fundamental ways from the original and that was strongly influenced by much later developments. Also, the ideas that began developing in Europe from the time of Fibonacci were increasingly integrated into presentations of ancient texts. This strongly contributed to the further dissemination of these mainly practical ideas within scholarly contexts.

One of the core elements deriving from the Greek tradition that gradually weakened as part of this process was the meticulous separation between continuous and discrete magnitudes. Together with the continued development of ever more efficient computational techniques, with the increased willingness to broaden the legitimacy of negative and imaginary numbers, and with the improved ability to handle flexible and rather abstract algebraic notations, the developments in mathematics in the late sixteenth century created a propitious background for consolidating, during the seventeenth century, a new idea of number as a totally abstract and general mathematical entity. We will devote the next two chapters to inspecting more closely these developments.

Appendix 6.1 Casting out nines

Readers who grew up as children in the electronic era are perhaps used to the idea that computations made "by hand" (assuming they ever did any such computing) can be immediately verified with the help of some electronic device that is always readily available, for example in their smartphones. However, before such pocket electronic devices were to be found everywhere, verifying computations was always a concern. Ever since the Renaissance, mathematicians have developed methods of verification to be used even in relatively complex calculations, especially in "long multiplications." The best known of these methods is the one called "casting out nines." In his treatise on the Hindu positional system of numeration, al-Khwārizmī taught the essentials of this method, and it continued to be taught at elementary schools until relatively recently. In this appendix, I explain the essentials of the method and its mathematical basis.

"Casting out nines" for a given number means adding up its digits and subtracting 9 every time this value is exceeded. If the given number is 190,871,823, for instance, then casting out nines is done as follows, beginning from the units:

- $3 + 2 = 5$
- $5 + 8 = 13$: we have exceeded 9, so we turn 13 into $13 - 9 = 4$
- $4 + 1 = 5$
- $5 + 7 = 12 \to 3$
- $3 + 8 = 11 \to 2$
- $2 + 0 = 2$
- $2 + 9 = 11 \to 2$
- $2 + 1 = 3$

In this way, 190,871,82 becomes, by way of casting out nines, 3.

Now, casting nines is used as a method for verifying multiplications as follows: multiply any two numbers, say 871,823 × 55,367, in the standard way:

```
            8 7 1 8 2 3
          ×   5 5 3 6 7
          ─────────────
            6 1 0 2 7 6 1
          5 2 3 0 9 3 8
        2 6 1 5 4 6 9
      4 3 5 9 1 1 5
    4 3 5 9 1 1 5
    ─────────────────────
    4 8 2 7 0 2 2 4 0 4 1
```

We cast out nines for the two factors as well as for the result:

- 871,823 → 2
- 55,367 → 8
- 48,270,224,041 → 7

We now arrange the obtained values around a big X as follows:

The values corresponding to the factors appear at the sides, and that of the result in the top space. Now, in the empty space at the bottom of the X, we write the result of multiplying the values on the sides, 2 × 8 (casting out nines, if necessary). The result is 16, which becomes 7 after subtracting 9. Thus, we obtain

If the multiplication was correct, then the value obtained, 7, is equal to the value in the top space, which corresponds to the result obtained by casting out nines in the multiplication itself. In this case, it is indeed 7. If we had obtained a different value in the bottom space, then we would have been certain that the multiplication performed had been wrong. But notice that this method provides a necessary but not a sufficient condition for the correctness of the operation. If we had obtained 48,270,224,014, rather than 48,270,224,041 (with the last two digits in the number flipped with each other), then after casting nines we would also have obtained 7, and the method would not have indicated that our result was indeed wrong.

Even though this method had been well known since at least the sixteenth century, and perhaps much earlier than that, we do not know at what point in time the underlying reason for its validity began to become clear to those using it. An important mathematical idea that was carefully worked out by Gauss in the early nineteenth century, however, help to explain this in very simple terms. This is the idea of "congruence." Given any two integers a and b, and a fixed number p, we say that a is congruent with b modulo p (written $a \equiv b \pmod{p}$), if and only if a and b leave the same reminder when divided by p. Equivalently, $a \equiv b \pmod{p}$ if and only if p divides (with no remainder) the difference $a - b$. Hence, the following expressions hold:

$$7 \equiv 2 \pmod{5}; \quad 113 \equiv 29 \pmod{7}; \quad 25 \equiv 1177 \pmod{2}.$$

Two simple, fundamental properties of congruences are the following:

(a) If $a \equiv b \pmod{p}$ and if $c \equiv d \pmod{p}$, then $a + c \equiv b + d \pmod{p}$.
(b) If $a \equiv b \pmod{p}$ and if $c \equiv d \pmod{p}$, then $a \times c \equiv b \times d \pmod{p}$.

Congruences modulo 9 satisfy some additional, simple properties that any reader can easily check. Thus, for instance,

$$1 \equiv 1 \pmod{9}; \quad 10 \equiv 1 \pmod{9}; \quad 100 \equiv 1 \pmod{9}; \quad \text{etc.}$$

More generally, we have the following property:

(c) For any integer value of n, $10^n \equiv 1 \pmod{9}$.

These three properties translate into a useful way of working with congruences module 9. Let us take one of the numbers already inspected above, 871,823 and write it as a sum of powers of tens:

$$871{,}823 = 8 \times 10^5 + 7 \times 10^4 + 1 \times 10^3 + 8 \times 10^2 + 2 \times 10^1 + 3 \times 10^0.$$

From properties (a)–(c) above, it follows that $10^5 \equiv 1$ (mod 9) and, likewise, $8 \times 10^5 \equiv 8$ (mod 9). By repeated application of this kind of reasoning, we obtain

$$871{,}823 \equiv 8 + 7 + 1 + 8 + 2 + 3 \text{ (mod 9)}.$$

But also $8 + 7 + 1 + 8 + 2 + 3 = 29$. Hence, for the same reason,

$$871{,}823 \equiv 8 + 7 + 1 + 8 + 2 + 3 \equiv 29 \equiv 2 \text{ (mod 9)}.$$

More generally stated, if we add the digits of any given number, then we obtain a second number that is congruent (mod 9) with the given number.

Now, moving to the context of the verification process for multiplication, in the sense explained above, we see that casting out nines of any given number is tantamount to adding up the digits of the given number. The next step in the process—namely, placing the values on the big ×—is equivalent to checking the multiplication of the congruences. In our example, the product of the congruences is 7 (mod 9). So, we need to check whether the product of the congruences modulo 9 of the multiplicands equals the congruence modulo 9 of the multiplication. If they are not equal, then the product is incorrect. As already indicated, if they happen to be equal, then we are on the right track, although it is possible that the outcome is right even though the product is incorrect.

A last interesting point to be noticed is that one can create additional proofs of this kind using a different basis, other than 9. Property (c) above makes the proof with 9 particularly convenient, but the principle can be applied with other bases. An interesting case is that of 11, because of the following property:

$1 \equiv 1$ (mod 11),
$10 \equiv -1$ (mod 11)   (because $10 - (-1) = 11$);
$100 \equiv 1$ (mod 11)   (because $100 - 1 = 99$);
$1000 \equiv -1$ (mod 11)   (because $1000 - (-1) = 1001$ and $1001 = 11 \times 29$);
etc.

Thus, taking once again the same example as above,

$$871{,}823 \equiv -8 + 7 - 1 + 8 - 2 + 3 \equiv -1 \text{ (mod 11)}$$

Casting out elevens works similarly to the procedure with nines, except that instead of adding up all digits, we alternatively add and subtract, beginning from the units. All the rest works as in the case of nines, namely using the big × as above. You are invited to try it.

CHAPTER 7

Number and Equations at the Beginning of the Scientific Revolution

We arrive now at the historical period of time usually known as the "scientific revolution." Historians use this term to refer to a series of processes and events that took place about the sixteenth and seventeenth centuries and that led to the abandonment of a world picture that had consolidated in the late Middle Ages. At the center of this older view stood what was called the Aristotelian philosophy of nature. In its stead, there arose a new way to see the world around us, natural phenomena, the fabric of the universe, and the human body. The climax of these processes came with the rise and consolidation of Newtonian science as the central paradigm of the new kind of knowledge that came to dominate the learned world.

Historians continue to hold intense debates around the questions of whether it is justified to use the term "revolution" in relation to these processes, of the causes that led to them, and of the main characteristic traits of the changes that took place. One way or another, there is wide consensus that this was a time of significant changes both within the body of knowledge of the various scientific disciplines and concerning the more general question of the aims and the legitimate methods of scientific knowledge. Moreover, there were certainly deep changes concerning the place of science in society and its perceived role. And there is little debate that one of the main traits of the new science that evolved over this period of time is the transformation of physics into a decidedly mathematical discipline.

Of particular interest for our account here is that, together with its changing role in science, mathematics itself underwent substantial transformations at the time of the "scientific revolution," mainly around the development of the infinitesimal calculus. Even more to the point here, current ideas about numbers and equations were also substantially transformed over this period of time. It is clear, though, that these transformations can be more adequately described, not as a revolution but rather as the culmination of long-term processes. Some of these processes were described in previous chapters. Their next stages will be discussed now.

The year 1543 is often singled out as a milestone on the way to the scientific revolution. That year, two important books were published that have since been identified as harbingers of the new period: *De revolutionibus orbium coelestium* ("On the Revolution of Celestial Bodies") by Nicolaus Copernicus (1473–1543) and *De humani corporis fabrica* ("On the Fabric of the Human Body") by Andreas Vesalius (1514–1564).

To put this in the context of our story here, recall that Cardano's *Ars Magna* was published in 1545, Bombelli's *Algebra* was published in 1569, and Clavius' edition of the *Elements* appeared in 1574. In Chapter 6, I described these three books as representing three high points in Renaissance mathematics. The works of some prominent mathematicians that worked in the period immediately thereafter will be at the focus of this and the next chapter.

As in previous historical periods, at this time there is more than one single kind of activity that falls under the heading of "mathematics" and more than one kind of practitioner. We find intense activity related to the creation of new methods in geometry and in arithmetic, alongside debates about numbers in various kinds of professional communities. One such community comprised followers of the abbacus and cossic traditions, occupied with developing new methods for solving problems with unknowns. Others were associated with university circles and focused on the study of scholarly texts (including, of course, Euclid's *Elements*). Universities continued to be leading institutions of knowledge where the classical traditions of Greek geometry were persistently cherished. Yet another group was more clearly oriented toward practical tasks and applied new calculation techniques to problems of construction, astronomy and naval navigation.

New mathematical ideas coming from various directions gave rise to questions of legitimation, against the background of the accepted canons commonly associated with the "ancients." The humanists continued to study outside the universities the classical sources, mathematical text being only a part of their scholarly world. For them, the works of Diophantus (rather than the acknowledged classical texts of geometry) and whatever could be derived therefrom were the main focus of attention.

Common to all these groups was that in their works, each in their own way and for different reasons, they took one step further the processes that called into question the strict classical separation between continuous and discrete quantities. Also, questions about the status of the various kinds of numbers (irrational, negative, rational) continued to attract attention, as did questions about the possibility of considering ratios as numbers. In addition, efficient symbolic techniques continued to develop both for writing fractions and for indicating unknowns. All of these converged into a process whereby a truly general, abstract concept of number gradually emerged in the actual *practice* of mathematics. An idea of "quantity," which is not itself either continuous or discrete and may be both of them at the same time, began to establish itself in all mathematical areas of activity.

This process, however, was seldom accompanied or followed by a scholarly, systematic and convincing discussion about the new understanding of the nature of number. Moreover, there was nothing like a consensus about these matters among the groups or inside them. But as the arithmetic and algebraic practices developed intensely over this time, the main related concepts gradually settled down and eventually became widely accepted, even in the absence of a commonly agreed, underlying theoretical definition that would support it. Once again, the tensions between, on the one hand, innovative and fruitful practices that led into unexplored directions and, on the other hand, the concepts and definitions that inevitably lingered behind are at the center of the history of numbers, and we will devote attention to them here.

7.1 Viète and the new art of analysis

I begin the discussion of this period by focusing on the important work of François Viète (1540–1603). In his work, we find the most important example of substantial progress in developing symbolic methods for solving problems with unknowns. His main innovation in this regard concerns the simple, but powerful idea that one can denote with letters not only the *unknown*, but also the *known quantities*. It is remarkable how long we had to wait in our history before someone made systematic use of such an apparently self-evident idea. More specifically, Viète suggested that a clear distinction can be achieved just by denoting the unknown quantities with consonants and the known ones with vowels. This move offers a compelling historical example of how a truly simple idea (at least simple in retrospect) can be groundbreaking in its consequences. A typical expression appearing in his texts is the following:

$$A \text{ cubus} + C \text{ planum in } A \quad \text{aequatus} \quad D \text{ solidum}.$$

In modern notation, this could be translated as $x^3 + cx = d$, or perhaps more precisely: $x^3 + c^2x = d^3$. Viète developed rules of formal manipulation that can be applied to an expression like this one, in order to yield a value for the unknown, A, in terms of the other terms appearing in the expression. Two examples of such operational rules are the following (to the right, in parentheses, I have translated into modern algebraic symbolism):

$$\frac{A \text{ planum}}{B} \text{ in } Z \quad \text{aeq.} \quad \frac{A \text{ planum in } Z}{B} \qquad \left(\frac{A}{B} \cdot Z = \frac{A \cdot Z}{B}\right),$$

$$\frac{Zpl.}{G} + \frac{Apl.}{B} \quad \text{aeq.} \quad \frac{G \text{ in } Apl. + B \text{ in } Zpl.}{B \text{ in } G} \qquad \left(\frac{Z}{G} + \frac{A}{B} = \frac{G \cdot A + B \cdot Z}{B \cdot G}\right).$$

The kind of problems that Viète solved using this kind of symbolic calculus appear to us as an interesting mixture of those appearing in Diophantus' work and those of the abbacus tradition. On the one hand, he was able to improve on Cardano's methods for solving cubic and quartic equations. On the other hand, he dealt with word problems such as the following:

Given the sum and the difference of two numbers, find the two numbers.

In this case, Viète would indicate the sum as D, the difference as B (these are the two known values), and the smaller, unknown number as A. Then he would proceed as follows:

- The larger number is $A + B = E$.
- The problem is translated into the equality $2A + B = D$.
- It follows that $2A = D - B$.
- The two numbers are found via the expressions $A = \frac{1}{2}D - \frac{1}{2}B$ and $E = \frac{1}{2}D + \frac{1}{2}B$.
- Finally, if we take $B = 40$ and $D = 100$, we get $A = 30$ and $E = 70$.

This is the first time in the story thus far that I have not needed to translate the expression found in the original text (typically in a purely rhetorical manner and occasionally with the help of abbreviations) into something that is close to our symbolic expression. What is written above is what Viète himself wrote. Moreover, the symbols used here have been manipulated according to formal rules given in advance. Thus, even at a purely technical level, one can easily see the progress involved in Viète's approach. His predecessors from the abbacus and coss traditions had made important advances on the way toward the fully fledged idea of an algebraic equation. They applied sophisticated methods to the handling of unknown quantities and their powers, aiding themselves with symbolic abbreviations and partial operations on them. Cardano and Stifel, as we saw, devised methods for working out some solutions to problems involving two (and even more) unknown quantities. Viète's step of denoting with letters both the unknown and the known quantities, however, was a crucial turning point.

But this *technical* aspect, fundamental as it was for subsequent developments in algebra, was only part of Viète's overall outlook. Viète always referred to his own work with a high degree of self-importance, and indeed he was much more ambitious than any of his predecessors concerning the aims of his innovations. Whereas in the treatises of both Diophantus and the cossists, we find compilations of large numbers of ad hoc tricks that could be conveniently used for solving different problems, each case corresponding to a specific situation, Viète attempted to develop a *general* method, based on a handful of well-defined rules, that would be fit for "leaving no problem unsolved" (*nullum non problema solvere*)—no less than that.

Viète's personality is interesting in many respects. He was, in the first place, representative of an important kind of contemporary practitioner of mathematics, and one with far-reaching contributions at that. After studying law at Poitiers and serving as tutor to the daughter of an aristocratic family, he moved to Paris in 1570. There he became involved in politics for the rest of his life, but at the same time he continued to develop his original mathematical ideas. These were made known via privately distributed as well as printed treatises, which were always dense with content and not at all easy to read.

Viète's mathematical abilities led him into diverse areas of activity. One of his known achievements is an improvement of Archimedes' method for calculating the value of π by approximation of polygons (see Section 3.6). He used a polygon of 6×2^{16} (that is, 393,216) sides to obtain the first ten fractional places. Then, in 1579, he introduced what is perhaps the first infinite product used in this context:

$$\frac{2}{\pi} = \frac{\sqrt{2}}{2} \cdot \frac{\sqrt{2+\sqrt{2}}}{2} \cdot \frac{\sqrt{2+\sqrt{2+\sqrt{2}}}}{2} \cdots$$

In a different direction, Viète worked as a cryptanalyst in the service of King Henry III. It has been speculated that some of his symbolic techniques for algebra developed in relation to his secret activity in cryptography. This is an intriguing hypothesis that would make the story much more interesting, but unfortunately there is no actual evidence that could support it. The only respect in which such a claim may seem to make sense (without ceasing to be conjectural) concerns the need, naturally arising in the art of code-breaking, to consider separately the frequency of vowels and of consonants, and the parallel distinction that lies at the heart of Viète's symbolism.

A more mathematically significant place where we can look for the sources of Viète's symbolic methods concerns the continued attention, and sometimes openly stated reverence that he, like many of his contemporaries, continued to pay to the "ancients." According to a view that was widely accepted at the time, the Greek sources offered more than enough evidence to assert the existence of some kind of secret "analytical method" that had served them well for finding solutions to many mathematical problems. The synthetic, austere mode of presentation typically found in the classical texts of Euclid or Apollonius was, according to this view, just a way to conceal the actual method with which they were found in the first place. Through the work of Pappus, the term "analysis" had come to be associated in the seventeenth century with a Greek method of discovery in which one started by assuming what was to be proved or constructed.

One of the classical references to the contemporary attitude toward the kind of putative techniques that the ancients had developed and then concealed is found in the fourth of Descartes' *Rules for the Direction of the Mind*. In a rather critical tone, he wrote the following:[1]

But when I afterwards bethought myself how it could be that the earliest pioneers of Philosophy in bygone ages refused to admit to the study of wisdom anyone who was not versed in Mathematics, evidently believing that this was the easiest and most indispensable mental exercise and preparation for laying hold of other more important sciences, I was confirmed in my suspicion that they had knowledge of a species of mathematics very different from that which passes current in our time. I do not indeed imagine that they had a perfect knowledge of it ... But I am convinced that certain primary germs of truth implanted by nature in human minds ... had a very great vitality in that rude and unsophisticated age of the ancient world ... Indeed I seem to recognize certain traces of this true Mathematics in Pappus and Diophantus ...

But my opinion is that these writers then with a sort of low cunning, deplorable indeed, suppressed this knowledge. Possibly they acted just as many inventors are known to have done in the case of their discoveries; i.e., they feared that their method being so easy and simple

---

[1] Quoted in (Mahoney 1994, p. 31).

would become cheapened on being divulged, and they preferred to exhibit in its place certain barren truths, deductively demonstrated with show enough of ingenuity, as the result of their art, in order to win from us our admiration for these achievements, rather than to disclose to us that method itself which would have wholly annulled the admiration accorded. Finally, there have been certain men of talent who in the present age have tried to revive this same art. For it seems to be precisely that science known by the barbarous name of Algebra, if only we could extricate it from that vast array of numbers and inexplicable figures by which it is overwhelmed, so that is might display the clearness and simplicity which we imagine ought to exist in a genuine Mathematics.

Now, when Descartes wrote about "certain men of talent who in the present age have tried to revive this same art," he was surely referring to Viète. The putative method of the ancients was assumed to have been similar to the kind of algebraic methods known in Renaissance Europe. Viète, for one thing, saw his own contribution as part of a more general effort to reconstruct the assumed Greek analytic methods that had been ruined by the "barbarians" (namely, the Arabs). In order to stress this attitude as an essential part of his work, he consistently declined to use the term "algebra" in favor of "analysis." He also introduced some new Greek-sounding terms of his own invention such as *poristic, exegetic* and *zetetic*.

In order to make better sense of the ideas about numbers that arise in Viète's work, it is necessary to stress that, for all of its innovative character, this work did not abandon many basic aspects of the traditions from which it emerged. At the most visible level, one can immediately notice that the symbolic language adopted by Viète preserves significant aspects of the cossic tradition: C represents a cube, Q a square, and R or N an unknown number or a root. Also, while in his symbolic expressions he used signs as + or –, Viète often described the operations verbally, as did mathematicians of previous generations. Moreover, Viète never used numerical notations for powers such as introduced recently by Bombelli and others, but rather more archaic expressions such as *A-quadratus* or *A-cubus*. This implied some obvious limitations on the kinds of general symbolic expressions that could be derived, and on the concomitant idea of where and how a fully abstract idea of number may be applied.

A typical example of this appears in the following sequence of expressions that we find in Viète's text (here expressed in modern notation):

$$(A - B) \cdot (A + B) = A^2 + B^2 \quad \text{or} \quad (A + B)^2 - (A - B)^2 = 4AB.$$

Viète also developed expressions that in modern symbolism are equivalent to

$(A + B)^2 = (A^2 + 2AB + B^2)$ and the equivalents for $(A + B)^3$ and up to $(A + B)^6$, $(A - B)^2 \cdot (A^2 + AB + B^2) = (A^3 - B^3)$ or $(A - B)^2 \cdot (A^3 + A^2B + AB^2 + B^3) = (A^4 - B^4)$,

and so forth up to the power of 6. Viète was able to develop with relative ease each of these cases separately, but he could not reach a general formula for *an arbitrary* power n. The reason for this is clear, and it relates to the lack of a general symbolism, flexible enough to denote a general power, rather than a specific one, rhetorically described. What I just wrote above in terms of exponents 2 or 3 does not appear in this way in Viète's original. Rather, Viète would write each case, like the cossists before him, by way of adding a specific word such as *planum, quadratum, cubus,* or *solidum* (or abbreviations thereof). He could likewise write *cubo-planum* or *cubo-cubum* (for the fifth and sixth powers, respectively), as in the abbacus tradition. But still, he had no clear and consistent way to write the general exponent.

This is just one aspect of a broader sense in which Viète's understanding of number remained close to that of his predecessors. If we return to the already-mentioned expression

$$A \text{ cubus} + C \text{ planum in } A \quad \text{aequatus} \quad D \text{ solidum}$$

($x^3 + cx = d$), we would say nowadays that the various letters involved in the expression, x, c and d, represent just "numbers," rather than a number of this or that kind. For Viète, as we clearly see from his notation, every quantity in the expression is clearly identified by its type. In this example, all the terms compared are solid magnitudes: "C planum" is a plane number and "A" is a length, so that the product "C planum in A" is a solid. To this, Viète added the solid "A cubus" and equated their sum with another solid magnitude, namely, "D solidum."

Without strictly paying attention to this dimensional homogeneity, the entire expression would be, for Viète, meaningless. Thus, Viète was using only magnitudes of given kinds and was also paying close attention to the classical requirement of homogeneity, but at the same time (as this example indicates), in addition to the classical operations of comparison and addition, his rules did allow for the innovative ability to multiply or divide magnitudes of different types. These operations yield, of course, a magnitude of yet a different type. Viète's adherence to homogeneity is strongly related to the central role that the use of the Greek classical theory of proportions (with all the concomitant view on number and magnitude) plays in his treatises.

It is also remarkable that the kind of homogeneity of magnitudes required by Viète had been abandoned in the symbolic techniques found in many texts of the abbacists and cossic tradition (which were less powerful in many other respects). In this sense at least, Viète's innovations were achieved at the cost of consciously reimposing a restriction that was much more strongly followed in classical Greek text that in those of his immediate predecessors.

We can also compare Viète's practices with those developed by his followers, and above all Descartes, in order to realize that some fundamental elements are still missing here before the full idea of an algebraic equation can be fully discerned. At the level of style, to take one interesting point, Viète often accompanied all his symbolic manipulations with a fully rhetorical account of each of the steps, perhaps with the aim of convincing the skeptics. Moreover, the use of "aeq." instead of our "=" symbol is also highly suggestive of the underlying views. Viète's symbolic rules of manipulation, such as illustrated in the previous examples, are in the first place *replacement* rules rather than rules for handling "an equation." What I mean by this is that if we are given an expression like

$$\frac{A \text{ planum}}{B} \text{ in } Z,$$

then we can write *in its stead* (i.e., we can *replace* it with)

$$\frac{A \text{ planum in } Z}{B}.$$

But this does not yet mean that we will want to go in the opposite direction, and when given the latter expression replace it by the former. To state this more generally, Viète's rules, while general in their aims, allow handling of the specific situation they describe, but not beyond that. We will not find in his text a simple, general rule for dealing with equations, such as "a number multiplying in one side of the equation passes to the other side as a divisor."

In spite of the limitations imposed by the demand for homogeneity, I want to emphasize once again that the importance of Viète's idea of representing symbolically the known, as well as the unknown, quantities can hardly be exaggerated. Recall that Cardano, for all his ability to solve problems with unknowns in the third and fourth degree, was not able to come up with a "formula" in which the relationship between the coefficients and the result is evident. Viète's approach to writing the known quantities opened the way precisely to taking this crucial step.

In the above examples, we can see that rather than describing the steps leading from a problem to its solution, we obtain a formula that connects, once and forever, any pair of numbers taken as sum and difference with the two numbers we are looking for. This is, of course, the approach we are used to following nowadays, but against the background of our account so far we can clearly pinpoint this bold step of Viète as being decisive for allowing this approach. It was decisive, moreover, toward completing the picture of modern symbolic algebra, whereby the general equation itself, rather than some specific question about relations among numbers, becomes the focus of enquiry. The very few obstacles remaining for the complete picture will be soon removed by Descartes, as we will see in the next chapter.

But the innovation promoted by Viète's methods was not limited to the technical level alone. Rather, it operated at the conceptual level as well. His analytic methods, conceived as a general tool, were meant to be applied to *abstract* magnitudes rather than specifically to numbers, line segments, or figures (although, remarkably, and in spite of the sweeping generality he was aiming at, Viète did not abandon the view that negative numbers cannot legitimately appear as solutions to the problems; the "abstract" magnitudes that he had in mind did not include anything like "negative" magnitudes).

This attempted generality, however, raised questions about, and had important implications for, the current conceptions of number, and the still ongoing process of blurring the distinction between continuous and discrete magnitudes. Indeed, when two numbers are multiplied, say 5 times 7, this means that the number 7 will be added to itself five times. But there is no single, natural translation for this idea when it comes to the multiplication of two lengths or two areas. Viète chose to identify multiplication of two lengths with constructing a rectangle having those lengths as sides. There is nothing particularly innovative to this. Similar ideas appeared at the time in widely used texts, and it was naturally incorporated into the standard body of knowledge, for instance in Clavius' edition of Euclid's *Elements*. But in the case of Viète, precisely because of the generality he sought to associate with his methods, this choice raised an issue when applied to areas multiplied by areas, for example, or to quantities of higher dimensions. The results thus obtained admit of no direct geometric interpretation, and yet Viète's methods allow them to be handled without the need for such an interpretation.

Fortunately, Viète does not seem to have been bothered by this apparent conceptual difficulty, and he just went on to work out the methods as part of his ambitious plan. Indeed, Viète's general analysis—aimed at extending to quantities of all kinds what in algebra had been thus far applied mainly to numbers, and in which preservation of dimensional homogeneity was still fundamental—provided a convenient framework against which the idea of abstract quantities could gradually evolve, which are neither numbers nor geometric magnitudes, and which represent the two species at the same time.

A revealing statement in this regard appears in the text where he presented his method for approximating the value of π. He wrote:[2]

Arithmetic is absolutely as much a science as geometry [is]. Rational magnitudes are conveniently designated by rational numbers, and irrational [magnitudes] by irrational [numbers]. If someone measures magnitudes with numbers and by his calculation gets them different from what they really are, it is not the reckoning's fault bu the reckoner's.

So, Viète's "new algebra" not only contributed to the intensifying trend towards eliminating the separation between discrete and continuous magnitudes, but was also instrumental in facilitating the soon-to-be-completed fusion between algebra and geometry in the framework of the analytic geometry that Fermat and Descartes would develop about forty years later. Viète had no direct interest in the kind of problems that would lead these two mathematicians to understand the benefits of applying symbolic methods to the study of curves. But whereas for all of his predecessors, Cardano and Bombelli included, algebraic methods required a justification that only geometry was seen to provide, Viète inverted the conceptual hierarchy and laid the priority on algebra. This was a step of momentous consequences. His most important book, *In Artem Analyticam Isagoge* ("Introduction to the Art of Analysis"), published in 1591, reached wide mathematical audiences all around Europe. It is easy to spot its direct influence in texts that appeared soon thereafter in the British Isles, Italy, Holland and France.

7.2 Stevin and decimal fractions

A second contemporary focus of influential activity related to numbers and arithmetical practice is found in the work of Simon Stevin (1548–1620), a Flemish mathematician and engineer. Stevin was a highly ingenious and original thinker, with many intellectual and practical contributions to various fields of knowledge ranging

---

[2] Quoted in (Berggren et al (eds.) 2004, p. 759).

from the construction of military fortifications and dams for flood control to the writing of theoretical books on geometry and physics.

In 1585, Stevin published two short but influential texts. The first, *L'arithmétique* developed an explicit call for the elimination of all distinctions between number and magnitude. We have already seen how this separation had been gradually weakened and then blurred in the works of the abbacists and cossists, and in a different way in the more recent work of Viète. Stevin was the first person to not just explicitly declare that this distinction is unnecessary and indeed detrimental, but also to come forward with a consistent and fruitful body of arithmetical knowledge that followed the spirit of this declaration.

The second text appeared in Dutch under the title *De Thiende* ("The Art of Tenths"). It was soon translated into French as *La Disme*, and later into other European languages. It was an innovative booklet that presented in a concise, coherent and systematic manner many ideas that had previously surfaced only implicitly or that had been discussed sporadically and in isolated texts. Of particular importance among this was the idea of decimal fractions.

Stevin defined number in *De Thiende* as "that by which the quantity of each thing is explained." Tacitly, Stevin was identifying proportions with numbers, and allowing the use of Eudoxus' theory of proportions as the basis for a new kind of generalized arithmetic of quantities. Explicitly, Stevin indicated that also "one" is a number. He wanted to leave no doubt that his definition of number was opposed to that of Euclid. Moreover, as the part and the whole are made out of the same elements, so the unit, being part of any number, is of the same nature as any other number. Therefore, one can divide the unit into parts, as small as we may want, and these parts will be numbers as well.

A number, then, is not a "discrete quantity," for the same reason that fractions and incommensurable lengths are numbers in every respect. The fraction $\frac{3}{4}$ has a common measure with 1, while $\sqrt{8}$ does not. But, on the other hand, $\sqrt{8}$ has a common measure with $\sqrt{2}$ or with $\sqrt{32}$, and hence in all cases we are speaking of legitimate numbers. For Stevin, the use of terms like "irrational," "absurd," "unexplained," or any similar terms for denoting numbers other than the natural ones, was utterly unjustified. And yet, interestingly, expressions containing square roots of negative numbers were still for him "useless," and he did not discuss at all the issue of imaginary or complex numbers.

Stevin had few connections with the scholarly institutions of knowledge of the time. He certainly did not depend on them for his ability to continue developing his own original ideas. Nevertheless, it is quite remarkable that even a person like him would try to ground and provide a sound justification of his views on the authority of the "ancients." In his case, however, these were not the ancient Greeks. Stevin expressed his conviction about the existence of an even more ancient time, *Wisentijt*, an era of sages where perfect order reigned and in comparison with which even the world of the Greeks with their ideas seemed as decline and corruption.

Ideas of this kind were quite common among scholars in the humanistic tradition, and they were still pervasive in the seventeenth century. According to Stevin, for instance, the origins of the Hindu–Arabic numeration and of algebra as a whole can be found in that era. The complex ideas on incommensurable magnitudes that are at the core of Book X of the *Elements*, he thought, had been previously known in that early period. They were developed in purely arithmetical terms, in a way similar to the one he was presenting now in his booklet. It was only later—he added—that these ideas were translated by the Greeks into a theory of ratios between geometric magnitudes. The different treatment accorded to such ratios and to ratios between numbers was for him the source of the complexity that characterized the Euclidean approach. This called for the new kind of approach he was now offering.

What made Stevin's arithmetical ideas valuable, important and widely accepted were not his somewhat weird scholarly analyses of the past (which were shared by many humanist intellectuals), but the way in which he used his ideas to considerably enlarge and enhance current arithmetical practice. Indeed, Stevin was working at an interesting historical crossroads where scholarly traditions met and mixed with practical traditions developed by engineers like himself. His theoretical consideration received far-reaching resonance because he ably translated them into concrete steps of efficient number writing and of calculations with numbers of all kinds. This is the case, above all, with the new system he introduced for writing decimal fractions.

Stevin's decimal fractions embodied a shorthand for writing what at the time would be typically presented as a concatenation or as a sum of fractions, such as the following:

$$27 + \frac{5}{10} + \frac{3}{100} + \frac{2}{1000} + \frac{8}{10000}.$$

This he suggested be written as a single number, as follows:

$$27①5①3②2③8④.$$

The small cipher inside the circle is in fact an abbreviation of the power of ten by which each of the digits is multiplied. At the same time, it also indicates the place of that digit within a sequence that follows the integer part of the number thus represented. This notation was developed under the explicitly acknowledged influence of Bombelli's powers of the unknown. Notice, also, that in Stevin's original approach there is a small zero, rather than a point separating the integer from the fractionary part.

After introducing the basic idea behind his notation, Stevin went on to explain simple rules for performing, on numbers written in this way, all the arithmetic operations as known for the integers. So, for example, if we wish to multiply two numbers, 3⓪7①5②7③ and 8⓪9①4②6③, then all we need to do is arrange them as follows:

```
        ⓪ ① ② ③
        3  7  5  7
        8  9  4  6
        ―――――――――
        2 2 5 4 2
      1 5 0 2 8
    3 3 8 1 3
  3 0 0 5 6
  ―――――――――――――
  3 3 6 1 0 1 2 2
```

The partial multiplications are done in the standard way for integers, and thus we obtain as a result 33610122. But then it is still necessary to state the exact value of this string of digits as a decimal fraction. This, Stevin explained, is done by adding up the two "last signs" of each factor, ③ and ③ in this case, thus obtaining ⑥. From here, we get the final result, which is 33⓪6①1②0③1④2⑤2⑥. Following a similar approach, Stevin explained how to perform all other standard arithmetic operations, including root extraction.

As with Viète's symbolic method, it is also difficult to exaggerate the importance of Stevin's simplifying step, at the basis of which there is an effort to unify the notational system (decimal–positional) and make it work simultaneously for integers and for fractions. No less important was his intention that this method be adopted for denoting, in a unified manner, units of measurement used at the time in diverse contexts. This, he thought, would considerably contribute to enhance their ability to perform demanding calculations. He explicitly referred to the obstacles faced by astronomers, land surveyors, sea navigators and some additional professions, who had to address complex and tiresome computational tasks. Their jobs had become plagued with errors stemming in the first place from the use of fractions of various kinds all mixed with each other. In astronomy and cartography, which were two important examples, the fractions mostly used were those of basis 60, because the degree is divided into 60 minutes and these, in turn, into 60 seconds. In matters related to land surveying, every region used its own units of measure, which were differently subdivided, typically not on a decimal basis: inches, yards, feet, etc. And the same was the case for matters related to local currencies and accounting problems.

Stevin addressed members of different professional communities and explained to them, with separate arguments relevant to each profession, why, in their particular field of occupation, the adoption of the decimal system together with his notational methods for fractions would be especially useful. Stevin's appendices are devoted to computations of surveying, of measuring of tapestry, of gauging and measuring of wine casks of various shapes and of volumes in general, of astronomy and of mintmasters and merchants. Astronomy was at the time the field where the most lengthy and complex computations were continually performed, and the interesting point is that Stevin's recommendation to use the decimal system has *not* been adopted to this day. Both in geometry and in astronomy, the sexagesimal basis continues to be the most commonly used. Also in the case of currencies, Stevin's suggestion was essentially overlooked and continued to be so in some countries until relatively recently.

Stevin was aware of the expected difficulty of introducing the decimal approach into the currency system, as every local administration has the authority and the will to decide on this matter, based on all kinds of considerations—not necessarily those that follow from serious mathematical thinking. But he also expressed his hope that if the system were not adopted in his own lifetime, then perhaps statesmen of the future would turn out to be wiser, as were those of the past (his *Wisentijt* sages), and would understand the importance of such a decision. Thus, natural as it may seem for us to work with the same decimal logic in the numbering system and in the various measurement systems, the cases of astronomy and of currencies show, each in its own way, and each for its own reasons, that historical circumstances may have a greater influence on the processes of development of ideas than purely logical considerations.

The decimal system became the standard in Europe (not including the British Isles) for measurement systems only in the late nineteenth century, and in matters of currency it had to wait even longer than that.

Although, as we saw in Section 5.7, mathematicians in Islamicate cultures, such as Al-Uqlīdisī, had made sporadic use of the idea of decimal fractions, no-one before Stevin stressed, as he did, the importance and the advantages of adopting decimal fractions as a way to turn the decimal system into much more than just a convenient system for writing numbers. His text was translated into several European languages, and it exerted a crucial influence on contemporary views on numbers, numeration techniques, and writing and calculation with decimal fractions. The latter were widely adopted and, as a matter of fact, several suggestions were put forward for improving Stevin's original way of writing them. The use of a period to separate the integer from the fractional part, as is commonly used nowadays, appeared with various authors, but it gained final acceptance only when it was adopted as the main way to write logarithms, beginning in 1614.

The massive use of logarithms as part of the day-to-day work of astronomers and cartographers, and the ability to write them in a simple and efficient way as decimal fractions, according to the line of thought introduced by Stevin, represented a new peak in the lengthy process that led to the definite adoption of the decimal system of numeration in all of its aspects. In the European context, this is a process that began, as we saw, in 1202 when Fibonacci published his *Liber Abbaci* and explained the basics of the decimal system as he had learnt them from Islamic mathematics. But additional ideas and substantial changes in the concept of number and in the notational systems were still needed in order to attain a fully fledged consolidation of the system. In order to complete the picture of this complex process, I devote the last part of this chapter to a brief description of the introduction of logarithms as an efficient and sophisticated computational tool in the hands of John Napier (1550–1617) and Henry Briggs (1561–1630).

7.3 Logarithms and the decimal system of numeration

Some of the mathematicians mentioned thus far addressed the question of the relationship between operations on powers and on exponents, which is the necessary link underlying the idea of logarithm. Chuquet, for instance, wrote down in 1484 tables of powers and exponents. Later, in 1544, Michael Stifel discussed in greater detail some of the interesting properties of such tables, with particular reference to the following:

0	1	2	3	4	5	6	7	8
1	2	4	8	16	32	64	128	256

The numbers in the first row of the table he called exponents and he emphasized that while they comprise an arithmetic sequence (i.e., each number is obtained by adding to the previous one a fixed factor, in this case 1), the numbers in the second row comprise a geometric one (i.e., each number is obtained by multiplying the previous one by a fixed factor, in this case 2). From here, it is easy to deduce some clear operational rules that depend on the table alone. The simplest rule is, of course, that a sum of elements in the arithmetic sequence corresponds to a product in the geometric one: 2 is associated with 4; 3 is associated with 8, and the sum of these factors, $2 + 3 = 5$, is associated in the table with the product of 4 and 8, namely, with 32. Likewise, division in the lower row can be translated into subtraction in the upper row, power of power into multiplication of exponents, and root extraction into division. For example, 6 is associated in the table with 64, and 8, which is the square root of 64, is associated with 3, which is 6 divided by 2.

Given that Stifel worked extensively with negative and irrational quantities in other contexts, he also wondered about the possibility of using these kinds of numbers as exponents, but in the final account he was unable to handle such issues systematically and successfully. The irrationals, for instance, did not for him and for others correspond to any "true" number, and therefore the possible role of exponents of this kind could not be properly conceived. The same held for negative numbers. Even though they did appear with increasing frequency in texts throughout the fifteenth and sixteenth centuries, there was no agreement about their possible significance as exponents.

Before going further into the historical account, it is worth noticing that the kinds of sequences that at this time attracted the attention of someone like Stifel can be analyzed from a more general and abstract perspective. Consider the following example:

2	4	6	8	10	12	...
3	9	27	81	243	729	...

In this table, the correspondence between the sequences constituting the two rows allows us to calculate with rules similar to those established by Stifel. For example, we can calculate the product of 9 and 81 via the sum of the corresponding elements in the upper row: $4 + 8 = 12$. Since 12 corresponds in the table to 729, we realize that $9 \cdot 81 = 729$. The calculation is based on relative location on the tables alone, provided that the first is an arithmetic sequence and the second a geometric sequence. So, it is not really necessary to check if the numbers in the second row are powers of some given base number. (But, I should add, for those readers interested in details, that if we really want to, we can see this as table of powers of $\sqrt{3}$: for example, $(\sqrt{3})^4 = 9$.) At any rate, if we now choose a similar table, in which the bottom row is one of powers of tens, which is a geometric series, then we obtain a table of logarithms without having to have spoken at all about a connection between power and exponent:

0	1	2	3	4	5	...
1	10	100	1,000	10,000	100,000	...

On the other hand, if this table is to provide a useful tool for anyone intent on actually performing long and complex calculations (e.g., an astronomer in the early seventeenth century), then the large gaps between successive powers appearing in the bottom row must be drastically reduced. And this was precisely the starting point of Napier's work on logarithms. The details of Napier's ingenious construction of logarithmic tables are given in Appendix 7.1. The remarkable point that I want to stress here is that he worked close to twenty years in preparing them. Indeed, it is rather amazing to realize how precise his calculations were, and how truly few errors they contained.

In their original form, Napier's logarithms do not satisfy exactly the same kinds of properties which are well known to us nowadays, but rather some other, similar ones. Because of the specific way in which he constructed them, the logarithm of 0 was 10^7, and the logarithm of 1 was 9,999,999.

A related difficulty was that his original approach led him to wrongly believe that the logarithm of x equals the logarithm of $-x$ (this belief continued to be accepted up until the time of Euler). Nevertheless, with time, the more he got involved with calculations of his tables, the better he understood that their efficiency as a practical tool could be easily improved with just some little modifications that turned the value of the logarithm of 1 to be 0, as we have it nowadays. Also, the usual rules followed easily:

$$\log x \cdot y = \log x + \log y \quad \text{and} \quad \log x/y = \log x - \log y.$$

More importantly, by setting the value of log 10 to be 1, then the logarithm of any number $a \cdot 10^n$ (with $1 \leq a < 10$) becomes $n + \log a$. This basic property makes the preparation of tables a much easier task, since, once we know the value of, say, log 5, then we automatically obtain the values of all its multiples with the powers of 10:

$$\log 50 = 1 + \log 5, \quad \log 0.5 = 1 - \log 5, \quad \log 500 = 2 + \log 5, \quad \ldots$$

Napier died before he could complete the computations for the full tables following this new approach. But then Briggs, professor of mathematics at Oxford, who was among the most enthusiastic early readers of Napier, was able to speak with him on these matters before his death. Briggs completed the job by computing new tables from scratch, starting with values such as

$$\log \sqrt{10} = 0.5, \quad \log \sqrt{\sqrt{10}} = (0.5)^2, \quad \log \sqrt{\sqrt{\sqrt{10}}} = (0.5)^3, \quad \ldots$$

In this way, he went all the way up and could also calculate values such as $\log 10^{1/2^{54}} = (0.5)^{54}$.

Briggs calculated each value up to the thirtieth decimal place, and he wrote them down following Stevin's then still novel notation, but already with a period separating the integer from the fractional part (a practice that Napier had initiated). With the help of the rules of multiplication and division of logarithms, Briggs constructed a table with very dense values. What he was able to calculate in his lifetime he published by 1624. This comprised logarithms of all numbers between 1 and 20,000 and then between 90,000 and 100,000, which he wrote as decimal fractions with 14 digits after the period. Throughout the years, about 1100 mistakes were found in the tables.

This amounts to no more than 0.04% of the total entries. In most cases, these were deviations of ±1 in the last digit, plus some simple typos.

The entire table was completed in 1629 by the Dutch Adriaan Vlacq (1600–1667). Briggs also published a highly influential book with explanations of possible uses of logarithms in computations related to astronomy and navigation, but also to financial calculations (for instance, future values of loans with compound interest) and to theoretical calculations with no immediate, practical use. Following the publication of Briggs' books, logarithms became a central tool for scientific activity in computation-intensive disciplines, and above all in astronomy.

Logarithm tables that did not essentially differ from those of Briggs continued to be in extensive use in the hands of engineers, chemists, artillery officers and many other kinds of specialists, up to the 1970s when the first hand-held calculators with logarithmic and trigonometric capabilities began to appear. What is truly important for our account here is the pervasive and long-ranging influence of the innovations just described on the further development and consolidation of the modern conception of number and on the practices related to it.

In the first place, it is important to notice that without the flexibility afforded by Stevin's approach to writing fractions, one cannot even begin to imagine such an exhausting and demanding enterprise as Napier and Briggs undertook. It is not just the matter of calculation, which is difficult enough in itself, but also all that refers to the material side of the enterprise, like printing, correcting and making the tables available to a wide readership that could easily make use of them. Without Stevin's decimal fractions, these demanding tasks would most likely have proved impracticable. In turn, the logarithmic tables were the ultimate vehicle that facilitated the widespread adoption of decimal fractions, including the use of a period to separate the integer part from the fractional part.

But it is also crucial to realize that not just a *technical* issue was at stake here. The use of Stevin's decimal fractions implied, or at least promoted, a view of numbers in which the distinction between discrete and continuous quantities became increasingly meaningless. Not that Napier or Briggs undertook to actively eliminate the distinction, but in practice this was an immediate and important consequence of their work. Indeed, a close inspection of Napier's texts as they evolved through time show an initial attempt to abide by the classical, Euclidean approach to numbers and then a gradual willingness to abandon it in favor of Stevin's approach under the weight of the actual mathematical work. Indeed, given the connection between the two sequences with which he built his tables, and the fact that one of them was geometric, Napier initially approached the computations using the ideas of ratio and proportion and the tools provided by propositions proved in Books II and V of the *Elements*. The actual source for the term *logarithm* is in the Greek *logos* (ratio) and *arithmos* (number). But the more Napier was dragged into complex calculations, the more he started to treat the ratios obtained simply as numbers and to write them as approximate values in decimal fractions of up to six digits after the decimal period.

Because of the practical orientation of their books, Napier and Briggs devoted very little attention to theoretical discussions about the concept of number as they saw it, or about the kinds of calculations they did with numbers. Moreover, one may speculate that the very few such discussions that did appear scattered through their texts most likely passed unnoticed by the many readers of their books and made little impression on them. One can imagine that the typical user of the tables, an astronomer or an artillery officer in need of aids for calculations, went directly to the contents, without asking many questions about the nature of discrete and continues quantities or about the possible distinctions between them. The possibility of writing decimal fractions with a separation point, in such an efficient way and with no need for further explanations, helped, no doubt, to prevent such questions from being asked and to focus on the wonderful practical opportunities opened by their use.

The works of Viète, Stevin, Napier and Briggs changed in a deep sense the overall map of arithmetic and algebra at the beginning of the seventeenth century. In the next chapter, we will examine some of the main mathematical works that were strongly influenced by the new views on numbers implied by their contributions, or reacted critically against them. By doing so, they completed the process whose main stages we have been discussing in this and the preceding chapters. They led, finally, to consolidating the early modern views on numbers and equations.

Appendix 7.1 Napier's construction of logarithmic tables

The basic idea of the logarithm arises from the correspondence between an arithmetic sequence of exponents and a geometric sequence of powers. This was noticed well before the time of Napier. This is the case with the following example of the table of powers of 10 already mentioned above:

0	1	2	3	4	5	...
1	10	100	1,000	10,000	100,000	...

As already stated, however, if the gaps between successive values are as large as in this table, then one cannot think of turning the general idea into a practical tool to be used in actual, complex computations. Napier addressed this difficulty by introducing another sequence with very small gaps, and by calculating the corresponding values with the help of an ingenious approach based on considering the numbers in both sequences as points moving along a straight line. In relation to the first sequence, the point is considered to move with constant velocity, while in relation to the second sequence, the point is considered to be moving with a velocity that decreases in inverse proportion to the distance covered. Following the details of Napier's original calculations requires considerable effort, but here I will present the basic ideas along the lines of a nice and simple account appearing in (Pierce 1977).

Let us imagine a straight line TS, as in Figure 7.1, and a point b located on T at time, $t = 0$, within a distance of 10^7 units from S. Point c is located on S. Now assume that point b starts to move along the straight line, with decreasing velocity. Specifically, we assume that within every unit of time, b advances a distance equal to $(1 - 1/10^7)$ times the current distance between b and c. In other words, since $(1-1/10^7)$ equals 0.9999999,

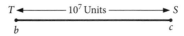

Figure 7.1 Point b moves over a straight line with decreasing velocity.

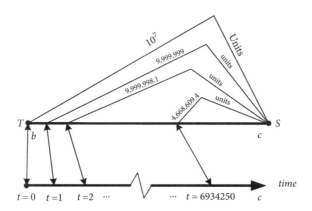

Figure 7.2 Defining logarithms by means of two points moving at different velocities.

after the first unit of time the distance between b and c is 9,999,999, and after the second unit of time this distance is 9,999,998.1, and so forth. If we concurrently think of a second point representing time, which is also moving along the straight line TS, but with uniform velocity, then, by looking at the relative positions of the two points, we obtain two sequences of values that we can make to correspond to each other as shown in Figure 7.2.

Alternatively, we can represent the corresponding values of times and distances with the help of a table, as follows:

0	1	2	3	...	6,934,250	...
10^7	9,999,999	9,999,998.1	9,999,997.1	...	4,998,609.4	...

In this table, the upper row displays the values of the arithmetic sequence of times, starting at 0: 0, 1, 2, 3, ... The sequence of distances in the bottom row, on the other hand, is a decreasing geometric sequence: $10^7 \cdot (0.9999999)^0$; $10^7 \cdot (0.9999999)^1$; $10^7 \cdot (0.9999999)^2$; $10^7 \cdot (0.9999999)^3$; ... $10^7 \cdot (0.9999999)^{6,934,250}$; ...

With the help of a table like this one, Napier set out to calculate the logarithms. He called the exponents 0, 1, 2, 3, ... "logarithms," and the numbers 9,999,999, 9,999,998.1, etc., he called "sines." The reason for this terminology is directly connected to the context of his investigations, namely, astronomical computations, in which trigonometry had played, ever since the time of Greek science and increasingly in the decades preceding the work of Napier, a very central role. Particularly lengthy multiplications that required much effort and were often the source of mistakes appeared in connection with the use of formulae such as

$$2 \cos a \cos b = \cos(a + b) + \cos(a - b).$$

The trigonometric context was also the reason for the special way in which the rows of Napier's tables were presented. In their original version, they comprised values for each minute of the degree, going from 0 to 45 degrees. In each row, there were seven values, as in the following example:

34°40' 5688011 5642242 367872 1954370 8224751 55°20'

The three values to the left represent, respectively from left to right, an angle α, the sine of the angle (that is, $\sin 34°40' = 0.5688011$), and the logarithm of the sine (the value here is somewhat different from the currently accepted value, but the principle is the same). The three values to the right represent the same values, but for the complementary angle $90° - \alpha$. The figure in the central column represents the value $\log \sin \alpha - \log \sin(90° - \alpha)$, which is equal to $\log \tan \alpha$.

CHAPTER 8

Number and Equations in the Works of Descartes, Newton and their Contemporaries

Let us complete now our overview of the world of numbers and equations as it emerged at the time of the scientific revolution, by looking at the all-important work of René Descartes (1596–1650) and some of his British contemporaries, including the giant figure of Isaac Newton (1642–1727). Before entering into details about Descartes' ideas, however, it is important to stress that his entire scientific enterprise, including his views about numbers, should best be understood in the framework of a broader discussion of his philosophical system. As a matter of fact, there are not many cases in the history of mathematics where the connection between a philosophical doctrine and the development of scientific ideas is as strong and unmistakable as in the case of Descartes. For the purposes of our account here, suffice it to say that the text whose contents we will examine more closely in order to understand his views on arithmetic and geometry, *La géométrie*, appeared in 1637 as one of three appendices to one of Descartes' well-known philosophical treatises, *Discours de la methode*. For Descartes, mathematics was in the first place an invaluable tool for educating the mind so that it could be fit for penetrating the secrets of nature and the true grounds of metaphysics. Above all, the mathematical ideas presented in the appendix to his philosophical book were intended as a well-focused and particularly important illustration of the philosophical system discussed in the main text.

8.1 Descartes' new approach to numbers and equations

A convenient way to understand Descartes' original ideas on numbers and equations is by comparison with Viète. First, like Viète, Descartes too saw his ideas as part of a general method for "solving any problem" in mathematics. Descartes also took for granted the same widespread assumption about the putative method of "analysis" that the ancient Greeks had maliciously hidden from us, and that should be renovated now. Descartes thought that Viète had already made an important contribution in that direction, but there was still need to do more.

Viète, as we saw, devoted himself to developing the old–new analysis from a hands-on mathematical perspective: he started from the algebraic methods known in his time and tried to extend the use of the symbolic methods to all possible quantities, both discrete and continuous. Descartes' starting point, in contrast, was a philosophical perspective that looked at scientific problems in general and attempted to systematically classify them so as to be able to determine beforehand the appropriate solution for each and every one of them. Algebra was his tool of choice for approaching this task. In general, he adhered to the widespread view that algebraic methods such as taught by Viète involved a rediscovery of techniques the ancients had systematically concealed. But in some places he was eager to emphasize the novelty implied by his own methods.

Both Viète and Descartes faced the need to translate geometric situations into symbolic language, but each made this translation in his own way. Viète defined the multiplication of two segments as a rectangle formation (i.e., an operation between two magnitudes of the same kind yielding a third magnitude of a different kind). But when it came to equations, he always adhered to strict dimensional homogeneity. Descartes, in contrast, defined multiplication of two lengths in way that yielded a third length, as we will immediately see. He did as much for the other algebraic operations: dividing one length by another to yield a length and extracting the root of a length to yield a length. In addition, he made the first steps in the newly conceived idea of analytic geometry, whereby he established a direct link between geometric figures, such as a straight line and a parabola, and a certain well-defined class of equations representing each type of figure.

In retrospect, one can find some of these ideas in the works of Descartes' predecessors (Bombelli defined a multiplication of lengths that yield a length; Fermat came up with some of the basic ideas of analytic geometry), but what appeared in previous works as hesitant or sporadic, as a first step that was not carried through, or perhaps as a passing comment, becomes with Descartes an organically interconnected whole, which was systematically pursued with far-reaching consequences. Descartes' views on the interconnection between algebra and geometry led to a more general and abstract understanding of the idea of an equation under which the traditional requirement of dimensional homogeneity would eventually (but, as always, only hesitantly) be abandoned. Descartes himself, for one, did try to preserve homogeneity in many of the equations he considered. But his approach allowed in principle, and in practice also led to, the possibility of ignoring, once and for all, this burdensome requirement that was a legacy of centuries and that hindered the

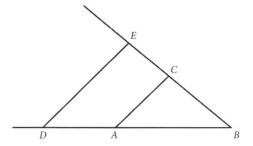

Figure 8.1 Descartes' multiplication of two segments.

full understanding of what is involved in working with a polynomial equation. Let us see some details of how he worked out these ideas.

Descartes' definition of the multiplication of two segments, *BD* and *BC*, is presented in Figure 8.1. The most important feature of this multiplication is that its outcome is not an area but rather a third segment, *BE*. The procedure is based on defining a certain length, *AB*, which is considered to be a "unit length," namely, a segment whose length is 1. The segment *BE* is constructed by placing the three segments *AB*, *BC* and *BD*, as indicated in the figure. Segment *AC* is first drawn, and then *DE* is drawn from *D* and parallel to *AC*. A simple consideration of similarity in the triangles yields the proportion *AB*:*BC* :: *BD*:*BE*. And since the length of *AB* is 1, then we obtain that the length *BE* equals the product of the lengths *BC* and *BD*, as requested. It is clear how this same diagram can be used to construct the segment *BC* as the division of two given segments *BD* and *BE*, with the help of the same unit length *AB*.

In a different example, shown in Figure 8.2, Descartes showed how to obtain a segment that is the square root of a given segment. Given a segment *GH*, we extend it to reach the point *F*, with *GF* being the unit-length segment. The segment *FH* is bisected at *K*, and we trace the circumference *FH* with center at *K* and radius *KF*. At *G*, we raise a perpendicular that cuts the circumference at *I*. A simple theorem about circles, which was known to Greek geometers, states that the square built on *GI* equals the rectangle built on the segments *GF* and *GH*. As the length of *GF* was set to be 1, we get $GI^2 = GH$, or, in other words, *GI* is the square root of the given segment *GH*.

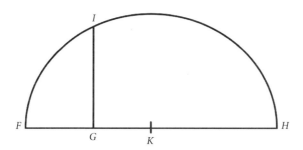

Figure 8.2 Descartes' extraction of a square root of a given segments.

To most readers of the present book, these two examples may seem utterly trivial and requiring no additional explanation. What have we done here, after all? No new knowledge seems to be involved, other than simple theorems about similarity of triangles and a property of the circle. And indeed, Descartes' *La géométrie* is the first among the texts we have examined up to this point in the book that a modern-day mathematical reader can approach with concepts, terminology, notation and methods that are essentially known to him. But this is precisely the point that I am trying to emphasize here and that highlights the striking innovations involved in Descartes' work. Descartes' approach to solving problems via geometric constructions, where magnitudes are considered as numbers without restrictions, and where one freely operates with these magnitudes in the framework of algebraic equations, is absolutely close to our understanding because it has left behind the bulk of the previous, more limited views on numbers. Descartes wrote explicitly that it is possible to find the appropriate construction to solving any geometric problem by finding the lengths of some segments. His definition of operations with segments was aimed precisely at fulfilling this task.

If Descartes defined the operations with lengths on the basis of theorems that were known to the Greeks, it is evident that it was not a technical difficulty that prevented any of his predecessors from taking the step he took in defining operations between segments the way he did. Rather, what was at stake were more fundamental questions of principle about mathematics, about the relationship between geometry and algebra, and about the question of what are numbers and what is their role in mathematics. At the technical level, the key for defining the various operations lies in one step that is almost imperceptible, and certainly almost insignificant from the point of view of modern mathematics, but which is the crucial one here: the use of a segment of "unit length" in each of the operations above (*AB* in the case of multiplication, *FG* in the case of root extraction). It is this unit segment that allows, at the bottom line, to finally overcome the need to distinguish between magnitudes of different dimensions and to abide by the homogeneity among terms appearing in an equation. Unit lengths, to be sure, had appeared in geometric texts from the time of Islamicate mathematics. They had also appeared more recently in texts that explored in a tentative manner the relations between geometry and algebra in more modern terms. But it was only the systematic use the unit length by Descartes in the framework of his innovative treatment of geometry—and of the operations with segments as part of it—that turned it into a fundamental piece of a new, overall conception of numbers and magnitudes.

The systematic introduction of the unit length afforded the possibility of abandoning the need to strictly abide by dimensional homogeneity, and from his explanations it is clear that Descartes was aware of this. Still, he did not immediate give up the habit of doing so. He used a symbolic language similar to that of Viète, with one stylistic difference that has remained in use up to our times: the first letters of the alphabet are used for the known quantities and the last ones for the unknown quantities. In this regard, he wrote in the opening passages of Book I of *La géométrie*:[1]

---

[1] (Descartes 1637 [1954], p. 5).

Often it is not necessary thus to draw the lines on paper, but it is sufficient to designate each by a single letter. Thus to add the lines BD and GH, I call one a and the other b, and write $a+b$. Then $a - b$ will indicate that b is subtracted from a; ab is that a is multiplied by b; a/b that a is divided by b; aa or a^2 that a is multiplied by itself; a^3 that this result is multiplied by a, and so on, indefinitely. Again, if I wish to extract the square root of $a^2 + b^2$, I write $\sqrt{a^2 + b^2}$, if I wish to extract the cube root of $a^3 - b^3 + abb$, I write $\sqrt{C.a^3 - b^3 + abb}$, and similarly for other roots.

Notice another interesting, if minor stylistic difference between Descartes and current usage, namely that Descartes indicates the cubic root with a letter C inside the root symbol, rather than as an index outside it. This follows from a fact that is interesting in itself in the context of the history of the concept of numbers, namely, that roots are not yet conceived as fractional powers, and as a matter of fact not as powers at all. The square root is taken here of a quadratic expression, whereas the cubic root is taken of an expression that is a sum of cubes. But, from the explanation, one can easily understand that this kind of homogeneity is not necessary thanks to the use of the unit length. And indeed, after the above passage, Descartes added the following:

Here it must be observed that by a^2, b^3 and similar expressions, I ordinarily mean only simple lines, which, however, I name squares, cubes, etc. so that I make use of the terms employed in algebra ... It should also be noted that all parts of a single line should as a rule be expressed by the same number of dimensions, when the unit is not determined in the problem. Thus a^3 contains as many dimensions as abb or b^3, these being the components of the line which I have called $\sqrt{C.a^3 - b^3 + abb}$. It is not, however, the same thing when the unit is determined, because it can always be understood, even where there are too many or too few dimensions; thus if be required to extract the cube root of $a^2b^2 - b$, we must consider the quantity a^2b^2 divided once by the unit, and the quantity b multiplied twice by the unit.

In order to understand the full significance of this new possibility of bypassing the traditional demand for homogeneity, thanks to the introduction of the unit length and the steps taken by Descartes based on it, you will find it relevant to revisit briefly some passages in Section 3.7 (especially those related to Figures 3.12 and 3.13), where we discussed the lack of length measurements in synthetic geometry as presented in Euclid's *Elements*. This lack of measurement continued to influence the mainstream of Greek geometry and thereafter, but all of this changed now with Descartes.

By adopting a thoroughly abstract algebraic approach in geometry, based on an appropriate symbolism and on the use of a unit length, Descartes came up with truly novel constructions that could be used in solving longstanding open problems, as well as in providing new solutions to problems that had previously been solved. A straightforward example is that of the quadratic equation (see Appendix 8.1).

A more complex example concerns Descartes' solution of the geometric locus of four lines. This problem had remained unsolved since the time of Pappus (see Figure 4.4). It was as a result of Descartes' efforts to solve this problem (as he understood it at the time) that he introduced the basic ideas of analytic geometry. The techniques he developed allowed him to tackle, with the same method he used for the 4-line locus, also the general, n-line locus problem. His solution represented a crucial milestone in the history of mathematics, but, for lack of space, it will not be possible to discuss it in this book.[2]

From his treatment of geometric constructions with the help of algebraic methods, Descartes was also led to focus on the investigation of equations and of polynomials as objects of intrinsic mathematical interest. In this context, he systematically developed some important ideas that had already incipiently surfaced in the works of Cardano and others. One of them is the relationship between the solutions of an equation and the possibility of factorizing the corresponding polynomial into elementary factors. Descartes, by the way, did not yet clearly distinguish between the polynomial (say, $x^2 - 5x + 6$) and the equation ($x^2 - 5x + 6 = 0$).

Descartes also analyzed the relationship between the signs of the solutions and those of the coefficients of the polynomial. When we replace the unknown x with a certain value a, and the value of the polynomial expression is zero, Descartes called that value a "root" (and we continue to do so). If the roots of a given polynomial are, say, 2 and 3, Descartes showed that the polynomial is obtained as a product of two factors (or, as he said, of two equations), $(x - 2)$ and $(x - 3)$, and hence the equation in question is $x^2 - 5x + 6 = 0$. If we want to add the root 4, then we need to multiply by $(x - 4) = 0$, thus obtaining the equation $x^3 - 9x^2 + 26x - 24 = 0$.

Descartes was very clear in his attitude toward negative solutions: he considered them as possible roots, but he called them "false" (*faux*) roots, as Cardano had also done much earlier. "If we suppose x to represent the defect [*defaut*] of a quantity 5—he said—we have $x + 5 = 0$, which multiplied by $x^3 - 9xx + 26x - 24 = 0$, gives $x^4 - 4x^3 - 19xx + 106x - 120 = 0$." This was for him an equation with four roots, three of which are "true roots" (2,3,4) and one is a "false" root, 5. In addition, the number of "false" roots equals the number of times that the signs of the coefficients of the equation remain the same when passing from a power of the unknown to the one immediately under it (in this case, this happens only once: $-4x^3 - 19x^2$). These are known nowadays as "Descartes' rules of signs." The number of "true" roots equals the number of times that the signs of the coefficients of the equation change when passing from a power of the unknown to the one immediately below it.

It was unavoidable that as part of this kind of investigation, Descartes would have to deal with equations leading to the appearance of roots of negative numbers. His position on this issue is interesting, and it turned out to be very influential. To see what it was, we need to comment on the so-called fundamental theorem of algebra, already mentioned in Chapter 1. It states that every polynomial equation (with real or complex coefficients) of degree n has exactly n roots, some or all of which may be complex numbers.

Descartes was aware, as we have just seen, of the relationship between the fact that a is a root of the polynomial and that the latter can be divided exactly by the factor $(x - a)$. It was surely natural for Descartes, then, to somehow come up with the basic idea behind the fundamental theorem. Indeed, the idea had already been hinted at in various ways by Descartes' predecessors, such as Cardano. It had appeared quite explicitly in a book that was well known at the time, *L'invention en algebra*, published in 1629 by Albert Girard (1595–1632).

[2] See (Bos 2001, pp. 273–331).

Descartes pointed out, in the first place, that the number of roots of a polynomial equation cannot be greater than the degree of the equation. Later, in Book III, he wrote explicitly:[3]

For the rest, neither the false nor the true roots of the equation are always real, sometimes they are only imaginary, that is to say that one may always imagine as many in any equation as I have said, but that sometimes there is no quantity corresponding to those one imagines.

In other words, Descartes asserted the existence, beyond "true" roots, also of roots that are false and also of others that are only "imaginary" or "imagined." Only if one takes these roots into consideration, he emphasized, will their number equal that of the degree of the polynomial. Even if at this point Descartes was not yet willing to bestow on all these kinds of roots equal status as *legitimate* kinds of numbers, the identical role they played from the point of the polynomial was now a major mathematical consideration that could not be easily overlooked.

Descartes' views on polynomials, including the important insight on the number of roots, implied a significant breakthrough, especially since it appeared in the framework of an influential book on geometry. From now on, mathematicians would refer to a new kind of entity, autonomous and abstract, the polynomial, to which much attention needed to be directed. Many questions that arose previously in different contexts, and that were investigated separately, would be treated now from a unified point of view based on the new knowledge developed in relation to the polynomials and their roots. Rather than speaking about specific questions with unknown quantities, each separately treated according to its type and to the degree in which the unknown appears in it, all of them were now seen as particular cases of a more general theory, the theory of polynomials. The availability of a flexible and efficient symbolism such as Descartes developed in his work, as a highpoint of a long and hesitant process that preceded it, played a crucial role in this development. Many important additional developments over the following centuries in mathematics in general, and in algebra in particular, derived from this new perspective whose clear origin was with Descartes.

But, at the same time, we should pay close attention to the interesting nuances that appear in Descartes' approach to numbers of the various kinds. When he spoke about "imaginary" roots, for example, Descartes meant it quite literally; that is to say, he saw them as numbers that are only in our imagination, and hence they neither represent a geometric quantity nor are similar to other kinds of numbers that appear as roots. Under the marked influence of this text, terms like "imaginary" and "real" were sweepingly adopted by mathematicians of the following generations. Nonetheless, this was not really of great help in giving some coherent meaning or in understanding the nature of expressions that comprised square roots of negative numbers. Even his attitude to the "false" roots cannot be seen as true progress toward a more systematic incorporation of the idea of negative numbers. Descartes did include the "false" roots under those that are "real," but he never considered the possibility of speaking about a negative number in isolation: the false roots appear as part of the expression $(x + a)$

---

[3] Quoted in (Bos 2001, p. 385).

that divides the polynomial, and in the polynomial one may find coefficients that are preceded by the sign "−", as we saw.

Evidently, this kind of practice involving the continued use of all such kinds of numbers as part of the theory of polynomials made it easier and more natural to accept them gradually as part of the general landscape of arithmetic. But the full acceptance was not part of Descartes' own view, as one may see in his approach to solving purely geometric problems. When explaining the geometric way to solve quadratic equations (as shown in Appendix 8.1), Descartes specifically refrained from dealing with the equation $z^2 + az + bb = 0$, precisely because a and b represent here lengths (that is, positive quantities), while the only solutions that one obtains are negative, which are devoid of geometrical significance. For the same reasons, in his analytic geometry also, we find only positive coordinates. The idea that coordinates may be either positive or negative appeared somewhat later, simultaneously with the increasing acceptance of the legitimacy of negative numbers.

8.2 Wallis and the primacy of algebra

The year 1631 was important in the history of algebra in the British Isles. It saw the publication of two books that marked the beginning of an increased pace in algebraic activity in the British context: *Clavis Mathematicae* ("The Key of Mathematics") by William Oughtred (1575–1660) and *Artis Analyticae Praxis* ("The Practice of the Analytic Art") by Thomas Harriot, posthumously published. This increase in algebraic activity was accompanied by debates about the question of the relationship between algebra and geometry, and in particular about the nature and role of numbers of various kinds. Following Oughtred and Harriot, no one played a more significant role in helping assimilate and develop algebraic ideas among British mathematicians than John Wallis (1616–1703).

Wallis contributed many of his own original ideas to the continued expansion of the concept of number and the continued blurring of the borderline between numbers and abstract quantities. Unlike many of his predecessors and many contemporaries who continued to abide by more classical views, Wallis unequivocally attributed full conceptual precedence to algebra over geometry. Likewise, he actively put forward many attempts to finding coherent ways to legitimize the use of negative and imaginary numbers. It is therefore important to devote some attention to his ideas as part of our account here.

Wallis' serious involvement with mathematics came at a relatively late age and in a rather non-systematic way. He was formally trained in the classical tradition, which included mainly the study of Aristotelian logic, theology, ethics, and metaphysics and in 1640 he was ordained priest. Like Viète, also Wallis developed a keen interest in cryptography. During the English Civil Wars of 1642–1651, he exercised his skills in decoding Royalist messages for the Parliamentarian party. It was only in 1647, at the age of 31, that he studied Oughtred's *Clavis* for the first time. This marked the beginning of his highly creative mathematical career. In 1649, he was appointed to the Savilian Chair of Geometry at Oxford.

Wallis' most original contributions relate to calculations of areas and volumes, as well as tangents. At the time, such calculations started to involve geometric situations of increasing complexity that, over the next few decades, became the core of the infinitesimal calculus. The problems discussed as part of this trend occupied the minds of the leading mathematicians of the period. But while most of them tried their best using methods that were essentially geometric and followed the Greek indirect method for dealing with the infinite (see Appendix 3.3), Wallis went his own way and introduced many new *arithmetical* methods for dealing with infinite sums and products. It was here that Wallis displayed the full power of his mathematical ingenuity and developed truly original methods.

One of his most stunning results, published in 1656 in his *Arithmetica Infinitorum*, involved an innovative method for approximating the value of π. Like Viète's calculation almost eighty years earlier, also Wallis' method involved an infinite product. However, this one was based on arithmetical considerations, rather than on geometric approximation, and hence it was much more powerful. It can be symbolically represented as follows:

$$\frac{\pi}{2} = \frac{2}{1} \cdot \frac{2}{3} \cdot \frac{4}{3} \cdot \frac{4}{5} \cdot \frac{6}{5} \cdot \frac{6}{7} \cdots$$

Wallis also expanded the concept of power to include negative and fractional exponents, being the first to work out useful insights such as $a^{1/2} = \sqrt{a}$ or $a^{-n} = \frac{1}{a^n}$ (even though his notation was somewhat different from ours).

One of the more impressive displays of the power of algebraic methods in the work of Wallis appeared in his treatment of conic sections. The recent development of Cartesian methods had helped characterize a parabola with the help of a quadratic equation $y = ax^2 + bx + c$. Wallis was the first to do something similar for the ellipse and the hyperbola. Because of his success in providing algebraic tools for dealing with a topic that, in the purely geometric treatment of Apollonius had traditionally been considered to be of extreme difficulty, Wallis saw himself as implanting simplicity in a field that had previously deterred many. Apollonius' original treatment was still the only one available at the time of Wallis, and Wallis was truly proud of the deep change he brought into the field of research on conic sections.

Wallis joined those who believed in the existence of a lost analytic "method of discovery," which "was in use of old among the *Grecians*; but studiously concealed as a Great Secret." Therefore, like Viète and Descartes before him, he saw his work as both a continuation and an improvement of that putative analytic method.

From the perspective afforded by his strongly arithmetic approach Wallis also advanced a further significant step in the direction of looking at proportions as no more than equalities between two fractions. In doing so, he simply dismissed off-hand, in a more decisive and unequivocal fashion than anyone before him, the age-old separation between ratios and numbers. Wallis spoke of a ratio between two numbers or two magnitudes simply as a division of the first by the second—plain and simple, as we would consider it nowadays. In this view, four magnitudes are said to be in proportion if the ratio of the first to the second, *seen as a number*, equals the ratio of the third to the fourth, *also seen as a number*. In other words, for Wallis, the proportion $a{:}b :: c{:}d$

was not different from the identity $a/b = c/d$. Remarkably, Wallis did not emphasize that he was changing an accepted view rooted in a centuries-old tradition and based on a completely different definition.

In considering arithmetic rather than geometry as the more solid conceptual basis for mathematics at large, Wallis had few constraints in using all kinds of numbers in various mathematical contexts. Nevertheless, he did not always promote a full acceptance of numbers of all kinds. Wallis' views on the legitimate use of negative, rational and irrational numbers were somewhat fluctuating. Negative numbers were for him a necessity, but he did not always consider them as legitimate in all situations, because it is not possible that a quantity "can be Less than Nothing, or any number fewer than None." A ratio between a positive and a negative number he initially considered as devoid of meaning, but later he came up with a strange argument that proved—so he thought—that dividing a positive by a negative number yields a result "greater than infinity."

And yet, since the idea of negative number was so useful and it is not "altogether absurd," Wallis suggested that these numbers should be given some kind of interpretation via a well-known physical analogy. More generally, for a mathematician like Wallis, it was imperative to provide some kind of underlying conceptual consistency to arithmetic and to avoid "impossible" situations that might arise in operating with natural numbers: subtracting a greater number from a smaller one, dividing a number by another number that is not a factor, extracting the square root of a non-square number or a cube root of a non-cubic number, or coming up with equations whose roots are square roots of negative numbers.

In spite of his own definition of ratio as a division of numbers, Wallis had doubts about the legitimate status of fractions and of irrationals. Still, given their practical usefulness in the solutions of many mathematical problems (including the kinds of innovative solutions that he himself had been developing with infinite series and the like), he did not limit their use. He chose to consider them as approximate values expressible in terms of decimal fractions as Stevin had taught. Concerning negative numbers, he was not able to come up with any definitive argument to justify their legitimacy, and all he was able to gather was a series of more-or-less convincing claims.

So, somewhat along the lines of Descartes, Wallis defined both negative numbers and their roots as "imaginary," in the sense that negative numbers represent a quantity that is "less than noting." With this definition, his intention was to promote the view that whoever accepts the legitimacy of the negative numbers has no real reason to reject that of their square roots. This was a wise move. He suggested extending to imaginary numbers the kinds of arguments typically used for providing legitimacy to the negatives, namely, some well-conceived physical analogy. The details of his argument are worthy of discussion here.

In his *Treatise on Algebra*, published in 1685, Wallis came up with the following original account of imaginary numbers (Figure 8.3): on a straight line where a starting

Figure 8.3 Wallis' graphical representation of negative numbers on a straight line.

point A is indicated, a man walks a distance of 5 yards in the direction of B, and then he retreats a distance of 2 yards in the direction of C. If asked what is the distance he has advanced, one will have no hesitation in answering 3 yards. But if the man retreats from B 8 yards to D, what is the answer to the same question? Clearly −3, and Wallis said "3 yards less than nothing."

This obvious argument is presented just in preparation for Wallis' original idea on how to interpret in a similar, graphical way the square roots of those same negative numbers. What happens, he then asked, if on a certain place on the seashore we gain from the sea an area of 26 units and lose to the sea, in some other place, and area of 10 units? How much have we gained, all in all? Clearly, an area of 16 units. If we assume this area to be a perfect square, then the side of this square is of 4 units of length (or −4 units, if we admit the negative roots of positive squares). Nothing special or new thus far. But now, what happens if we gain from the sea 10 units and lose in some other place 26 units. In analogy with the previous case, we may say that we lost 16 units, or gained −16 units, and if the area lost is a perfect square, what is then its side? It is, he concluded, the square root of −16. This was Wallis' point: negative numbers and imaginary numbers are equally legitimate or equally illegitimate, and there is no reason to accept the former and reject the latter in mathematics.

Wallis tried yet some other, possible geometric interpretations of the imaginary numbers. One of his ideas was based on the construction of the geometric mean of two (positive) magnitudes b and c (we can denote it here as $\sqrt{bc}$). A classical construction of this means is embodied in the diagram of Figure 8.4. This construction is based on an elementary theorem on circles (already mentioned in relation to Descartes), stating that if AC is the diameter and PB is orthogonal at any point B on the diameter, then the square built on PB equals in area the rectangle built on AB,BC. Wallis suggested to consider the square root of a negative number as the geometric mean of two lengths, one of them positive, the other negative: for instance, $-b, c$ or $b, -c$. Graphically, this is represented in Figure 8.5. Indeed, if we take the quantity b to the left of A, so that $AB = -b$, and then the quantity c to the right of B, so that $BC = c$, and hence $AC = -b+c$,

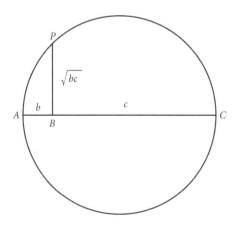

Figure 8.4 The geometric mean of two quantities, b, c.

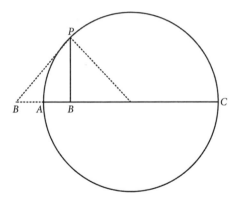

Figure 8.5 Wallis' representation of an imaginary number as the geometric mean of a positive and a negative quantity. This diagram is reproduced directly from (Smith (ed.) 1951 [1931], p. 49). The next two are modified versions from the original.

then it is easy to see by a simple geometric argument that PB, the tangent to the circle on P, represents the geometric mean $\sqrt{-bc}$.

Wallis was not really satisfied with this, and he went one step further, suggesting yet another geometric interpretation, which comes very close to the one that will eventually turn into the accepted interpretation of complex numbers. We will speak in greater detail in Chapter 9 about this later interpretation and its origins in the eighteenth century, but, as a brief reminder for readers at this point, I just want to emphasize that the geometric interpretation is based on extrapolating the representation of real numbers on a straight line into the entire plane as a way of representing the complex numbers (see Figure 1.5). Wallis' last and very original attempt to interpret imaginary numbers geometrically also extended the representation from the straight line to the plane, but the way he followed turned out to have serious limitations. It appeared as part of an explanation of the geometric meaning of solutions to quadratic equations $x^2 + 2bx + c^2 = 0$, where b and c are positive quantities. The solutions are obtained, of course, through the formula

$$x = -b \pm \sqrt{b^2 - c^2}.$$

Wallis drew up a diagram in which the solutions appear as two points P_1 and P_2, and in which one sees that real solutions may exist only when $b \geq c$ (Figure 8.6). But what happens here when $c > b$? From the algebraic point of view, the formula says that the solution would involve square roots of negative numbers. In terms of the diagram, what we see is that the points P_1 and P_2 would lie outside the line that was chosen to represent the numbers, and yet they lay on the same plane (Figure 8.7). This appears indeed as a possible representation of these roots of negative numbers. But a significant problem arises immediately: if b is taken to decrease continually, then P_1 and P_2 will approach each other on the plane. If b finally becomes zero, then $P_1 = P_2$. The meaning of this is that $\sqrt{-1} = -\sqrt{-1}$, which is clearly unacceptable.

Seen against the background of contemporary debates on the nature of number, and in particular of Wallis' own uncertain views about the negative numbers, one is not

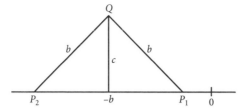

Figure 8.6 Wallis' graphical representation of two solutions to the quadratic equation. The sides of the triangle are of length b, and the same length is taken on the horizontal axis to the right of O.

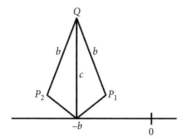

Figure 8.7 Wallis' graphical representation of imaginary solutions to the quadratic equation on the plane.

surprised to realize that he tried hard and had some brilliant starts, but that he did not succeed in forming for himself a coherent view of a possible geometric representation of imaginary numbers. As I have already pointed out, analytic geometry started its way more or less at that time in the work of Descartes (and independently also in the work of Fermat). Also, the more general implications of the relationship between geometric forms and algebraic expressions embodied in this mathematical discipline took time to be fully worked out. Negative coordinates, for example, did not appear from the beginning, among other things because of the uncertain status of the negative numbers. The work of Wallis, precisely because of his acknowledged accomplishments in successfully applying innovative and powerful arithmetic methods, is of special interest in highlighting the difficulties still encountered at this time in dealing with such concepts that we deem nowadays so simple and straightforward. It also highlights the influence, still pervasive at that time, of views on numbers stemming from the ancient Greek tradition.

8.3 Barrow and the opposition to the primacy of algebra

In the last part of the seventeenth century, there developed in England a trend that took a more restrained attitude toward the rising tide of Viète's and Descartes' kind of algebra. This trend mistrusted algebra as a possible source of certainty in mathematics and sought to restore primacy to synthetic geometry seen along the lines of the classical Greek tradition. Two main figures in this trend were Thomas Hobbes (1588–1679) and Isaac Barrow (1630–1677). The transparent structure of geometry was in their view the perfect paradigm of simplicity, certainty and clarity. Nothing like this could be found in either arithmetic or algebra, in their view. They opposed the use of algebraic arguments for solving geometric problems, but they cared to state that their opposition did not imply a more general, negative attitude to the spirit of the new science and the mathematics of their times. Rather, they had very clear and specific arguments against the use of algebra in certain situations. Accordingly, some of the algebraic ideas and methods introduced by Viète and Descartes, and by their British followers, did find a way into their works and were naturally incorporated therein in spite of their declared opposition. This gave rise to an interesting and original synthesis involving an underlying, inherent tension, which is of particular interest for our story. Let us see some details as they are manifest in the work of Barrow.

Barrow was a profound scholar with a very broad background in classical and modern languages and a deep interest in divinity studies. He was a professor of Greek until 1663, when he was appointed the first Lucasian Professor of Mathematics at Cambridge. A few years thereafter, he would renounce the chair on behalf of Newton, whose outstanding talents he was among the first to recognize while Newton was still a student.

One of Barrow's earliest mathematical publications was an abridged and commented Latin edition of Euclid's *Elements*, published in 1655. In its English translation of 1660, it became a widely used text in the British context up until the eighteenth century. Barrow incorporated into this text some clearly algebraic elements and combined them into his purist approach to geometry. At the same time, he explicitly emphasized that in his presentation he was not deviating in any sense from the original. It is likely that he was sincere in this belief, even though in historical perspective the deviation is more than obvious.

In many of the propositions discussed in his edition, instead of the classical accompanying diagrams, Barrow preferred to write the geometric property to be proved in an idiosyncratic symbolic language. This did not translate the property into an algebraic equation, to be sure, and his symbols were not mean to be manipulated. But Barrow's symbolism allowed, if it did not actively suggest, a reading of Euclid in which the geometric magnitudes could also be seen as abstract quantities. He mixed without much constraint classical geometric constructions, symbolic expressions and numerical examples. Still, he kept stressing that he followed this approach just in order to present the proofs (which he characterized as fully geometric in spirit) in a more condensed manner. This point is better understood by looking at a detailed example, which I have presented in Appendix 8.2.

This unique blend of a declared attitude that promoted the classical standards of Greek geometry, on the one hand, and, on the other hand, favored the adoption of a symbolic language as a way to allow for a clearer presentation of geometric results stands also in the background of Barrow's attitudes towards numbers. His views are known to us via the texts of his lectures in Cambridge, beginning in 1664. Arguing for the primacy of geometry over algebra, Barrow put forward some philosophical statements, not always very convincing, about the way in which the objects of geometry are perceived through the senses. Quantities, he said, appear in nature only as "continuous magnitudes," and these are the only true objects of mathematics. Numbers, as opposed to magnitudes, are devoid of an independent existence of their own, and they are nothing other than names or signs with the help of which we refer to some magnitudes.

Barrow explicitly criticized Wallis' views on numbers and algebra. If for Wallis the formula "2+2 = 4" was true independently and previous to any geometric embodiment of it, for Barrow it was arbitrary and devoid of autonomous meaning. Indeed, for him, it was constrained by the ability to apply it in some specific geometric situation. For example, when we add a line of length 2 feet to another line of the same length, then we obtain a line of length 4 feet. But when we add a line of length 2 feet to a line of length 2 inches, we obtain a line neither of length 4 feet nor of length 4 inches, nor of any other 4 known units. So, in his view, the meaning of the sign 2 depended directly on the geometric context to which it is applied.

But if natural numbers are no more than signs for magnitudes, what can then be said about irrational, negative, or imaginary numbers? Irrational numbers were the easiest to adapt to Barrow's views, and indeed he used them to strengthen his position as opposed to that of Wallis. His claim was that there is no number, either integer or fractional, that when multiplied by itself yields 2, and from his own point of view there is no need to understand $\sqrt{2}$ in terms of natural or fractional numbers, or even approximations of decimal fractions (as was the case of Wallis). For Barrow, $\sqrt{2}$ was nothing but a name, or a sign, that indicates a certain geometric magnitude, namely, in this case, the length of the diagonal of the square with side 2. And from here he also derived an additional criticism of Wallis, namely, of the latter's arithmetical interpretation of ratios and proportions, which so strongly deviated from the classical Greek tradition, as we have just seen above. Barrow admitted that certain ratios, *but by no means all of them*, can be expressed as fractions. The classical case of the diagonal of a square was for him the indisputable instance to think about in this regard. Ratios, in Barrow's views, could in no way be conceived as numbers, since numbers represent only magnitudes.

Negative numbers—Barrow suggested very much like Wallis—should be seen as differences between a smaller and a larger natural number. But then, how can one interpret the number –1, when 1 is no more than a sign to indicate a magnitude? Well, here Barrow admitted the difficulty of thinking about a number that is "less than nothing," but he illustrated the idea with the same kinds of physical–geometric analogies adduced by Wallis. And concerning the roots of negative numbers, interestingly, Barrow did not mention them at all.

Wallis and Barrow are emblematic representatives of two trends in mid-seventeenth-century British mathematics that laid their stress on different aspects of mathematical practice. These trends, however, were not diametrically opposed, and they complemented each other in various respects. Wallis' concerns with algebraic methods as new instruments for discovery, for example, did not imply a disregard for Barrow's insistence on classical rigor. On the other hand, Barrow's preference for geometry should not be seen just as a stubborn refusal to adopt "modern" methods. At the time, only a rather limited kind of curves could be treated with the help of algebraic methods (curves that we call nowadays "algebraic curves"). Other kinds of curves, such as spirals and cycloids (which we refer to nowadays as "transcendental curves"), could not be covered by algebra. A mathematician like Barrow aimed at developing mathematical methods of a clarity and generality that algebra could not deliver at the time the way geometry did.

8.4 Newton's *Universal Arithmetick*

The trends of ideas embodied in the works of Wallis and Barrow interacted at the heart of a process where innovative views on the relationship between algebra and geometry gradually consolidated. A modern conception of number was among the outcomes of this process. The intellectual stature of Wallis and Barrow and their acknowledged status within the British mathematical community turn their contrasting views and debates, as well as their points of convergence, into a highly visible milestone from which to analyze this significant crossroads in the history of mathematics. But, at the bottom line, these processes, and indeed all significant processes that shaped British contemporary ideas in the exact sciences, crystallized under the towering shadow of Isaac Newton and his pervasive influence.

I devote the last section of this chapter to a brief description of Newton's views on numbers. When examining his work, however, one must always keep in mind the complexity of the task involved. The entire seventeenth century, as we have seen thus far, is a truly transitional period in all what concerns the disciplinary identity of mathematics. While in the sixteenth century, Euclidean methods provided a stable reference model, and in the eighteenth century, mathematicians will refer to the calculus as the language and method that will provide an underlying unity to their field, Newton's time is precisely that of passage from the former to the latter. Questions about the interrelation between geometry and arithmetic, and related questions about the nature of magnitudes and numbers, arose in this context alongside new questions about the applicability of mathematics to the study of the natural world.

But on top of the difficulty generated by the originality and intrinsic depth of Newton's mathematical ideas in a time of deep changes, one cannot overlook the variety of methodological, institutional and personal considerations that keep affecting his work at different periods of his lifetime. We find interesting tensions between his declared intentions, his practice and his method. We must examine linguistic and publication choices related to the various dialogues and confrontations that he entertained with his contemporaries (and of particular interest are those with Descartes and Gottfried Wilhelm Leibniz (1646–1716)). We need also to consider the different kinds of intended readers he addressed in different texts that he wrote. In short, we should not assume that Newton's ideas on any topic, the idea of number included, can be summarized under a simple, coherent formula or description.

As a young student in Cambridge, Newton immersed himself in the study of the mathematical works of Viète and Descartes, as well as those of Oughtred, Wallis and Barrow. He came up with an efficient and thoroughgoing synthesis of concepts and symbols introduced earlier in all active fields of mathematics. He also went on to develop many new fields of research while introducing highly innovative methodologies, the most important of which comprised the techniques of "fluxions and fluents," which later would become part of infinitesimal calculus (about which we will not speak here).[4] Later in his life, Newton became increasingly critical of Descartes' methods and views, and he devoted efforts to reconsidering his own earlier achievements against the principles of the classical tradition. Newton sought to consolidate a unified view of mathematics in which the calculus of fluxions could be reconciled with Euclid's *Elements* or with Apollonius' *conics*.

In 1669, Newton was appointed to the Lucasian Chair of Mathematics at Cambridge, following Barrow's resignation to take a position as chaplain to the King. Newton's lecture notes indicate that he devoted great energies between 1673 and 1683 to algebra, the field of knowledge that Barrow had described a few years earlier as "not yet a science." It is not completely certain that the dates retrospectively added to the notes reflect the actual teaching of Newton during those years, but it is quite clear that the notes underwent many transformations before being published as a Latin book in 1707. Several English editions of the book, *Universal Arithmetick*, were published over the following decades, and they were widely read and highly influential in eighteenth-century England.

From the point of view of its intrinsic mathematical value, *Universal Arithmetick* is far from being one of Newton's most important texts. As a matter of fact, he did not really mean to publish his notes. Retrospectively, he even manifested his discontent when the book was published thanks to the efforts of William Whiston (1667–1752), Newton's successor in the Lucasian chair. Between 1684 and 1687, most of Newton's efforts were devoted to the writing of *Philosophiae Naturalis Principia Mathematica* ("Mathematical Principles of Natural Philosophy"), the real climax of his scientific opus (and it must be said that, also in the case of this epoch-making book, Newton was not at all enthusiastic about its publication for fear of criticism that it might attract. Publication became possible in the end thanks to the continued intercession of the famous astronomer Edmond Halley (1656–1742)).

In the following years, Newton was very busy with debates that arose in the wake of the publication of the *Principia*. At this time, all plans for a possible publication of his algebraic notes remained unattended. But then, in the 1705 elections to the British Parliament, when Newton presented his candidacy but his campaign did not show signs of taking off, some of his colleagues at Cambridge promised their support in exchange for a considerable donation on his side to Trinity College, and a final permission to publish the notes, after these had been revised and edited by Whiston.

I take the trouble to tell all these details in the background to the publication of *Universal Arithmetick*, just in order to stress the almost incidental character of its appearance. If we compare the published version with some of the manuscripts found in Newton's scientific legacy, it is easy to recognize the many hesitations and continued changes throughout the years. This should come as no surprise, of course, given that these were drafts and teaching notes, rather than a text prepared carefully for publication. But those who prepared the various editions for print did not always pay close attention to all the nuances and changes that they contained. Accordingly,

---

[4] See (Guicciardini 2003).

different points of emphasis are noticeable within the published texts, as well as ideas that conflict with those appearing in earlier versions.

The point is that whatever the background to its publication, the many readers of the book saw its contents as expressing, in all respects and without qualifications, the ideas of the great Newton. No doubt, beyond the intrinsic mathematical assets or setbacks of the ideas exposed in the book, the very authority of Newton as their perceived supporter gave them an enormous legitimation that would help in disseminating and assimilating them as part of the mainstream of ideas about algebra and arithmetic in Europe.

The focus of *Universal Arithmetick* was on algebraic practice, and there was little room in the text for debates on the foundations of the discipline. The central concepts were only briefly explained, and the rules of calculation were presented without any kind of comments or arguments for legitimation. By contrast, every technique that was explained was accompanied by many examples that were worked out to the details. The influence of Viète is clearly visible throughout the text, but even more pervasive is the presence of Cartesian algebraic methods for problem-solving. While the book as a whole is an implicit way to fully legitimize the methods of algebra and its use as a tool for solving geometric problems, Newton used every available opportunity to stress his own preference for the classical methods of synthetic geometry, and he continued to praise its virtues and to support it as the example to be followed everywhere in mathematics.

Newton's attitude toward Descartes' ideas was complex and ambivalent at best. In the margins of Newton's copy of *La géométrie*, we find many critical annotations: "*Error*," "*Non probo*," "*Non Geom*," "*Imperf*." They may have been written while Newton was still a student at Cambridge, and they refer to Descartes' use of algebra in a geometric context. Later on, however, as he himself began to teach algebra, and after having been exposed to the kind of ideas developed by Wallis, Newton was more open to admit the advantages of applying algebraic methods to geometry.

In the opening chapter of the book, Newton provided a concise definition of number, combining together ideas that had appeared in the various traditions from which he was taking inspiration:[5]

By *Number* we understand, not so much a *Multitude of Unities*, as the abstracted ratio of any Quantity, to another Quantity of the same Kind, which we take for Unity. And this is threefold; integer, fracted and surd: An *Integer*, is what is measured by Unity; a *fraction*, that which a submultiple Part of Unity measures; and a *Surd*, to which Unity is incommensurable.

This synthesis is extremely interesting. On the one hand, following Barrow, Newton also tended to eliminate the separation between continuous and discrete magnitudes. On the other hand, like Wallis, he identified ratios with numbers, but he abode by the classical demand that the ratio be between "quantities of the same kind." The unit, the

---

[5] There are various editions of this book. I cite here from the 1769 edition: *Universal arithmetick: or, A treatise of arithmetical composition and resolution. Written in Latin by Sir Isaac Newton. Translated by the late Mr. Ralphson; and rev. and cor. by Mr. Cunn. To which is added, a treatise upon the measures of ratios, by James Maguire, A.M. The whole illustrated and explained, in a series of notes, by the Rev. Theaker Wilder*, London: W. Johnston. This passage is on p. 2.

integers, the fractions, and the irrational numbers appear here—perhaps for the first time and certainly in an influential text in such clear-cut terms—all as mathematical entities of one and the same kind, the differences between them being circumscribed to a single feature clearly discernible in terms of a property of ratios: either the ratio with unity is exact (integer), or the ratio with a part of unity is exact (fraction), or there is no common measure between the two quantities in the ratio (surd). Moreover, and very importantly, numbers are *abstract* entities: themselves they are not quantities, but they may represent either a quantity or a ratio between quantities.

The influence of Newton's definition is clearly visible in many eighteenth-century books throughout Europe, in which it is sometimes repeated verbatim. But the remarkable fact is that this definition is not put to use within Newton's own book. As already said, Newton focused in the book on the practice of problem-solving, and he gave little attention to philosophical or methodological questions concerning the central concepts of algebra and arithmetic.

Newton also introduced the negative numbers without much comment or philosophical considerations, while indicating what is it that characterizes them as quantities: quantities may be "affirmative," that is, larger than nothing, or "negative," that is, less than nothing. Newton did not adopt the terminology of Wallis, who had called them "fictions" or "imaginary quantities." Rather, he relied on analogies, relating the negative numbers to "debts" or "subtraction of a larger number from a smaller one." Where Wallis had spoken of "impossible subtraction," Newton just spoke of a subtraction to whose outcome we anticipate a "−" sign, and without further distinguishing between positives and negatives. He also presented the rules of multiplication with signs without further ado, with no explanations or justifications, and simply providing numerous examples of their use.

There was no new element in Newton's presentation that had not previously appeared in some British book on algebra, but the systematic and simple picture arising from the book, and—perhaps more importantly—the fact that this picture carried with it the authoritative legitimation stamp of the great Newton, endowed it with a special status that helped turning it into the standard point of necessary reference for both concepts and terminology all around Europe over the decades to come.

Newton's attitude to imaginary numbers is of particular interest because of the hesitation and lack of final decision that arises from the published text. This attitude reflects the remaining weaknesses in the concept of number, seen either as a quantity or as a ratio of quantities. Newton's inability to take a final stance on this matter derived from the difficulty involved in considering square roots of negative numbers as quantities of some specific kind, like the rest of the numbers. Imaginary numbers appear in Newton's book in the section where he discusses Descartes' rules for counting roots of polynomials (explained above) and the relationship between roots and coefficients.

In his early lectures in Cambridge, Newton spoke—following Descartes—about the possibility that a polynomial equation may have roots that exist "only in our imagination," but to which no quantity can be associated. In this sense, the term "imaginary" described quite literally the way these roots were conceived. The manuscripts of the lectures show that he gradually changed this view and the associated terminology. In one place, Newton formulated a version of the fundamental theorem of algebra as the assertion that the number of roots of a polynomial equation cannot surpass the highest order of the unknown in the equation, but these roots may be either positive, or negative or "impossible" (rather than "imaginary"). And what he meant by "impossible" he explained by reference to the solution of the equation

$$x^2 - 2ax + b^2 = 0.$$

Here, we obtain two roots, namely,

$$a + \sqrt{a^2 - b^2} \quad \text{and} \quad a - \sqrt{a^2 - b^2}.$$

Now, when a^2 is greater than b^2—Newton wrote—the roots are "real." In the opposite case, when b^2 is greater than a^2, of course, the root is "impossible." But, interestingly, Newton nevertheless went on to stress that both expressions are roots of the polynomial, for the simple reason that when they are introduced in the equation in place of the unknowns, the equation is satisfied because "their factors eliminate each other." In other words, a square root of a negative number is an impossibility and hence does not represent a number in the proper sense of the word, but expressions containing such impossible entities are legitimate roots of an equation and allow for an appealing formulation of the fundamental theorem of algebra, as Newton conceived of it.

We have already seen these kinds of ambiguous attitudes appearing in algebraic texts at least since the time of Cardano and Bombelli. The fact that by the time of Newton the ambiguity has not been fully bridged is highly indicative of the pervasiveness of certain basic ideas that in retrospective we see as completely inadequate. Newton's formulations make patent the continued tension between what the existing concepts of number implied and what the actual practice required. More than a century of intense mathematical activity would still be needed before truly satisfactory definitions of imaginary numbers would appear, as we will see in the following chapters.

No less confusing for the reader could be Newton's remarks on the relationship between algebra and geometry. As I have already said, Newton made extended use of Cartesian methods that combine algebra and geometry, but, nevertheless, in some places he specifically refrained from using algebra, while stressing that even in apparently difficult geometric problems algebra is not an adequate tool for finding the solution. In the printed edition of *Universal Arithmetick*, there is a well-known passage in which Newton declared that Cartesian methods endanger the purity of geometry. He wrote:[6]

Equations are Expressions of Arithmetical Computation and properly have no Place in Geometry, except as far as Quantities truly Geometrical (that is, Lines, Surfaces, Solids and proportions) may be said to be some equal to others. Multiplications, Divisions and such sorts of Computations, are newly received in geometry, and that unwarily, and contrary to the first Design of this Science ... Therefore these two Sciences ought not to be confounded. The Ancients did so industriously distinguish them from one another, that they never introduced Arithmetical Terms into Geometry. And the Moderns, by confounding both, have lost that Simplicity in which all the Elegancy of Geometry consists.

[6] *Universal Arithmetick*, p. 470.

This passage was repeatedly cited in many European mathematical texts over the following decades. The mathematicians who cited it were actually those who sought to preserve the primacy of geometry over algebra. It is quite ironic to contrast what Newton wrote with the approach that is dominant in *Universal Arithmetick*, where the prominence of algebra *in practice* is so blatant. From reading the manuscripts of the various versions of the book, one readily realizes that Newton continually hesitated and changed his views on this important point. The clear-cut statement cited above is what in the end was included in all editions of the book, and this is what readers came to associate with the name of Newton.

I want to stress that in the late 1670s—the time when he was involved with these texts—Newton had just begun reading Pappus. He came to the conclusion that the putative method of discovery of the ancients (the "analysis" that I have already mentioned) was superior to Cartesian algebra. At this time, Newton began conceiving Descartes and Cartesians of all sorts as his personal enemies, while at the same time, he began also to conceive himself as a direct heir of the ancients. The view of algebra that emerged in this context saw it as a heuristic method that could be used for discovery but was not adequate for publication. Algebra—Newton thought at this time, in line with Barrow's views—lacks the clarity of geometry, and it is also philosophically misleading since it makes us believe that non-existent things actually exist.

The underlying relationship between algebra and geometry is even more complex in the case of Newton's most famous and influential book, the *Principia*. From the vantage point of later developments in mathematics, it appears to us convenient to present this revolutionary book in the language of the calculus, a language whose initial stages Newton himself was instrumental in helping to shape in some of his other important works. But a reader of the original text of the *Principia* will find its style, on the face of it, more reminiscent of classical Greek geometry than of a nineteenth-century treatise on classical mechanics. This is, however, a kind of "classical façade" that Newton worked hard to bestow upon his text. Behind it, one can find in several places a wide variety of recently developed mathematical methods underlying the classical surface: infinite series, infinitesimals, quadratures, limit procedures and also algebraic methods. There was plenty of evidence in Newton's text to create an image of him as an uncompromising champion of classical views about the primacy of geometry over algebra. In truth, however, especially when it came to developing a mathematical practice in relation to numbers and algebra, he followed a more flexible and variegated attitude.

The end of the seventeenth century marks a significant inflection point in our story. The face of science had profoundly changed, as had changed the place of science in society in most of Europe. The consolidation of the new symbolic algebra, especially in the works of Viète and Descartes, and the rise of the infinitesimal calculus in the works of Newton and Leibniz, were a truly significant turning point in the history of mathematics. Many of the topics that we have been discussing thus far receded into the background. The influence of Euclid's *Elements* in the mainstream of advanced mathematical research declined. The importance of the Eudoxian theory of proportions almost disappeared. The divide between continuous magnitudes and discrete numbers lost its interest. The concept of number had undergone deep changes, and the door was now open to a new stage in which significant additional changes would completely reshape this concept over the next two centuries. These changes will be described in the remaining chapters of the book.

Appendix 8.1 The quadratic equation. Descartes' geometric solution

A particularly illuminating perspective from which to understand the innovation implied by Descartes' approach to the relationship between algebra and geometry is afforded by a detailed examination of his treatment of quadratic equations and the geometric solutions he suggested for them. In this appendix, I bring some direct quotations from *La géométrie*. The reader may thus get a first-hand grasp of the way Descartes handled the various kinds of numbers and geometric magnitudes that appear in his equations. Of particular interest is the unhesitant transition from algebraic expressions to geometric interpretations. We are of course used to such transitions, but at the time they implied a far-reaching innovation, even if Descartes did not particularly emphasized this in the book.

The following is quoted from the original text:[7]

For example, if I have $z^2 = az + bb$, I construct a right triangle NLM with one side LM, equal to b, the square root of the known quantity bb, and the other side, LN, equal to $\frac{1}{2}a$, that is to half the other known quantity which was multiplied by z, which I suppose to be the unknown line. Then prolonging MN, the hypothenuse of this triangle, to O, so that NO is equal to NL, the whole line OM is the required line z.

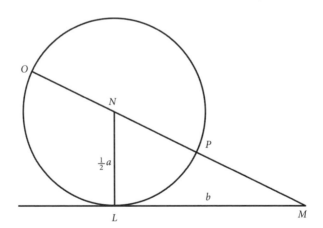

[7] (Descartes 1637 [1954], pp. 13–14).

This is expressed as follows:

$$z = \frac{1}{2}a + \sqrt{\frac{1}{4}aa + bb}.$$

But if I have $y^2 = -ay + bb$, where y is the quantity whose value is desired, I construct the same right triangle *NLM*, and on the hypotenuse *MN* lay off *NP* equal to *NL*, and the remainder *PM* is the desired root. Thus I have:

$$y = -\frac{1}{2}a + \sqrt{\frac{1}{4}aa + bb}.$$

In the same way if I had $x^4 = -ax^2 + bb$, *PM* would be x^2 and I should have

$$x = \sqrt{-\frac{1}{2}a + \sqrt{\frac{1}{4}aa + bb}}.$$

And so for the other cases.

Finally, if I have $z^2 = az - bb$, I make *NL* equal to ½*a* and *LM* equal to *b* as before, then instead of joining the points *M* and *N*, I draw *MGR* parallel to *LN*, and with *N* as center describe a circle through *L* cutting *MGR* in the points *G* and *R*;

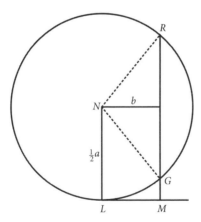

then *z*, the line sought is either *MG* or *MR*, for in this case it can be expressed in two ways, namely,

$$z = \frac{1}{2}a + \sqrt{\frac{1}{4}aa - bb} \quad \text{and} \quad z = \frac{1}{2}a - \sqrt{\frac{1}{4}aa - bb}.$$

And if the circle described about *N* and passing through *L* neither cuts nor touches the line *MGR*, the equation has no roots, so that we may say that the construction of the problem is impossible.

The very fact that Descartes refrained from dealing with the equation $z^2 = -az - bb$ stresses the then still strongly geometric perspective from which he addressed this problem. The said equation has no positive roots (since a, being a length, is a positive quantity), and hence, from Descartes' point of view, it makes no sense to include its treatment in this part of the book, where he is presenting the possible solutions to a quadratic equation.

Appendix 8.2 Between geometry and algebra in the seventeenth century: The case of Euclid's *Elements*

In Chapter 7, I explained some of the important developments that took place in the seventeenth century concerning numbers and the changing interrelations between algebra and geometry. In this appendix, I want to focus more narrowly on one specific mathematical result that affords an illuminating perspective on these issues, namely, Proposition II.5 of Euclid's *Elements*.

Euclid's II.5 is not a particularly profound mathematical result. On the contrary, it is a rather simple technical, auxiliary result that is used in the *Elements* for proving more substantial theorems. Seen in its changing versions throughout time, however, it affords interesting insights into some of the historical processes that we have been discussing so far. Specifically, it provides a good perspective from which to consider the "geometric algebra" interpretation of Greek geometry already mentioned in Section 4.2 and in Appendix 3.1. According to this interpretation, Book II of the *Elements* is devoted to developing a collection of algebraic relations that can be used, among other things, to solve quadratic equations. These relations appeared in the *Elements* in geometric garb, just because Euclid had no proper symbolic language at hand, such as we currently associate with algebraic thought. But other than this difference in "language," so the interpretation goes, the book is algebra through and through.

As I have already indicated, this approach incurs many historiographic problems, but one specific point that I have been trying to stress throughout this book is that the putative separation that it postulates, between "mathematical ideas," in a very abstract and perhaps ethereal sense, and the language used to expressed these ideas is artificial and untenable. Speaking about an abstract idea of "number" and of "equation" as separate from the way in which we are able to write them does not make much historical sense. Euclid's II.5 provides a useful vantage point to stress this point even further.

I want to examine here two versions of II.5: one of Wallis, the other of Barrow. Their diverging views on the relationship between geometry and arithmetic are interestingly reflected in their respective versions of the proposition. Before presenting them, however, and before seeing how they relate to the issue of "geometric algebra," I need first to introduce the proposition itself as originally formulated by Euclid. The English version of the proposition, as it appears in Heath's edition of the *Elements*, reads as follows:

If a straight line be cut into equal and unequal segments, the rectangle contained by the unequal segments of the whole together with the square on the straight line between the points of section is equal to the square on the half.

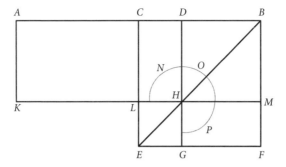

Figure 8.8 Euclid's diagram for Proposition II.5.

For let a straight line AB be cut into equal segments at C and into unequal segments at D; I say that the rectangle contained by AD, DB together with the square on CD is equal to the square on CB.

In the diagram of the proof (Figure 8.8), a figure such as that obtained when joining the rectangles HF and CH together with the square DM is what the Greeks called a "gnomon," and in this case "gnomon NOP." It plays a central role in the proof. The proof is quite simple, but it is interesting to read the details in order to come to terms with its purely geometric character. It involves only straightforward geometric constructions and comparison. It involves no arithmetical operations, and surely no algebraic manipulation of symbols representing the magnitudes involved. It reads as follows:

For let the square CEFB be described on CB, and let BE be joined; through D let DG be drawn parallel to either CE or BF, through H again let KM be drawn parallel to either AB or EF, and again through A let AK be drawn parallel to either CL or BM. Then, since the complement CH is equal to the complement HF, let DM be added to each; therefore the whole of CM is equal to the whole of DF. But CM is equal to AL, since AC is equal to CB; therefore AL is equal to DF. Let CH be added to each; therefore the whole AH is equal to the gnomon NOP. But AH is the rectangle AD, DB for DH is equal to DB, therefore the gnomon NOP is also equal to the rectangle AD, DB. Let LG, which is equal to the square CD, be added to each; therefore the gnomon NOP and LG are equal to the rectangle contained by AD, DB and the square on CD. But the gnomon NOP and LG are the whole square CEFB, which is described on CB; therefore the rectangle contained by AD, DB together with the square on CD is equal to the square on CB.

Notice how the entire deduction relies on basic properties of the figures that derive from the initial construction or that were proved in previous theorems (which in turn were proved purely geometrically). Thus, for instance, the claim that "the complement CH is equal to the complement HF" is Proposition I.43 of the *Elements*. The gnomon NOP is a geometric figure built out of other figures, and similar gnomons appear in many other proofs in Greek geometry. It is clear that while one might claim (ahistorically, but at least with some mathematical justification) that rectangle formation is a geometric equivalent of arithmetic multiplication, no such direct arithmetic equivalent may even be suggested for "gnomon formation."

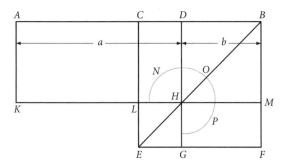

Figure 8.9 An algebraic interpretation of Euclid's II.5.

Let us look now at Heath's comments on the proposition. They embody the emblematic example of the "geometric algebra" approach. Heath translated this proposition into an algebraic identity by setting $AD = a$ and $DB = b$, as in Figure 8.9. He thus obtained the following identity:

$$ab + \left(\frac{a+b}{2} - b\right)^2 = \left(\frac{a+b}{2}\right)^2.$$

By manipulating this identity, Heath deduced that II.5 can be stated in the following algebraic form:

$$\left(\frac{a+b}{2}\right)^2 - \left(\frac{a-b}{2}\right)^2 = ab.$$

This is now, of course, visibly removed from a direct translation of II.5. But for someone who wants to look at the proposition as if it expresses an underlying algebraic idea, all the equivalent formulations of the same identity are actually one and the same. Indeed, Heath went on to claim that this way of stating II.5 "could hardly have escaped the Pythagoreans," because the expression may be further elaborated into some arithmetic identities usually assumed to have been known to them.

Heath also pointed out that the precise form of the algebraic identity underlying the proposition may change if we chose alternative ways to denote the segments. One might for example write $AC = BC = a$ and $CD = b$, and the proposition would then translate into: $(a + b)(a - b) + b^2 = a^2$. Heath himself chose another translation, from which he deduced that the proposition is actually a solution of the quadratic equation: set $AB = a$, $DB = x$ (and hence $AD = a - x$, and $AC = (a + x)/2$), and then focus on the step of the proof where the gnomon NOP is equated with the rectangle AH. This yields

$$\text{rectangle } AH = ax - x^2 = \text{gnomon } NOP.$$

By assigning a value to the gnomon, say b^2, it turns out that II.5 affords what Heath sees as a geometric method for solving the equation $ax - x^2 = b^2$.

Beyond the more general historiographic concerns that arise from Heath's interpretation, we can use this example to indicate a particularly significant difficulty: if we interpret II.5 algebraically as Heath does, it is necessary to explain what are the putative arithmetical operations underlying the algebra whose use he is assuming. According to Heath, every number is represented with a segment; the sum of two numbers is represented by concatenating two segments, whereas their difference is represented by subtracting the smaller from the larger. Multiplication can be likewise naturally defined by the construction of a rectangle with the given numbers as sides. So far so good. But what about division? In his view, to divide two segments one simply sets a ratio between them. But, as we have seen in the foregoing chapters, a ratio was not, for the Greeks, a kind of number, but rather something completely different. Moreover, in Greek mathematics in general and in Book II of the *Elements* in particular, there is no place where we find ratio formation as the inverse operation to rectangle formation (or vice versa). Hence, the attempt to interpret ratio creation as an adequate translation of geometric division gives rise to many difficulties. As a matter of fact, Heath himself was aware of the interpretive overweight that his explanations added to the original text. He thus wrote:

The algebraic method has been preferred to Euclid's by some English editors; but it should not find favour with those who wish to preserve the essential features of Greek geometry as presented by its greatest exponents, or to appreciate their point of view.

Heath's algebraic interpretation of what Euclid had in mind when writing Book II can be criticized on historical grounds, as I have done here. But one cannot deny that it is mathematically plausible, and in indeed that it is quite appealing. At any rate, the mathematical plausibility of the interpretation raises some interesting historical questions concerning the actual process whereby the algebraic (or some kind of arithmetic) interpretation of the propositions in Book II started to gain ground and to permeate the existing editions of the *Elements*. The historical process that afforded the mathematical possibility of interpreting the propositions of Book II in algebraic terms was hesitant and complex. Hence, it has a lot to teach us about the long-term changing interrelations between geometry and arithmetic, beginning with the Greeks and up until the seventeenth century. I will not go here into the details of that complex process.[8] But the cases of Wallis and Barrow, which I want to present here, are highly illustrative of the variety of ways in which a proposition like II.5 could be reformulated according to the changing points of view of the mathematicians involved in the process.

Before describing the versions of Wallis and Barrow, however, I just want to remind the reader that we have already come across Proposition II.5 in Section 5.8. Al-Khayyām mentioned two results, Euclid's II.5 and II.6, as results that are purely geometric, but which, from the perspective of Arabic algebra, were already seen by some of his contemporaries (erroneously, in his opinion) as embodying something other than "Those who believe that algebra is an artifice devoted to the determination of unknown numbers—he said while referring to the way in which these propositions provide the underlying justification to the procedures of al-Khwārizmī—believe the impossible."

---

[8] See (Corry 2013).

I also want to bring a brief passage from a text that I mentioned in Section 6.5, namely, Clavius' 1574 edition of the *Elements*. It illustrates very nicely the ways in which, already by the mid sixteenth century, the proposition could be formulated as a purely arithmetical result. Clavius explicitly stated that the best way to understand Proposition II.5 (and he meant the same for the other propositions in Book II) is by giving a *numerical* example of what it states. His formulation was roughly the following:

If we divide a given number, for example 10, into two equal parts, 5 and 5, and into non-equal parts, 3 and 7, the difference between the two cuts is of length 3 (that is: 5 – 3). 21 is the product of 7 times 3, and if we add to that a square of the length between them, which equals 4, then we obtain 25, which is indeed equal to the square of the half.

Clearly, then, those readers who came to study the *Elements* via the Clavius edition (and there were many such, given the importance of this edition in the very well-developed educational system of the Jesuits), rather than from anything closer to the original Euclidean one, could by all means reach the understanding that all propositions of Book II were no more than a series of simple arithmetical exercises.

Wallis' version of II.5 appeared in 1657 as part of a well-known, lengthy elementary treatise called *Mathesis Universalis*. Very likely, the text was based on his lectures as Savilian Professor at Oxford, and it bears the clear stamp of his views on the conceptual primacy of arithmetic in mathematics. It is here that he introduced many of his notational innovations, and he also provided a detailed overview of various numeration systems known at the time. He explained all the arithmetic operations with integers, while providing a large number of examples, and introduced the symbolic techniques of Viète, while indicating how to interpret geometric operations, such as area formation, in terms of number multiplication. He stressed the disadvantages of associating the powers of a number (or an unknown) with geometric dimensions, and he suggested that they should be best seen simply as operations with numbers.

Wallis devoted special attention to Euclid's Book II, which he described as a kind of "failed arithmetic." In line with the general spirit of his treatise, it was natural that Wallis would consider Euclid's propositions as "easily demonstrated in arithmetical terms." His formulation of II.5 is not accompanied by a geometric construction or a diagram of any kind. Rather, he denoted with letters the various segments that appear in the construction, gave a numerical example, and then proved the proposition in a strongly algebraic fashion, with the help of letters for indicating the various magnitudes. He argued both rhetorically and by symbol manipulation, as follows:[9]

If the straight line (Z) is cut into equal segments (S, S) and unequal segments (A, E or $S + V$, $S - V$), then the rectangle comprehended between the unequal segments (AE) together with the square on the segment between the sections (Vq) equals the square on the half (Sq). That

---

[9] (Both the text and the symbolic argument are reproduced in Neal 2002, p. 145).

is, $Z = 2S = A + E$, then it will be that $S + V = A$, $S - V = E$ (just as $A - S = V = S - E$.) It will be that $AE + Vq = Sq$. Or (in numbers) if $12 = 6 + 6 = 8 + 4$ just as $8 - 6 = 2 = 6 - 4$. It will be that $8 \times 4 + 2 \times 2 = 6 \times 6$.

$$Z \begin{array}{c} S \overbrace{V+E} \\ \underbrace{S+V}\ E \\ A \end{array} \qquad \begin{array}{c} 6 \\ 6\ \overbrace{2+4} \\ \underbrace{6+2}\ \ 4=6-2 \\ 8 \end{array}$$

$$\begin{array}{r} S + V = A \\ S - V = E \\ \hline Sq + SV \\ - SV - Vq \\ \hline Sq - Vq = AE \\ Sq = AE + Vq \end{array} \qquad \begin{array}{r} 6 + 2 = 8 \\ 6 - 2 = 4 \\ \hline 36 + 12 \\ -12 - 4 \\ \hline 36 - 4 = 32 \\ 36 = 32 + 4 \end{array}$$

Wallis expected his readers to follow the steps of his argumentation (algebraically rendered on the left-hand side, arithmetically illustrated on the right-hand side), in a self-explained manner, and to realize that this formal manipulation of symbols embodies a proof of the general statement. As in all of Wallis' mathematics, the primacy of arithmetic (and algebra) over geometry is clearly manifest in this proof.

It is particularly illuminating to compare Wallis' version of II.5 with that of Barrow, which appeared in an edition of Euclid's *Elements* that the latter published in 1655. Barrow was among the seventeenth-century mathematicians for whom reverence to the ancients and their works was a guiding principle of his work. As part of this, geometry continued to be the source of certainty and legitimacy in mathematics, while algebra could be seen, at best, as an ancillary tool. In his own work, he strove to remain as close as possible to the classical tradition of synthetic geometry, and he certainly considered his presentation of Euclid to be faithful to the original spirit of Greek mathematics. However, in spite of this genuine commitment, he did introduce changes in geometry that included some traits of algebraic thinking. This is nicely illustrated in his treatment of Proposition II.5.

Like Wallis, also Barrow couched his proof in a symbolic language, but his symbolism was not abstract–algebraic in the sense of Wallis. Rather, Barrow's symbols were intended mainly as a way to make the purely geometric argument of the proof shorter and easier to express. Unlike Wallis' proof, Barrow's remained geometric in the original sprit of Euclid. Still, the use of symbols in a proof by a mathematician who claimed and made actual efforts to preserve the primacy of synthetic geometry, definitely helped make a possible algebraic interpretation of this proposition more amenable to any reader of the *Elements*. Let us take, then, a look at this proof.

Instead of Euclid's original diagram, Barrow accompanied the proposition with the following, simpler one:

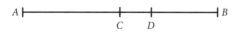

Instead of Euclid's original, purely rhetorical formulation, Barrow formulated the result to be proved in the following, symbolic terms:

$$CB_q = ADB + DC.$$

Denoting the rectangle on AD, BD as ADB was not uncommon in Greek geometric texts. Denoting the square on CB as CBq was in use at the time (Wallis, for one, used it), and in any case it was not totally foreign to the purely geometric spirit of Euclid's formulation. But then, the way in which Barrow used these notations in his proof was unlike anything that he could have found in the classical Greek texts that he liked so much to praise as paradigm of mathematical transparency and certainty. It read as follows:[10]

For thefe are all equal		
	CBq.	
	$aCDq + CDB + DBq + CDB.$	a 4.2.
	$CDq + bCBD(cAC \times BD) + CBD.$	b 3.2.
	$CDq + dADB.$	c hyp.
		d 1.2.

This is all. Four consecutive formulae that one must understand as being derived from each other, in a simple and evident way that requires no verbose explanations whatsoever. The passages from one step to the next are justified with the help of propositions that are indicated with the small letters a, b, c, d both within the formulae and in the column to the right. The argument is quite simple, but what is of interest for us here is the symbolic shorthand that Barrow adopted as legitimate and useful to represent succinctly the steps of his proof. In doing so, he was not actually moving into a fully algebraic interpretation of the proposition. The symbols were not formally manipulated according to abstract rules. Moreover, the only way to understand what Barrow was doing in his proof is by figuring out the successive steps (which are only symbolically indicated) as embodying a fully geometric situation. Barrow did not include any accompanying diagram in his proof, but it is evident that he was thinking geometrically. We must simply imagine the diagram ourselves. I suggest to do this as indicated in Figure 8.10.

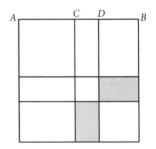

Figure 8.10 A possible diagram for Barrow's proof of II.5.

---

[10] As reproduced in (Neal 2002, p. 122).

With a view to such a possible diagrammatic representation, we can easily figure out the steps of the proof, as follows:

1. We start with the square on CB (CBq), and we want to see how it can be made to equal the sum of a rectangle and another square (i.e., ADB + DCq), as required by Proposition II.5.
2. The said square CBq equals the sum of two squares, CDq and DBq, together with twice the rectangle CDB. The letter *a* indicates that this step is justified by Proposition II.4 (of the *Elements*, of course), which handles precisely cases like this one ("If a straight line be cut at random, the square on the whole is equal to the squares on the segments together with the rectangle contained by the segments"). Translated into modern algebraic terms, II.4 corresponds to the identity $(x + y)^2 = x^2 + y^2 + 2xy$, but in Euclid it has a purely geometric meaning, namely, the decomposition represented in the bottom-right corner of the diagram. This is also how Barrow saw it—i.e., geometrically.
3. The rectangle CDB together with the square DBq equal the rectangle on CB, DB (and this is justified, as indicated by *b*, with Euclid's II.3).
4. In addition, instead of taking the side CB, one may take AC (a step warranted by hypothesis, as indicated by the letter *c*).
5. Finally, the sum of rectangles on AC, DB and on CB, BD equals the rectangle on AD, DB (warranted by II.1, as indicated by the letter *d*).
6. The previous two steps together yield the desired identity, CBq = ADB + DCq.

In spite of my insistence that Barrow's is a fully geometric poof, and that his symbols act as shorthand rather than as terms in an abstract language that can be formally manipulated, it is interesting to read a comment that Barrow added to the proof. In this comment, he essentially states that the actual mathematical core of the proposition is algebraic, not geometric. Thus, he wrote:[11]

This theorem is somewhat differently expressed and more easily demonstrated thus: A rectangle made of the summe and the difference of two right lines A, E, is equal to the difference out of them [i.e., the difference of their squares]. For if A + E be multiply'd into A − E there ariseth Aq + AE − EA − Eq = Aq − Eq, which was to be Dem.

---

[11] Cited in (Neal 2002, p. 122).

CHAPTER 9

New Definitions of Complex Numbers in the Early Nineteenth Century

The seventeenth century was a significant turning point in the history of science. Physics turned into a mathematically based discipline and at the same time into a paradigm that other disciplines aspired very often to imitate. Together with the internal dynamics of ideas within mathematics itself, the role of mathematics as the basic language of science became a powerful motivation that led to the development of novel, far-reaching techniques, concepts and ideas. The infinitesimal calculus had started its way, independently and following different approaches, in the work of Newton and of Leibniz. At the turn of the eighteenth century, in the wake of their work, this new calculus took center stage, with the concepts of limit, derivative and integral, with the increased attention paid to infinite series, and with all the new mathematical approaches that developed around them and around their use in physics.

All of these are aspects of momentous importance and great interest in the development of mathematics. All of them continued to be at the center of attention of mathematicians throughout the eighteenth century and a considerable part of the nineteenth. And within this powerful stream of ideas and innovation, old debates about the essence and nature of numbers of various kinds, about the legitimacy of using them, and about the relationship between geometry, arithmetic and algebra became somewhat marginal, but never fully disappeared. In this chapter, we explore some of the arithmetical ideas that were common during the eighteenth and early nineteenth centuries and particularly the way in which a completely new and highly influential approach to defining complex numbers appeared in the last part of this period.

9.1 Numbers and ratios: giving up metaphysics

An always revealing source of information about the ideas that were dominant in the mid eighteenth century is the famous *Encyclopédie, ou dictionnaire raisonné des sciences, des arts et des métiers*, the flagship of European Enlightenment culture, edited by Denis Diderot (1713–1784) and Jean le Rond d'Alembert (1717–1783). This is also a good place to look for evidence on contemporary conceptions about numbers. D'Alembert was a mathematician with important contributions to many active fields of research and with an intimate knowledge of the latest advances in many scientific disciplines. He wrote himself the lion's share of the mathematical entries of the *Encyclopédie* and was actively involved in the writing of most of the remaining ones as well. If collected together, these entries add up to a lengthy elementary summary that presents a panoramic overview of mathematical ideas as conceived at the time.

The entry on proportions, for instance, shows in an interesting way how the classical Greek conception had not yet been fully abandoned, despite having lost, during the seventeenth century, its centrality and efficiency as a main tool for proof and calculation in geometric contexts. Very much as the comparison of two quantities—d'Alembert wrote—can be expressed in terms of "ratios" or of "quotients," so can be the comparison of two ratios expressed in terms of "proportions." Given four quantities a, b, c, d such that $\frac{a}{b} = \frac{c}{d}$, d'Alembert explained that it is possible also to write them as a proportion $a:b :: c:d$.

For d'Alembert, then, a proportion and a quotient were still two different kinds of entities, but he understood that the difference was not really important. In a proportion, two ratios are compared, and the second quantity of each pair contains the first, "in exactly the same way." D'Alembert did not need to indicate anymore, like Eudoxus, that it is necessary for the two quantities compared in a ratio to be "of the same kind."

Of course, after Descartes' work, there is little significance to this requirement of sameness that had been so central ever since the Greek introduced it in their definitions. The quantities that d'Alembert's had in mind throughout (and which most contemporary mathematicians shared) were now "abstract" quantities, of the kind that had appeared in Newton's definition, mentioned in Chapter 8. Indeed in the entry "number" of the *Encyclopédie*, d'Alembert simply cited, almost word by word, the definition given in Newton's *Universal Arithmetick*.

D'Alembert's definition of proportion is therefore highly significant as representative of a crucial turning point. At this time, older ideas about numbers were still around, and they were respected on the basis of some kind of implicit reverence with the past. Yet, at the same time, some of these ideas had already lost most of their meaning when applied in the context of a completely new mathematical environment.

D'Alembert's definition of proportionis provides a good example of this, because it was of little practical use in any of the main concerns of mathematicians at the time, but it was used in the *Encyclopédie* for the sake of discussing the negative numbers. So, in attempting to oppose the view according to which negative numbers represent "quantities smaller than nothing," d'Alembert focused on the proportion $1:-1::-1:1$ and followed an argument introduced by Leibniz (and which was probably known earlier than that). He reasoned along the following lines:

On the one hand this proportion is valid on the basis of the rules of division with signs, because the following identity between fractions holds: $\frac{1}{-1} = \frac{-1}{1}$. On the other hand, according to the definition cited above this means that −1 contains 1 "in exactly the same way" that 1 contains −1. But if −1 were smaller than nothing (and hence smaller than 1), then it would follow that −1 is both greater than 1 (because it contains it) and smaller than it (because it is contained by it), thus yielding a contradiction.

D'Alembert did not indicate, however, that defining the negatives as quantities that are "smaller than nothing", the proportion 1:−1::−1:1 would not make any sense from the point of view of the classical definition, in the first place, for the simple reason that such quantities, if added together with each other, cannot add up to a quantity that becomes "greater than nothing" (as Eudoxus had stipulated). Quantities that are "smaller than nothing," then, if one is willing to speak about them at all, are not of "the same type" as quantities greater than zero. They simply cannot be compared in terms of the classical Greek definition. D'Alembert's conceptions of number and of proportion, then, were based on an uneasy mixture of old and new definitions that very often led to weird mathematical results.

If negative numbers are not quantities "smaller than nothing," what are they then? Here d'Alembert came up with a geometric explanation very similar to that previously advanced by Wallis: negative quantities differ from positive ones only with respect to their relative position to a fixed point on a straight line. The basic idea behind this definition is an attempt to "neutralize" value judgments implicitly attached to the adjectives commonly used for the various kinds of numbers, such as "negative" or "imaginary." The definition, however, creates an inherent tension with the basic idea of natural numbers seen as collections of units. This classical idea was so simple and straightforward that there was little willingness to abandon it. Still, neutralizing the value-laden adjectives attached to other kinds of numbers provided a reasonable approach that was to be conveniently pursued over the following decades for defining complex numbers, as we will see right below. It is strongly related to a more general attitude that was then becoming dominant in mathematics at large, and which implied a consistent estrangement from metaphysical interpretations of the essence of basic concepts (be they numbers, geometric entities, limits, or whatever).

So, instead of trying to find out "what is the real essence of numbers" in general, or what is the essence of this or that kind of numbers in particular, mathematicians gradually focused on the more concrete question "how numbers work." The clarification of the basic underlying principles that stipulate how each kind of numbers *works* were to gradually become the focus of interest of those (relatively few) mathematicians pursuing questions related to the foundations of arithmetic. Metaphysically oriented questions about essences and natures of numbers and of other mathematical entities were to become less and less common in mathematical texts.

9.2 Euler, Gauss and the ubiquity of complex numbers

In d'Alembert's time and immediately thereafter, with or without satisfactory explanations about the nature of numbers, mathematics continued to develop at an accelerated pace in all of its subdisciplines. In doing so, it used, almost with no restrictions or constraints, all kinds of numbers wherever required in order to solve problems or to develop new techniques. Among the most interesting developments during the eighteenth and nineteenth centuries were those connected with complex numbers. On the one hand, they started to be used with great success in all kinds of new and unexpected contexts. On the other hand, in spite of this extended use, such numbers continued to be considered as strange beasts, and debates about their legitimacy never disappeared.

This dual attitude could be found in one and the same mathematician at a given time, the most salient example of this interesting situation appearing in the case of the most influential figure of mathematics and physics in the eighteenth century, Leonhard Euler (1707–1783). One remarkable achievement of Euler in this context was his extension of some basic and well-known arithmetical concepts so that they could be used in a mathematically consistent and meaningful way, in the cases of negative and of complex numbers. A most interesting example of this was the extended definition of the exponential and logarithmic functions for negative and complex numbers. Previous attempts to come up with such definitions had led to considerable controversy. Euler helped settle and clarify them. He developed his ideas in the context of the continued efforts to elaborate the new field of the infinitesimal calculus. The classical example that amply illustrates the power embodied in them is the so-called Euler formula,

$$e^{i\pi} + 1 = 0.$$

This formula, whether or not understood by the reader, has always aroused a sense of awe and fascination because of the surprising combination, within one and the same expression, of these five "important" numbers $(0, 1, \pi, i, e)$ together with the two most basic arithmetic signs $(+, =)$. Incidentally, it was Euler himself who introduced the symbol i to indicate the number $\sqrt{-1}$, even though he was never fully consistent in its use. The symbol became standard only after Carl Friedrich Gauss (1777–1855) adopted it in his epoch-making book *Disquisitiones Arithmeticae* published in 1801 (more on which later). The still suspicious "imaginary" number $\sqrt{-1}$ appeared in this interesting result of Euler together with the other four $(0, 1, \pi, e)$, which were beyond any question of legitimacy. This alone was sure evidence that it made little sense to continue excluding $\sqrt{-1}$ from legitimate mathematical discourse. The very elegant and fascinating formula that connected the five numbers bestowed upon $\sqrt{-1}$ a mathematical status similar to those of the other four.

But this was not the only important place where Euler introduced complex numbers in a surprising mathematical context. Euler also applied them in a different field of research, number theory (or higher arithmetic, as it was called at the time). Euler solved a score of important open problems in this discipline with the help of the surprising new technique of factorizing integers into factors that contained imaginary numbers, such as in the following product (that you may want to check up):

$$5 = (1 + 2i)(1 - 2i).$$

By so doing, he broke new ground in a direction that was fully exploited by Gauss some decades later, as we will see below. (Again, incidentally, another place in which Gauss completed a task originally initiated by Euler was in the proof of the fundamental theorem of algebra. In 1751, Euler suggested a new proof that eventually was seen to fail for polynomial equations of degree greater than 4. Following in his footsteps, Gauss proved the theorem for the first time in 1799, at the age of 22, and later he came up with two additional and substantially different proofs.)

Euler's incorporation of complex numbers as a powerful, legitimate tool for solving problems in two of the most central branches of advanced mathematics, namely, calculus and number theory, was by all means a definite sign that they were mathematical entities that no one can further ignore or call into question. This did not mean, however, that anyone, let alone Euler himself, had a consistent and well-elaborated idea of how to satisfactorily define these numbers.

And, indeed, Euler himself, the great master of the use of complex numbers in innovative contexts, continued to formulate the basic definitions of the systems of numbers in ways that did not deviate significantly from those of his predecessors. In an introductory book on algebra published in 1770, *Vollständige Anleitung zur Algebra*, which in its various editions and translations was tremendously influential in all of Europe, Euler defined mathematics as the science that deals with "quantities in general," namely, anything that "can be either increased or diminished." On the one hand, he wrote, there are concrete quantities, such as money, length, area and velocity. On the other hand, in general, we speak about "abstract quantities," which we indicate with letters and calculate with in various ways. Among the latter, he included entities such as $\sqrt{-1}$, for the simple reason that there are algebraic expressions where they are used for various kinds of calculations.

But when he explained in his book the rules of calculations with numbers having different signs (similar to what had been done in other books on algebra beginning from the sixteenth century and up until Newton), he devoted some separate lines to the imaginary numbers and to the question of their essence. What he wrote is strongly reminiscent of the problematic statements of Descartes in this regard. He thus wrote:[1]

And, since all numbers which it is possible to conceive are either greater or less than 0, or are 0 itself, it is evident that we cannot rank the square root of a negative number amongst possible numbers, and we must therefore say that it is an impossible quantity. In this manner we are led to the idea of numbers, which from their nature are impossible; and therefore they are usually called imaginary quantities because they exist merely in the imagination. . . .

But notwithstanding this . . . we still have a sufficient idea of them; since we know that by $\sqrt{-4}$ is meant a number which, multiplied by itself, produces -4; for this reason also, nothing prevents us from making use of these imaginary numbers and employing them in calculation.

One can hardly say that this a well-formulated or illuminating statement about complex numbers. As in the case of Descartes, it seems to raise more questions than it answers. For one thing, if such numbers exists only in our imagination, where do all other numbers actually exist? In the empirical world? In the imagination of others? But this quotation by Euler is very symptomatic of the widening gap between the dazzling ability to use these numbers where they proved to be useful and the lack of progress in

---

[1] Quoted in (Martinez 2014, p. 31).

the ability to come to terms with their nature and to *define* them correctly. Of course, this is not disconnected from the fact that the corresponding definition of *natural* number was itself highly problematic, even though no one felt this at the time. For a mathematician like Euler, this much is clear—it was far less important to be able to *define* correctly the complex numbers than to *use* them advantageously in his research.

9.3 Geometric interpretations of the complex numbers

Continued efforts to better understanding the nature of negative and complex numbers persisted all the while. One important direction developed out of the geometric kind of interpretation of the negative numbers, which we have already met in the work of Wallis. Its starting point was in the suggestion that the signs +/− should be interpreted as no more than an indication of opposed directions. In this sense, the natural numbers are no more "natural" than the negative ones, and the latter are not to be explained in terms of the former, but both are to be accounted for together. Wallis attempted, without much success, to extend this same idea as a way to possibly account, on a geometric basis, for the idea of the number $\sqrt{-1}$.

The same idea was pursued more successfully at the turn of the nineteenth century in the works of various mathematicians. The first of these was Caspar Wessel (1745–1818), a Norwegian-born land surveyor whose work, first published in the Danish language in 1799 and later in French translation only in 1897, had little impact on others. Among those who developed similar ideas, one can count Jean-Robert Argand (1768–1822), Jacques Français (1775–1833), and several others. A brilliant formulation of the basic idea of geometrically interpreting complex numbers appeared in a text of 1806 by the French priest Adrien-Quentin Buée (1748–1826), who connected the meaning of $\sqrt{-1}$ with those of the signs +/− in the following words:[2]

> I speak of the sign $\sqrt{-1}$ rather than of the quantity or of the imaginary unit $\sqrt{-1}$. Because $\sqrt{-1}$ is a specific sign that we adjoin to the real unit and not a specific quantity in itself. This is a new adjective that is adjoined to an ordinary and well-known substantive, and not a new substantive. This sign does not indicate either addition or subtraction ... A quantity appearing with the sign $\sqrt{-1}$ is not added, is not subtracted, and it is not equal to nothing. The property described by $\sqrt{-1}$ is not opposed to that described by + nor to that described by − ... $\sqrt{-1}$ simply indicates a direction which is perpendicular to those indicated by + and by −.

Closely connected with d'Alembert's discussion on the status of negatives, the geometric interpretation of the complex numbers was initially motivated by a desire not to interpret the negatives as quantities less than nothing. Interestingly, it relied on the classical definition of proportion: $\sqrt{-1}$ was described as the geometric mean between 1 and −1, that is, a quantity x that satisfies the proportion $1:x :: x:-1$. Along the same line of thought, any given complex number $a + ib$ was interpreted, in the works of Argand and others, as a pair (r, α) representing a segment of length r and lying on the plane in

---

[2] Adrien Quentin Buée, "Mémoire sur les quantités imaginaires", *Philosophical Transactions of the Royal Society of London*, 1806, Part I, 23–88. Here on p. 28.

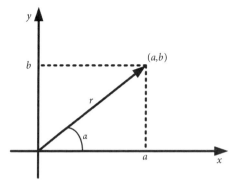

Figure 9.1 Argand's geometric interpretation of a complex number $a + ib = r(\cos\alpha + i\sin\alpha)$.

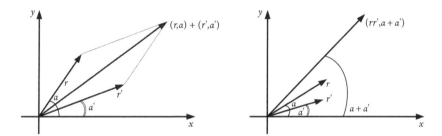

Figure 9.2 Addition and multiplication of complex numbers.

a direction indicated by the angle α. As can be seen in Figure 9.1, the values of r and α satisfy the relation $a + ib = r(\cos\alpha + i\sin\alpha)$, where $r^2 = a^2 + b^2$.

Argand also defined the basic arithmetic operations on complex numbers: addition corresponds to the "law of parallelogram," whereas multiplication corresponds to adding angles and multiplying lengths of the segment, as indicated in Figure 9.2. Division and root extraction can be defined in similar ways. In these terms, the number $a\sqrt{-1}$ is interpreted as a segment of length a with direction perpendicular to the horizontal axis, which is precisely what Buée had suggested.

It is important to stress that all the names just mentioned, Wessel, Argand, Français and Buée, were essentially unknown in the mathematical community during their lifetimes. They worked outside the mainstream of contemporary advanced mathematical research, and they did not publish important research other than the one related to the definition of complex numbers. The same is the case with some additional names in France, the British Isles and Italy, of people who were involved at the time in attempts to find a satisfactory geometric representation of the complex numbers. By contrast, we have already seen how Euler devoted intense efforts to incorporate complex numbers into the overall picture of mathematics as a powerful tool for advanced research, but much less so to the question of clarifying the proper foundations, geometric or other, of this system of numbers. Euler's attitude is truly representative in this regard of the mathematical community and of its leading figures at the time.

The link connecting these two groups—the leading mathematicians who used imaginary numbers in innovative ways and those outsiders who came up with original ideas on how to define them on a geometric basis—appeared in the figure of Gauss. Even before 1799, when he worked on his first proof of the fundamental theorem of algebra, Gauss had developed a geometric interpretation of the complex numbers. The technical details are similar to those of Argand's definition. But, unlike Argand and the other mathematicians just mentioned, Gauss made substantial use of his definition in order to attain deeper insights into the proof that he was then developing for a truly important mathematical result.

I have already mentioned Gauss's famous book of 1801, *Disquisitiones Arithmeticae*. This book laid down the main research agenda, the standard techniques, the basic concepts, and even the accepted notations (including the sign i for $\sqrt{-1}$) for much of the most important activity in number theory over the nineteenth century, and well beyond that. One of the important topics that Gauss dealt with in his own research after 1830 was the so-called higher reciprocity problem.

We cannot delve here into the intricacies of higher reciprocity,[3] but I would like to say some words about a main idea introduced by Gauss as part of his research on this problem. This was a truly new kind of numbers, the so-called Gaussian integers (also called "complex integer numbers" and typically denoted as $\mathbb{Z}[i]$). These are complex numbers $a + ib$ where both a and b are *integer numbers*. Gauss showed that these numbers satisfy many of the most important properties of the ordinary integers. He identified, for example, certain numbers that play within the new system the role that the primes play in the system of ordinary integers as its basic building blocks. Moreover, using these prime numbers, it is possible to prove a modified version of the fundamental theorem of arithmetic:

Every Gaussian integer can be represented as a multiplication of factors that are Gaussian prime integers and this in a unique way (except for the order of the factors, and multiplication by one of the numbers 1, –1, i or $-i$).

A simple, but highly illustrative example of the meaning of this result is the following: the number 5, seen in the context of ordinary integers, is of course a prime number; in the context of the Gaussian integers, however, it is not. The reason is that it can be written as a product:

$$5 = (1 + 2i)(1 - 2i).$$

An additional example of the ways in which Gauss explored the basic properties of this new kind of number is his definition of the "greatest common divisor" (G.C.D.). He showed that given any two Gaussian integers, their G.C.D. can be found using the same algorithm that is used for calculating it in the case of ordinary integers (the so-called Euclidean algorithm).

Gauss understood very quickly the power of the Gaussian integers, and they soon proved to be a highly efficient tool for proving theorems about *ordinary* integers as well. Indeed, using them, Gauss accomplished his intended aim of proving results

[3] See (Corry 1996 [2004], Section 2.2.1).

related to "higher reciprocity." Still, in spite of the fact that these ideas originated in his mind in close connection to his geometric interpretation of the complex numbers, he did not publish anything on the latter before 1832.

This long delay is not in itself very surprising, because this is the way Gauss acted in other, similar cases. Best known is the case of the so-called non-Euclidean geometries, where he refrained from publishing his discoveries and later claimed that he did so to avoid unnecessary criticism coming from philosophers and conservative mathematicians ("the Boeotians," he deridingly called them). It seems that also in relation to the complex numbers, Gauss feared that his innovative ideas would be received with hostility, since they implied the use of mathematical entities that were poorly understood or outright rejected (the complex numbers) in order to investigate the most straightforward and seemingly transparent kind of mathematical entities (the integer numbers).

When Gauss finally published his fascinating results on reciprocity, he decided to anticipate possible criticism with clarifications. He stated that in his view, complex numbers are not only unproblematic in any sense and that, moreover, their geometric representation over the Cartesian plane completely clarifies "their true metaphysics." He thought that the accepted views on the various number systems had been affected thus far by a kind of completely dispensable "mysterious obscurity," which derived in the first place from an unfortunate choice of wording. Gauss believed that some of these absurd attitudes would never had arisen if instead of calling the numbers $1, -1, i$ or $-i$ with the loaded terms positive, negative and imaginary (and even impossible), some more neutral terms had been adopted such as direct, contrary, or lateral.

9.4 Hamilton's formal definition of complex numbers

The geometric interpretation of the complex numbers had many advantages, but it also left open some significant issues. One consequence of this interpretation was that arithmetic and algebra remained conceptually subordinate to geometry. This was in contraposition to the view that was gradually gaining ground among most mathematicians of the time and that was becoming the most naturally accepted one. Indeed, the rise of non-Euclidean geometries, already mentioned above, and the increasing interest in the foundations of the infinitesimal calculus (as will be seen in the following chapters) involved fundamental processes of change that had been taking place in the central branches of mathematics, including arithmetic and algebra, and continued to shake the primacy of geometry as the main source of mathematical certainty.

The primacy of synthetic geometry had already been challenged by the early nineteenth century as part of a slow and complex, but uninterrupted, process. Ironically, then, the geometric interpretation of the complex numbers became consolidated at a time when it could be seen much more as a good illustration of their behavior than as a way of providing a satisfactory legitimation for the system, given that the foundations of geometry were showing signs of being more shaky than previously believed. There can be no doubt that these changes help explain the lack of a more unanimous acceptance of this interpretation among contemporary mathematicians, and the fact that alternative directions were proposed. One of these, about which we will talk now, represents the next and decisive stage in this long historical process of understanding and correctly defining the complex numbers. This stage came in 1837 in the work of the Irish mathematician, Sir William Rowan Hamilton (1805–1865).

Hamilton approached this issue from a novel, thoroughly formalist point of view that renounced any attempt to understand the "*essence*" or the "*nature*" of the complex numbers as the basis to account for their properties. Quite the contrary, Hamilton started from establishing the *rules* of the desired behavior of this system of numbers, and built the system out of these rules. He took for granted the system of real numbers and its arithmetic, and saw no need to explain or justify them in any way. The next generation of mathematicians, as we will see in the next chapter, would come to question even this starting point and would see a pressing need to clarify the foundations of the system of real numbers. But at this stage let us stay with Hamilton and see the way in which he constructed the system of complex numbers.

Hamilton started by considering ordered pairs (a, b) of real numbers and by trying to imitate with their help the desired properties of the complex numbers. Thus, he defined operations on these pairs based on operations with the real numbers. The desired properties of the complex numbers, of course, were those known to mathematicians since the time of Bombelli. So, the most important such property is the one related to a number, called here i and representing $\sqrt{-1}$, that when multiplied by itself yields the number -1. Besides, by conceiving a complex number as a binomial with a real and an imaginary part, $a + ib$, operations on them could be defined as follows:

$$(a_1 + ib_1) + (a_2 + ib_2) = (a_1 + a_2) + i(b_1 + b_2), \tag{9.1}$$

$$(a_1 + ib_1) \times (a_2 + ib_2) = (a_1 a_2 - b_1 b_2) + i(a_1 b_2 + a_2 b_1). \tag{9.2}$$

Defining division of two complex numbers requires some extra effort, but it can be done in analogy with the case of the integers. Thus, to divide an integer p by another integer q is equivalent to multiplying p by the "inverse" of q, namely by $\frac{1}{q}$. And $\frac{1}{q}$ is the inverse of q, because $q \cdot \frac{1}{q} = 1$. In this way, when we divide p by q, we obtain $\frac{p}{q}$. Something similar can be done for defining division of complex numbers: dividing any given complex number by $a + ib$ is equivalent to multiplying the given number by $a' + ib'$, where $a' + ib'$ is the "inverse" of $a + ib$. And this inverse, $a' + ib'$, is, in the framework of the system of complex numbers, another complex number that satisfies the condition $(a + ib) \times (a' + ib') = 1$. Some straightforward algebraic manipulation shows that this number ($a' + ib'$) has to be

$$a' + ib' = \left(\frac{a}{a^2 + b^2}\right) + i\left(\frac{-b}{a^2 + b^2}\right).$$

Thus, performing the division $(a_1 + ib_1) \div (a_2 + ib_2)$ is equivalent to performing a multiplication by the inverse of the second number, in the following way:

$$(a_1 + ib_1) \div (a_2 + ib_2) = (a_1 + ib_1) \times \left(\frac{a_2}{a_2^2 + b_2^2}\right) + i\left(\frac{-b_2}{a_2^2 + b_2^2}\right). \tag{9.3}$$

By the time when Hamilton began to work out his formalistic approach to defining the complex numbers, it was well known—I emphasize once again—how to perform these operations. But the point is that debates continued to be held around the question of the "essence" of the number i. It was still unclear whether the operations above, including division (or root extraction, which I have not mentioned here) were somehow justified on the basis of that putative essence. Hamilton suggested to bypass all those debates and to simply postulate, in formal terms and without further explanations, operations on ordered pairs of real numbers, in such a way that they would imitate those already known for complex numbers. These operations are

$$(a_1, b_1) + (a_2, b_2) = (a_1 + a_2, b_1 + b_2), \tag{9.4}$$

$$(a_1, b_1) \times (a_2, b_2) = (a_1 a_2 - b_1 b_2, a_1 b_2 + a_2 b_1), \tag{9.5}$$

$$(a_1, b_1) \div (a_2, b_2) = (a_1, b_1) \times \left(\frac{a_2}{a_2^2 + b_2^2}, \frac{-b_2}{a_2^2 + b_2^2}\right). \tag{9.6}$$

Clearly, operations (9.4)–(9.6) are defined so as to yield exactly the same results as those of (9.1)–(9.3). But notice that now the mysterious entity i has simply evaporated. Moreover, within this formal system, we can easily define a "real" number as a pair $(a, 0)$ and an "imaginary" number as a pair $(0, b)$. In this way, all the terms have become completely neutral, and they do not express any kind of "external" property besides what their formal appearance involves. Certainly, "real" and "imaginary" numbers are seen to have a completely symmetrical status, and neither of them is seen as existing either in the imagination or in the real world, in any sense, more than the other.

The direct, most important feature of this definition is that if we take the pair $(0, 1)$ and multiply it by itself using rule (9.5), then we obtain $(0, 1) \times (0, 1) = (-1, 0)$ (try it at home!). In other words, if we think of $(-1, 0)$ as representing the real number -1, we have just found an "imaginary" number $(0, 1)$ that stands, in the most basic sense of the arithmetic operations involved, for $\sqrt{-1}$ (or for i, if you wish). We need not, under this approach, give any special meaning to $(0, 1)$, much the same as we do not give any special meaning to $(1, 0)$, for that matter. We need just to know how the operations in the system work. This is indeed a good example of removing the dispensable "mysterious obscurity," as Gauss had wanted it to happen.

9.5 Beyond complex numbers

Hamilton's basic idea is as simple as one can wish, at least in appearance. However, as we will see below, it had some consequences that required further clarification and that led to further, important developments. Still, in order to get the picture right, it is important to keep in mind the broader context within which these ideas arose. They were closely connected with contemporary debates about the foundations of algebra, on the one hand, and with other fields of knowledge that attracted Hamilton's attention throughout his career, on the other hand.

In the early nineteenth century, a new mathematical tradition developed in the British Isles, the so-called symbolic algebra tradition, at the basis of which there was a fundamental distrust of the assumption that the symbols appearing in mathematical formulae necessarily represent some numbers or quantities of a certain kind. This assumption had naturally evolved through the centuries as part of the developments we have been describing in the previous chapters. By this time, there was a rather commonly accepted view of mathematics as "the science of quantities in general" (a view that Euler had helped promote in his writings). This view was an attempt to overcome narrower, previously held beliefs in which geometry alone was seen as the source of certainty in mathematics.

But the new views now developing in the British mathematical context aimed at an even broader and more abstract perspective. The idea was to look for the foundations of algebra in a set of abstract rules provided in advance, without any a priori constraints derived from the assumed nature of the entities involved in the operations. British symbolic algebra developed against the background of those far-reaching processes I mentioned above, whereby the rise of non-Euclidean geometry and the new debates on the foundations of the infinitesimal calculus contributed to the questioning of the primacy of geometry as the ultimate source of mathematical certainty.

Additional input came from recent progress in the "algebraization" of logic, particularly with the work of George Boole (1815–1864). The complex and multi-faceted processes related to these changes involve original and sometimes idiosyncratic mathematicians, such as George Peacock (1791–1858), Duncan Farquharson Gregory (1813–1844), Augustus De Morgan (1806–1871), and of course Hamilton himself. What interests us here, however, is just to stress one aspect of their work that is common to all of them, namely, the willingness to explore new directions and to broaden the scope of ideas related to possible algebraic systems defined on the basis of abstract operations defined by sets of rules stipulated in advance. In the view that they promoted, *the rules are not justified by the nature of the objects they apply to,* but it is rather the other way round: *the nature of the systems of numbers is determined by rules controlling the operations that underlie them.* This is a truly far-reaching idea that would have momentous implications for the development of mathematics in the decades to follow.

This general background to the new symbolic approaches that developed in British mathematics at the time combined in the case of Hamilton, in an especially interesting fashion, with his additional involvement in fields such as astronomy and mathematical physics (particularly optics and mechanics). He became aware, from very early on in his career, of the fruitful way in which Euler had used complex numbers in his work on infinitesimal calculus. At the same time, however, he was not satisfied by the existing definitions of negative and complex numbers. Moreover, unlike Euler, he thought he could contribute his time and efforts to clarifying these issues.

Around 1828, Hamilton came across Argand's geometric interpretation. He immediately came up with the original idea of possibly applying Argand's insights to his own ongoing research in mathematical physics, by extending Argand's approach from the plane to space. That is, whereas with complex numbers the basic operations of addition, multiplication and division had been defined for numbers $a + ib = (a, b)$, Hamilton now looked for similar definitions applying to triplets (a, b, c). One way that immediately comes to mind when attempting to do so is to add a new "imaginary" element j, which plays a similar role to i and allows one to represent the triplet as follows: $a + ib + jc$. It is, of course, easy to define addition of such triplets, as follows:

$$(a_1 + ib_1 + jc_1) + (a_2 + ib_2 + jc_2) = (a_1 + a_2) + i(b_1 + b_2) + j(c_1 + c_2).$$

But when Hamilton attempted to define multiplication and division in a somewhat similar way for the triplets, he encountered serious problems that would occupy his attention for many years to come. Defining these two operations properly requires that for any given triplet $a + ib + jc$, one should be able to find an "inverse" triplet $a' + ib' + jc'$ such that when these two are multiplied by each other the result will be $1 + i0 + j0$, that is to say, 1 (remember that this is the way we defined the inverse and, accordingly, division in the cases of integers and complex numbers).

After having succeeded in doing this for complex numbers, Hamilton did not think that he would face significant difficulties in trying to extend the same idea to the case of triplets. But, in spite of years of great effort, he did not succeed in this case, and in retrospect it became clear that he *could not* have succeeded. Indeed, in 1878, the German mathematician Georg Ferdinand Frobenius (1849–1917) proved a very important (and not easy to prove) theorem that implies that such a system of triplets with arithmetic operations as described above is an impossibility. But there were also good news. Although he failed in his attempt with triplets, Hamilton was led in 1843 to a surprising and highly important discovery that would forever remain associated with his name, namely, the system of quaternions. This is a system similar to the one Hamilton was looking for, and that extends the idea of complex numbers, but extends it directly into 4-tuples of real numbers, rather than into triplets (as it happens, Frobenius' theorem does not rule out the possibility of such a system of 4-tuples).

The process of discovery of quaternions and its immediate implications constitute a highly interesting chapter in the history of mathematics and they add an important dimension to our story here. I will devote some lines to discuss it in Section 9.6. But before doing that, I want to refer to an additional relevant aspect of Hamilton's work. In the period of time previous to the discovery of quaternions, Hamilton continued to work intensely on his research in mathematical physics, and at the same time he devoted serious effort to study the works of philosophers. His main focus of philosophical interest was on the influential work of Immanuel Kant (1724–1804).

While he continued to think out the issue of the foundations of the various number systems, Kant's ideas (and the entire web of philosophical considerations that Hamilton nurtured at the time) acted in combination with the technical tools he was using in this attempt. In 1833, he presented at the Royal Irish Academy his successful definition of complex numbers as ordered pairs of reals. Then, in the following years, he attempted to come up with a novel definition of numbers in general, based on a similar principle of ordered pairs of various kinds. In 1837, he published an article where he explained his ideas on this topic, but the mathematical arguments appeared embedded as part of a complex and rather obscure philosophical discourse that actually blurred some interesting insights he was presenting. Hamilton proposed that the foundations of algebra should be based on "the pure intuition of time," parallel to the way in which the foundations of geometry are based—in Kant's philosophy—on "the pure intuition of space."

Kant scholars usually relate to Hamilton's article on the pure intuition of time as a fundamental misunderstanding of the philosopher's doctrines. Mathematicians, in turn, typically react with an immediate estrangement from the metaphysical language and speculative tone of the article. And nevertheless, this article did contain some ideas that are worthy of attention and that pertain to the possibility of representing with the help of ordered pairs (this time, pairs of "instants" in time) the basic idea of the system of natural numbers as a "continuous" and "one-dimensional" sequence.

Irrespective of the actual meaning and the degree of success of this particular attempt by Hamilton, the whole thrust of his work in this context (including his definition of the complex numbers as pairs, the attempt to define the reals as pairs of instants, the failure to define an arithmetic of triplets, and the discovery of quaternions) definitely brought to the fore the need for clearer definitions of all number systems then in use, from the naturals up to the reals. It also suggested a general approach, later called "genetic" or "constructive," according to which each number system arises from a more elementary one via a process of very precisely defined construction (exemplified by the construction of complex numbers from the reals, in the way that Hamilton devised). In the next chapter, I will return to this important issue, and we will see how this approach was successfully implemented some decades later in the work of Dedekind.

9.6 Hamilton's discovery of quaternions

Let me then conclude this chapter by returning to the story of quaternions. Hamilton wrote a 4-tuple (a, b, c, d) with the help of three special symbols, i, j, k that generalize the idea of i as a symbol for $\sqrt{-1}$. Thus, the 4-tuple is written as $a+ib+jc+kd$. Hamilton could not know that his failure with the triplets was due to an inherent reason (as showed later by Frobenius), but once he began to work out the idea for the 4-tuples, he defined in a purely formalistic fashion one basic rule of multiplication, namely, $i^2 = j^2 = k^2 = -1$. On the basis of this, any two quaternions may now be multiplied, as follows:

$$(a_1 + ib_1 + jc_1 + kd_1) \times (a_2 + ib_2 + jc_2 + kd_2) = A + iB + jC + kD, \quad (9.7)$$

where

$A = a_1 a_2 - b_1 b_2 - c_1 c_2 - d_1 d_2;$ $\quad B = a_1 b_2 + b_1 a_2 + c_1 d_2 - c_2 d_1;$
$C = a_1 c_2 + c_1 a_2 + d_1 b_2 - d_2 b_1;$ $\quad D = a_1 d_2 + d_1 a_2 + b_1 c_2 - b_2 c_1.$

This appears, at first sight, to be a rather cumbersome definition. But, after some reflection, it is easy to see that the four expressions for A, B, C, D reflect a very obvious symmetry. The really important point that Hamilton was able to achieve with this way of defining the multiplication is that—as he had previously done with the complex numbers—one could define its inverse, i.e., a division of quaternions. And yet, this nice arithmetical feature came at a truly surprising price: the multiplication thus defined was not commutative! This is easily seen by multiplying two quaternions, for example,

$1 + i + j + k$ and $1 - i + j + k$, in different orders. Multiplying according to the rule stipulated in (9.7), you can readily check the following results:

$$(1 + i + j + k) \times (1 - i + j + k) = 4k, \quad \text{but} \quad (1 - i + j + k) \times (1 + i + j + k) = 4j.$$

And, at an even simpler level, if you multiply all possible pairs that can be formed with i, j, k, then you will realize that changing the order of the factors yields a change in the sign of the result, as follows:

$$ij = k = -ji, \quad jk = i = -kj, \quad ki = j = -ik.$$

Multiplication of quaternions thus provides us with a refreshing and insightful novelty: in spite of being a well-defined operation that satisfies the basic algebraic property of associativity (i.e., $X \times (Y \times Z) = (X \times Y) \times Z$), and in spite of the fact that one can satisfactorily define its inverse operation (i.e., division), the operation itself is not commutative (i.e., $X \times Y \neq Y \times X$). And this unprecedented situation had far-reaching consequences! Hamilton understood that it is possible to construct a system of 4-tuples with algebraic operations such as he was looking for, but for a price: he should give up a property that was taken for granted as self-evident for any system of numbers that one could possibly think of. This new system, the system of quaternions, broadened in an unexpected direction the idea of what is a number and what it could be.

One of the most famous anecdotes in the history of mathematics concerns this discovery. It was told by Hamilton himself in a famous and often quoted letter to his friend Peter Guthrie Tait (1831–1901). It describes the precise moment when the discovery popped up in his mind in October 16, 1843, like a bolt out of the blue, when he was pleasantly walking with Lady Hamilton on the way to a meeting of the Royal Irish Academy in Dublin:

I then and there felt the galvanic circuit of thought close; and the sparks which fell from it were the fundamental equations between i, j, k; exactly such as I have used them ever since.

Hamilton could not resist the impulse—"unphilosophical as it may have been"—to cut with a knife on the stone of Brougham Bridge the relations $i^2 = j^2 = k^2 = ijk = -1$.

The invention (or discovery?) of quaternions as an extension of the idea of complex numbers was not an isolated event. Hermann Günther Grassmann (1809–1877), for instance, investigated abstract, generalized systems of numbers from an original perspective. Because of a very untidy and complicate presentation of his ideas, however, his work received little attention at the time. It was only later that it was brought to the attention of a broader circle of mathematicians, who were then inspired to go on and develop them into important contributions to modern algebra.

Also, some leading British mathematicians of the mid nineteenth century investigated other kinds of generalized systems of numbers to which they give all kinds of strange names. In retrospect, these can all be identified as instances of what nowadays is covered by the somewhat broad term "hypercomplex numbers." Such systems appeared in the works of Arthur Cayley (1821–1895), James Joseph Sylvester (1814–1897), William Kingdon Clifford (1845–1879), and several others. Seen in the short run, however, it was Hamilton's quaternions that attracted the most attention, and this for a completely unexpected, and most welcome reason: it turned out that quaternions had direct applications in physics! Quaternions provided a very efficient and precise language for formulating mainstream physical theories being developed at the time, above all electromagnetism. Hamilton's 1853 *Lectures on Quaternions* presented in a systematic way his formulation of complex numbers as ordered pairs of real numbers, as well as the new theory of quaternions. But it was through two popular textbooks written by Tait that quaternions became a well-known idea, widely used by physicists.

As it happened, however, at the turn of the twentieth century, new concepts and techniques, particularly those related to the new vector spaces, proved to be even more efficient and general, and were preferred in most physical theories where quaternions had been playing an important role for a period of more than three decades.

CHAPTER 10

"What Are Numbers and What Should They Be?" Understanding Numbers in the Late Nineteenth Century

Hamilton's *Lectures on Quaternions* found one particularly enthusiastic reader in the young German mathematician Richard Dedekind. Investigating systems of numbers and their properties was always at the focus of Dedekind's interest, and he pursued this interest in innovative and original ways. On the one hand, he contributed seminal ideas to advanced research in the theory of numbers, and created sophisticated tools that made him at the turn of the twentieth century one of the most important figures in this field. On the other hand, he maintained a lifelong interest in foundational questions pertaining the nature and significance of the various kinds of number systems, and he came up with the first coherent, systematic and exhaustive picture of the entire hierarchy of the system of numbers.

Dedekind was one of the prominent followers of Gauss in all that concerns the attempt to understanding the world of numbers. Whereas before Dedekind foundational questions about numbers were mainly debated as part of philosophical discourse, or in introductions to both advanced and elementary mathematical books, or even as passing marginal remarks in such texts, he was the first to combine in a serious way these two, hitherto separate, levels of consideration. He dealt with foundational questions about numbers while always keeping an open eye on advanced research issues, and vice versa.

10.1 What are numbers?

"What are numbers and what should they be?" is an acceptable, non-literal translation of the title of one of Dedekind's most famous books, published in 1888 under the title *Was sind und was sollen die Zahlen?* This title reflects in a most appropriate manner the central concern that passes as a unifying thread throughout Dedekind's entire mathematical career, and at the same time it duly stresses the point that no one before him approached this question while attempting to provide a full, systematic, and *purely mathematical* answer to it.

Hamilton's proposal for defining the complex numbers as ordered pairs of real numbers, which appeared in detail in the book that Dedekind read, fitted neatly into the young Dedekind's early explorations of this big question. The proposal pointed to an interesting direction that Dedekind would follow in building the overall picture of the world of numbers that he then started to pursue. Hamilton's ideas, however, were not the only ones in the background to Dedekind's quest. Additional important insights originating in the works of Gauss and some of his followers, were part of it. There were also some well-conceived philosophical and methodological principles. Dedekind's contributions became crucial for the final consolidation of the modern concept of number at the end of the nineteenth century, but their influence went well beyond this relatively restricted issue, and it left a far-reaching imprint on the overall image of what the discipline of mathematics became at the turn of the twentieth century and throughout its first decades.

In this chapter and in Chapters 11 and 12, we will take a glimpse at some of Dedekind's most important work, together with those of some additional contemporary mathematicians involved with similar concerns. This chapter will be, of necessity, somewhat more technically challenging than the previous ones, but I will try to explain, as clearly and simply as possible, the main concepts involved here. I hope that readers who have been bearing-up up to this point will not raise their hands too quickly and will go along with me up to the end. Understanding Dedekind's ideas and his work is of fundamental importance for understanding the modern conception of numbers and its place within the overall disciplinary picture of mathematics.

Dedekind completed his doctoral degree at the University of Göttingen in 1851, and his advisor was none other than Gauss. Gauss was not particularly keen on advising research students, and also, in the case of Dedekind, their personal contact and interaction were quite limited. Nevertheless, Dedekind was no doubt one of the persons with the broadest and deepest understanding of Gauss's work on number theory and particularly on those topics that Gauss developed in *Disquisitiones Arithmeticae*. Moreover, Gauss's influence on Dedekind was not limited to the technical aspects of his work, and it touched also upon Gauss's views on the scope and aims of mathematical research and upon the best ways to pursue it.

Gauss's influence was even more strongly felt because of the mediation of a second crucial figure whom Dedekind met at Göttingen, Bernhard Riemann (1826–1866). Riemann was a taciturn and extremely introverted person, but a mathematician of an astoundingly broad knowledge and responsible for original breakthroughs in many fields. Dedekind learnt much and took inspiration from Riemann in significant ways. The third important person who helped shape Dedekind's mathematical world was

Peter Lejeune-Dirichlet (1805–1859), a prominent contributor to the establishment of the discipline of number theory as it was practiced in the early nineteenth century, who arrived in Göttingen following the death of Gauss.

A central methodological principle that Dedekind developed under the influence of these three Göttingen giants was the search for general abstract concepts around which broad and systematic theories are to be built. At the center of these theories, one must then look for a few fundamental theorems on the basis of which a vast number of specific, open problems can be solved with relative ease. According to this methodological principle, one must minimize the amount of specific calculations with particular cases. At the same time, one must maximize the number of general concepts and sweeping theorems that provide unified explanations and bring to light deeper, common reasons for a wealth of mathematical phenomena that have little apparent connection with each other. In order not to leave this somewhat broad formulation hanging in thin air as an empty statement, let me briefly introduce an important example of how Dedekind implemented this methodological principle as part of his research on the world of numbers. This example, which I discuss in the next section, is in my view both deep and beautiful. It may require some extra attention on the side of the reader, but, as with other difficult parts of this book, it will be rewarding.

10.2 Kummer's ideal numbers

In Section 9.3, we spoke about Gaussian integers $\mathbb{Z}[i]$, namely, complex numbers $a + ib$ with both a and b being integers. The collection of all Gaussian integers is a very specific collection within the system of all complex numbers, $\mathbb{C}$. As I already explained, Gauss introduced the Gaussian integers in the context of his attempts to deal with an advanced issue in number theory (the so-called higher reciprocity problem). As part of his research, he proved that the system $\mathbb{Z}[i]$ satisfies all the basic properties of the system of ordinary integer numbers, $\mathbb{Z}$, such as a corresponding version of the fundamental theorem of arithmetic.

Once we have thought about the system of Gaussian integers, it is easy to think about additional collections of specific kinds of complex numbers that generalize the same basic underlying idea. For example, consider numbers $a + \rho b$, with a and b integers and with ρ representing a complex number other than $\sqrt{-1}$. One possibility is to take ρ to represent a root of a different negative number, such as $\sqrt{-3}$, $\sqrt{-5}$ or $\sqrt{-19}$.

Another, more mathematically important, possibility is to take ρ to represent a "root of unity," namely a complex number satisfying the property $\rho^n = 1$, where n is any natural number. We already know the case $n = 4$, when ρ is simply i. Indeed, since $\rho^2 = -1$, it follows that $\rho^4 = i^4 = 1$. We can generalize this idea to other values of n. Notice that the roots of unity can be thought of as the solutions to the equation $x^n - 1 = 0$, and the fundamental theorem of algebra stipulates, as already explained, that there are exactly n complex roots to such an equation. If we think of the complex numbers in their graphical representation as points on a plane, as explained in Section 9.3 (see especially Figures 9.1 and 9.2), then it is easy to see that these roots of unity can all be located on a circle of radius 1 around the origin of the coordinate

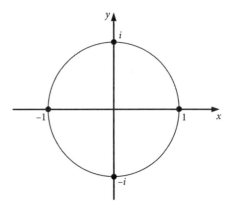

Figure 10.1 Four roots of unity represented graphically as complex numbers lying on a unit circle.

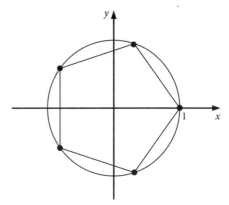

Figure 10.2 Five roots of unity represented graphically as complex numbers lying on a unit circle.

system. In the case $n = 4$, for example, we have four roots of unity, which are well known: $1, -1, i, -i$. This is represented graphically in Figure 10.1.

In the case $n = 5$, we have five roots of unity and they all lie on the unit circle, forming the vertices of a regular pentagon as represented in Figure 10.2.

Imagining the roots of unity in this way allows seeing them as "dividing the circle" into equal parts (in this example, five equal parts). For this reason, the equation $x^n - 1 = 0$ is typically known as the "cyclotomic equation" (the Greek term "cyclotomy" signifying "division of the circle"). Finally, we can introduce the idea of "cyclotomic integer numbers," which gives us a good feeling as to the ways in which mathematicians can begin to generalize into very abstract directions ideas that arise in the framework of their attempts to solve specific mathematical problems. A cyclotomic integer of order five, for example, is an expression of the form

$$a_0 + a_1\rho + a_2\rho^2 + a_3\rho^3 + a_4\rho^4,$$

involving the powers ρ^k of ρ, with $k = 0, 1, 2, 3, 4$ and with the coefficients a_k all being integers.

Now, Gauss himself suggested that cyclotomic numbers might be useful for the kind of research he was involved with, but he never really pursued this direction seriously.

The one to do so, beginning a few decades later, was the leading Berlin number theorist Eduard Ernst Kummer (1810–1893). Kummer tried to prove that in systems that generalize the idea of $\mathbb{Z}[i]$, the basic laws of ordinary arithmetic continue to hold. In particular, he assumed that the fundamental theorem of arithmetic would also hold in these systems. On the face of it, there was no reason to assume that what was the case for $\mathbb{Z}$, and what had been also proved to be the case for the more general system $\mathbb{Z}[i]$, would not also be true for even more general cases, including cyclotomic integers. Kummer, a mathematician never weary of working out in detail even the most daunting computations, studied many cases that seemed to consistently confirm the assumption.

But then a big surprise came up. After having checked many lower cases by endless, painstaking calculations, Kummer found out to his astonishment that cyclotomic numbers of degree 23 (!) provide a counterexample for the fundamental theorem of unique factorization. You have to be a great mathematician in order to start noticing that what seems to be a self-evident regularity, which is found as a matter of course in so many similar arithmetic contexts, will fail to hold in one specific case. But very few great mathematicians would take the necessary pains to calculate in detail, as Kummer did, so many cases before striking right on the one that indicated that there is a problem.

We can get a glimpse of the situation faced by Kummer, by looking at the following, much simpler, example that illustrates the kind of problem that he came across when studying the domains of cyclotomic numbers. Consider the domain of numbers $a + b\sqrt{-5}$, with a and b integers, and the following factorization of the number 21 into factors in that domain:

$$3 \cdot 7 = (4 + \sqrt{-5}) \cdot (4 - \sqrt{-5}) = 21.$$

Notice that we have here two different ways to factorize. It can be shown, though it is not a straightforward matter, that the four factors involved (3, 7, $4 + \sqrt{-5}$ and $4 - \sqrt{-5}$) are all "prime" in a very well-defined sense, namely, that each of them is divisible only by itself and by the "units" in this system (such as 1 or –1). In other words, in this domain, it is possible to factorize a number, 21, into prime factors in more than one way. The fundamental theorem of arithmetic does not seem to hold here!

This example is not just surprising in itself, but it also brings to light another related and interesting phenomenon. In the above equation, the prime number 3 divides the number 21 without, however, dividing either of the factors in the product $(4 + \sqrt{-5}) \cdot (4 - \sqrt{-5})$. Proposition VII.30 of the *Elements*, which I mentioned in Section 4.1, states that if a prime number p divides a product of two numbers, $a \cdot b$, then p either divides a or divides b. This proposition continues to hold true when we move from $\mathbb{Z}$ to $\mathbb{Z}[i]$, as well as when we consider some of the generalized domains that Kummer investigated, but it no longer holds in the specific example of the system we are considering here. Thus, whereas since the time of Euclid it had been self-evident to identify the statement of VII.30 as an alternative, equivalent definition of what is a prime number, it now turned out that in certain domains, numbers that cannot be further factorized into smaller factors (i.e., "indivisible" numbers) still do not satisfy VII.30.

After Kummer's surprising discovery, then, a new distinction had to be adopted between two properties that were previously considered to be one and the same. What was previously taken to be a *property* of primes, namely, that expressed by VII.30, now became the *definition*, while being indivisible into further factors became a property that all primes satisfy but that is, in some domains, weaker than being prime. In the domain of numbers $a + b\sqrt{-5}$, as we have just seen, 3 is actually indivisible, but it is not prime, after all. Since the time of Kummer on, then, the preferred definition of a prime number p is not via the property of indivisibility (i.e., that it cannot be divided other than by unity and by itself), but rather via the property that if p divides $a \cdot b$ then p either divides a or it divides b. This definition is adequate for the straightforward case of the ordinary integers $\mathbb{Z}$, and it remains equally adequate and useful for more abstract and intricate systems of generalized integer numbers.

Kummer now faced the following question: can we redefine some of the central concepts of arithmetic so that the main theorems (such as the fundamental theorem), perhaps in modified versions, will continue to hold not just in $\mathbb{Z}$ and $\mathbb{Z}[i]$, but also in all generalized domains of complex integers, including the cyclotomic integers? He developed some basic ideas on the way to answering this question, but he was only partially successful. It was Dedekind who undertook to give a full answer to this question, and, in so doing, he not only addressed the specific question left open by Kummer, but also laid the foundations for a completely new field of mathematical research, namely, the theory of fields of algebraic numbers.

10.3 Fields of algebraic numbers

The new theory of fields of algebraic numbers—inspired also by the important contributions of Leopold Kronecker (1823–1891)—was to become a fundamental domain of mathematical research in the decades to come, and it continues to be so until this day. But what is more important for our account here is that Dedekind's seminal contribution implied a consistent and inspiring implementation of the methodological principles that he preached. This crucial point may be illustrated more concretely by briefly discussing the concept of "ideal." This is the main (but not only) abstract concept that Dedekind developed in order to approach the question of unique factorization in generalized domains of integer numbers, from a broadest possible, unified perspective. It was just the kind of concept that he insisted should be applied in order to successfully address open problems in all fields of mathematics.

Consider first of all the system of rational numbers $\mathbb{Q}$. For any number $\frac{p}{q}$ in this system (with p and q integers, $q \neq 0$) there is a well-defined inverse for both the operations of sum and product, namely $-\frac{p}{q}$ and $\frac{q}{p}$, respectively. Indeed, $\frac{p}{q} + -\frac{p}{q} = 0$ and $\frac{p}{q} \cdot \frac{q}{p} = 1$. Something similar happens in the system of complex numbers, $\mathbb{C}$, where, as we already know, there exists a well-defined operation of division. In the system $\mathbb{Z}$ of integers, however, there is no operation that is the inverse of multiplication. If we want to multiply 3 by some number so as to obtain 1, we need to multiply it by $\frac{1}{3}$, which is not itself an integer. Dedekind introduced the term *Zahlkörper* ("field of numbers," is the English term that eventually became accepted for this concept) to indicate systems such as $\mathbb{Q}$ and $\mathbb{C}$ with inverses for both operations, addition and multiplication. Accordingly, $\mathbb{Z}$ is *not* a field.

Recall now the system of algebraic numbers $\mathbb{A}$, which I mentioned in Chapter 1. It comprises every real number that is a solution of a polynomial equation with rational coefficients. The generalized systems that Kummer investigated, including that of numbers such as $a + b\sqrt{-5}$, and similar ones, comprise specific kinds of algebraic numbers. More interestingly, if we consider all such numbers with a and b rational (rather than only integers as I defined them above), then the systems thus obtained happen to be "fields" of algebraic numbers, in the sense that all the inverses of these numbers for both addition and multiplication are also numbers of this kind. Once again, the system $\mathbb{Z}[i]$ of Gaussian integers is *not* a field, since there are no inverses for multiplication within the system.

Pioneering what would become a main area of mathematical research throughout the twentieth century, Dedekind studied in some detail the properties of these "fields of numbers," or *Zahlkörper*. Among other things, he proved that if, from within the system of algebraic numbers $\mathbb{A}$, we choose any particular set K that itself forms a field (i.e., a set of numbers closed with respect to addition and multiplication as well as to their inverse operations), then there is always a subset D of K that plays within K the same role that $\mathbb{Z}$ plays within $\mathbb{Q}$.

In other words, Dedekind proved that each field of numbers within $\mathbb{A}$ has its own subset of "integer numbers." Within the field of rational numbers $\mathbb{Q}$, for example, it is $\mathbb{Z}$ that plays the role of, well . . . "integer numbers" (and hence the right way to call them is "rational integer numbers," rather than just "integer numbers"). In exactly the same way, within each field K of algebraic numbers, we can find its own set D of "algebraic integer numbers." This is a result of fundamental importance, even though I cannot really explain here why it is so.[1] I must limit myself to state that it is at the heart of Dedekind's successful attempt to extend, to the most general systems of numbers that Kummer had helped introduce, all the basic concepts and theorems of arithmetic (including the fundamental theorem). Moreover, in Dedekind's systematic treatment, the original theorems of ordinary arithmetic reappear as particular cases of these more general ones.

Fields were not the only important concept in Dedekind's theory. In order to complete the task, he needed to add yet another one, which focuses on certain, special collections of algebraic integers and which he called "ideals." Again for reasons of space I cannot explain here what these "ideals" are and what the more general theorems look like,[2] but I want to stress one important point: Dedekind's generalized theorems of factorization are always formulated in terms of *collections of numbers* and of *operations among these collections* (especially among the ideals), rather than in terms of *individual numbers* and of *operations among individual numbers within the*

---

[1] See (Corry 1996 [2004], Section 2.2.2).
[2] See (Corry 1996 [2004], Section 2.2.3).

collections. It is the *collections and their properties as collections* that he considered, and *not the properties of the individual numbers* within the collections. This was the key to Dedekind's approach. According to this approach, then, if one wants to investigate in depth the underlying reasons for the behavior of specific kinds of numbers (for example, the reasons behind the behavior of primes as building blocks of the systems of natural numbers), it is not an individual number that one must consider, but rather a certain *collection of numbers* to which this number belongs.

Dedekind thus suggested that the law of unique factorization may be reformulated, no longer in terms of representing a number in a unique way as a product of prime numbers, but rather in terms of representing an ideal to which that number belongs as a product of "prime ideals." And, as I have already said, the classical version of the fundamental theorem of arithmetic appeared in Dedekind's approach as a particular case of the generalized version. The fundamental theorem of arithmetic in its classic formulation was nothing but the particular instance of Dedekind's generalized theorem of unique factorization, namely, the case of the field of rational numbers (rather than *a general field* of algebraic numbers, as in Dedekind's formulation) and of the ordinary integers (rather than of the algebraic integers of the given field).

Dedekind's approach to solving problems related to numbers by finding adequate, very general, and abstract concepts based on choosing certain sets of numbers was *not* easily accepted among mathematicians in the last third of the nineteenth century. Very few of them were willing, initially, to pay any attention to Dedekind's ideas. But, toward the end of the century, the situation changed dramatically when his approach was adopted as the guideline of a new book in number theory that was to set the agenda for research in the field for decades to come. The book was the *Zahlbericht*, or *Report on Numbers*, published in 1897 by David Hilbert.

David Hilbert (1862–1943) was the most influential mathematician of the early twentieth century, and, like many other prominent mathematicians mentioned thus far, he was (beginning in 1895) active in Göttingen. The theory of numbers was one of the fields where Hilbert left a lasting imprint. *Zahlbericht* was a summary of sorts of the current state of the art in the discipline as seen from the point of view of the algebraic fields of numbers, roughly 100 years after Gauss's *Disquisitiones* had opened this important (then mostly German) research tradition. Under the sweeping influence of Hilbert's book, not only Dedekind's approach and style became dominant in the discipline, but also his basic concepts and his methodological principles. They became a strong focus for many mathematicians, especially the younger ones who began their careers in the early twentieth century.

A peak in this process was achieved in the work of Emmy Amalie Noether (1882–1935), who also worked at Göttingen. In spite of the many difficulties she encountered in her attempt to build an academic career as a woman in early twentieth-century Germany, Noether is considered the main founder and promoter of the new, "structural" approach to algebra. Her ideas were instrumental in changing the face of research in many central fields of mathematics beginning in the 1920s. Noether always cared to stress the decisive importance of Dedekind's influence on her own research, and when praised for her achievements she would invariably repeat with sincere humbleness: "*Es steht alles schon bei Dedekind*" ("All of this is already found in Dedekind").

10.4 What should numbers be?

The theory of the fields of algebraic number is one of Dedekind's most important achievements, but in terms of its content (and certainly of its technical details) it is not *directly* related to the issue of the foundations of the concept of number and its essence. Nevertheless, the overall methodological approach embodied in the theory is indeed highly relevant to our discussion here, since it strongly reflects his entire view of the world of numbers. Let us then go back to Hamilton's idea of defining the complex numbers in terms of ordered pairs of real numbers, which, as already said, around 1857 made a strong impression on the young Dedekind.

In 1854, a couple of years after obtaining his doctorate, Dedekind had presented in Göttingen, as was customary at that time in the German university system, a *Habilitation* lecture, needed to obtain the *venia legendi*, or "right to teach." He chose to discuss the unique character of progress in mathematics, which he described as the mixed product of free creations of the human spirit and the constraints imposed by logical necessity. New concepts and mathematical objects—he said in the talk—continually appear as part of this process, but they must always arise in a natural way from the current state of mathematical knowledge at a given point in time. This is particularly the case with the numbers and the concepts related to them, he stressed. The constraints of developing new concepts, when it comes to numbers, are strongly tied to the need to allow unlimited completion of all arithmetic operations in the known domains of numbers (recall that this is the way in which I introduced the various system of numbers in Chapter 1, so that here we begin to close some open circles). The passage to the irrational and complex numbers had been particularly problematic, and no-one thus far—so Dedekind stated—had yet found the right way to construct them properly, so that all arithmetic operations would be adequately defined.

Against the background of these assertions, it is easy to see why Hamilton's text captured Dedekind's attention so strongly. Hamilton's definition not only answered Dedekind's specific concern about the system of complex numbers. The overall approach implied in it for defining a given system in terms of a previously defined one became the basis for Dedekind's own program for putting the entire world of numbers on stronger foundations than anyone else had done before him. Let us see now in greater detail what were his concerns and how he addressed them.

I begin with the rationals. The idea of a fraction does not seem to raise any particular conceptual problem, and also in our historical review we have seen that its use was never accompanied by the kind of concerns that negative and complex numbers aroused. It is easy to explain the significance of a fraction, say $\frac{p}{q}$ (with $q \neq 0$), in terms of dividing the unit (or a cake) into q equal pieces and then picking p out of the q pieces (here we are assuming that $p < q$, but if $p > q$, then it is easy to extend this idea to that case). It was also known how to multiply two fractions, $\frac{p}{q} \cdot \frac{r}{s} = \frac{p \cdot r}{q \cdot s}$, or to add them, $\frac{p}{q} + \frac{r}{s} = \frac{p \cdot s + q \cdot r}{q \cdot s}$. Justifying the correctness of these operations based on the idea of what is a fraction is also a straightforward matter.

But Dedekind was an extremely careful mathematicians (many of his colleagues, as a matter of fact, considered his work to be unnecessary prolix). Even a seemingly unproblematic issue like this one was for him a matter of concern, because it involved an idea that for him was far from clear. How do we "divide" the given unit? What is, indeed, the unit that we divide? How do we pick the p parts? And many additional questions that did not seem to bother other, more mainstream mathematicians. Moreover, this way of defining fractions based on the idea of rational numbers was for Dedekind too specific for the case at hand. He wanted one single general idea that would apply to defining any new system of numbers based on an existing one.

Using Hamilton's approach for constructing the complex numbers beginning from the real ones, Dedekind suggested to define the rationals based on the existence of the integers, on the use of pairs of such numbers, and on our knowledge of how to add, subtract and multiply them. These three operations are well defined within the integers, but division is not. In some cases, e.g., 7 divided by 3, the result is not an integer. This is the reason why rational numbers need to be defined in the first place, namely, to allow this operation to be consistently defined.

Following the direction taken by Hamilton, then, Dedekind defined a rational number simply as an ordered pair (p, q) of integers (with $q \neq 0$). Neither a unit (or a cake) has to be divided here, nor parts of it have to be picked. A meaningless pair of integers (p, q) is all that is involved. In addition, it is also easy to define in a formal way operations on the system of all such pairs, without having to appeal to any particular meaning of the fraction, of the unit, or of diving the unit into parts. So, addition and multiplication of pairs may be defined, in purely formal terms, as follows:

$$(p, q) + (r, s) = (ps + qr, q \cdot s), \quad (p, q) \cdot (r, s) = (p \cdot r, q \cdot s).$$

Obviously, while on the one hand I may stress that this definition is formal and that it is not based on giving any meaning to the pairs, this particular way of defining is meant to imitate exactly what we already know about operations with fractions (simply rewrite the pairs, e.g., (p, q) as $\frac{p}{q}$, and the translation is evident). Notice, for example, that the pair $(0, 1)$ has an interesting property

$$(p, q) + (0, 1) = (p, q).$$

In other words, the pair $(0, 1)$ plays the same role as the number 0 played in the rationals. We knew this before Dedekind's definition, because the pair $(0, 1)$ is parallel to the fraction $\frac{0}{1}$. Moreover, every pair (p, q) has an "inverse" with respect to the operation thus defined, namely, the pair $(-p, q)$, and this because if one applies the rule for addition, one gets $(p, q) + (-p, q) = (0, 1)$. And something similar happens with the multiplication of pairs as formally defined by Dedekind: $(1, 1)$ plays the role of "1" as a "neutral" element for the multiplication. In addition, for every pair (p, q), one can check that there is an inverse with respect to multiplication, (q, p), because $(p, q) \cdot (q, p) = (1, 1)$ (you may want to check the validity of all these claims simply by applying the definitions of the operations; but notice that when speaking of inverses of rational numbers with respect to multiplication, one always means inverses of *non-zero* rational numbers).

In this way, Dedekind was able to build the entire system of rationals $\mathbb{Q}$ out of pairs of integers in a purely formal way that imitated what Hamilton did in defining $\mathbb{C}$ out of pairs of real numbers. Hamilton's construction allowed the performance in the new system of an operation (extracting a root of a negative number) that could not always be performed in the previously existing one. So did Dedekind: division could not be fully performed within $\mathbb{Z}$, but it always could within $\mathbb{Q}$. And, like Hamilton with the imaginary numbers, Dedekind also did not need to say a word about the *nature* or the *essence* of the rational numbers thus defined, since we need not understand this essence in order to justify the rules for operating arithmetically on the rationals. Rather, it is the other way round: the system of rational numbers *is* the system of pairs of integers, and its arithmetic is fully defined by the formal rules stipulated in the construction. So, while the question how to define the rationals was not really a pressing one (even for Dedekind), this construction provided an useful clue about how to proceed in other cases.

So, the next question was: What about the integers that we have assumed in constructing the rationals? How are they constructed? Dedekind showed how they can easily be constructed just from the assumption that we have the natural numbers at hand and that we know how to add or multiply any two given such numbers. To the contrary, we do not always know how to subtract or divide any given two naturals within the system of natural numbers. We cannot, for instance, subtract 7 from 3 or divide 3 by 2. The task at this point is to construct a new system in which subtraction may be performed without limitations. Recall at this point the many analogies suggested earlier by mathematicians for explaining negative numbers (debts in an account, walking a distance in an opposite direction, etc.). These were obviously devoid of mathematical meaning for Dedekind, and he proceeded to define the new system, once again just with the help of ordered pairs and formally defined operations on them (this time pairs of natural numbers, since this is the system we are assuming as our starting point in this step).

Dedekind simply defined an integer as a pair (m, n) of natural numbers m, n on which the following operations are also defined:

$$(m, n) + (o, p) = (m + o, n + p), \quad (m, n) \cdot (o, p) = (m \cdot o + n \cdot p, m \cdot p + n \cdot o).$$

What he had in mind when defining the pairs in this way is even simpler (though not always immediately transparent) than what he did for the rationals. The pair (m, n) is intended to represent the difference $m - n$. For example, the pair $(7, 3)$ represents what we know as the integer 4, whereas the pair $(3, 7)$ represents the integer –4. In this way Dedekind introduced for every natural number another number which is its "inverse" with respect to addition, but without having to say a word about the meaning of being "negative." All we have are some pairs (m, n) with $m > n$ and some others with $m < n$. Moreover, addition and subtraction between the pairs can be now performed without limitations. Notice, however, that not all pairs have inverses with respect to multiplication. This is indeed what we expect to be the case for integers.

You can check that multiplication of integers as defined above also works correctly with respect to the question of "multiplication with signs." If we take the example of two pairs (m, n) and (m', n'), with $m < n$ and $m' < n'$ (i.e., two "negative numbers") and if we multiply them according to the rule above, then we will obtain a pair where the first element is larger than the second, namely, a "positive" number. (You can check this, for example, by taking a concrete example of two such pairs.) And this is exactly what Dedekind was looking for: to produce, following a way analogous to that of Hamilton for the complex numbers, not just a system embodying the idea of the negative numbers, but also the knowledge that *all of its arithmetic* is correctly defined and all of this without having to appeal to external metaphysical, pseudo-empirical, or analogical explanations, besides those inherently provided by the formal rules of definition.

Before we go on with the construction of the other systems as Dedekind did it, it is worth pointing out an interesting difference between the two constructions presented thus far. We know that it is possible to represent a fraction in more than one way. So, for instance, $\frac{3}{5} = \frac{6}{10} = \frac{-18}{-30} = \ldots$ Only one of these, however, is a *reduced* fraction. Dedekind did not actually speak about "pairs of integers," but rather of "classes of pairs of integers," meaning by this that each rational number equivalently appears as more than one pair, and we only take one representative in a class of equivalent pairs. However, he could not speak of "factoring" fractions as we commonly do, precisely because he did not want to assume anything about the fractions as representing divisions of a unit, etc. He therefore defined the equivalence of two pairs (or of two fractions) in purely formal terms: two pairs (p, q) and (r, s) are equivalent if and only if $r \cdot q = s \cdot p$. For instance, $(3, 5)$ and $(6, 10)$ represent the same rational number, not because $\frac{3}{5}$ and $\frac{6}{10}$ represent the same fraction or because the second can be factorized to obtain the first, but simply because $6 \cdot 5 = 10 \cdot 3$. The pair (p, q) represents all other pairs that are equivalent to it according to this definition.

As for the integers, it is also necessary to define a similar kind of equivalence, since, following what was said above, we could represent the number –4 as either $(3, 7)$ or $(10, 14)$, or $(110, 114)$. The formal rule is that two pairs (m, n) and (o, p) will be considered to be equivalent if and only if $m + p = n + o$. Notice that equivalence of pairs of integers is defined in terms of multiplications with integers alone (because division of two integers is not always defined), whereas equivalence of pairs of naturals is defined in terms of sums of naturals (because the difference of a pair of naturals is not defined *within the naturals* if the first number in the pair is smaller than the second).

10.5 Numbers and the foundations of calculus

At this point, we are in a position where we know, thanks to Dedekind, how to construct the integers $\mathbb{Z}$ out of the naturals, and how to build the rationals $\mathbb{Q}$ out of the integers. We also know, thanks to Hamilton, how to construct the complex numbers $\mathbb{C}$ out of the reals $\mathbb{R}$. So, in order to construct the entire system of numbers, we need to complete only two stages: (1) to construct $\mathbb{R}$ out of $\mathbb{Q}$ and (2) to define the natural numbers in some way as the starting point of all of this edifice (or perhaps to construct them out of something else—although it is not clear at the moment what that could be). This situation is summarized graphically in the following diagram:

$$? \to \mathbb{N} \xrightarrow{(m,n)} \mathbb{Z} \xrightarrow{(p,q)} \mathbb{Q} \xrightarrow{??} \mathbb{R} \xrightarrow{(a,b)} \mathbb{C}$$

The meaning of the pairs over the arrows in this diagram, in the steps that have already been clarified, varies from stage to stage:

$$(m, n) \Leftrightarrow m - n, \quad (p, q) \Leftrightarrow \frac{p}{q} \, (q \neq 0), \quad (a, b) \Leftrightarrow a + bi.$$

These two missing steps turned out for Dedekind to be essentially different from those already performed (and also different from each other), and much more mathematically challenging. They are also the most significant in the story. Still, in sorting them out, Dedekind abided by similar methodological principles as he had done for the easier ones. The "easy" stages that I mentioned above Dedekind just drafted sketchily in some papers later found in his legacy. For the two "difficult" stages of constructing the reals and the naturals, he wrote two separate booklets, which were to become two of his most famous and influential publications. We will return to Dedekind's construction of the natural numbers in the next chapter. The remaining part of this chapter I devote to discussing Dedekind's construction of the irrational numbers, and to the strong link between this construction and the attempts to deal with the foundations of the infinitesimal calculus.

The question of the foundations of calculus had remained open since the seventeenth century. It concerned the ability to define in a precise way the two main concepts and tools of the new calculus created by Newton and by Leibniz, the derivative and the integral, as well as the nature of the relationship between them. The integral embodied a generalized tool to perform calculations related to areas, volumes, lengths of arc and other related issues. The derivative, in turn, allowed addressing in a very efficient way questions related to tangents, rate of change, maxima and minima, and many others. Taken together, both tools had opened since the last part of the seventeenth century the path to a completely new mathematical world and allowed the solution of a great number of important problems in mathematics as well as in physics. Of course, the predecessors of Newton and Leibniz had developed interesting and sophisticated methods to deal with problems related to areas or tangents, maxima and minima, etc., but this was typically done with techniques that concerned individual cases.

The far-reaching innovation implied by the new calculus concerned the ability to see all this enormous variety of problems and mathematical situations as part of a single, large family that could be addressed from a unified perspective, and with far greater success, using the new tools. And in a similar way as with arithmetic and algebra, in the calculus also there is one "fundamental theorem," whose proof becomes a major theoretical challenge. The fundamental theorem of calculus, which is at the center of Newton's and Leibniz's contribution, states that the operations of finding the derivative (or differentiation) and of integration are mutually inverse. This fundamental and far from evident relationship is at the heart of calculus, but, curiously (or should we say "not so curiously," given that we have already seen similar situations in our account here), the fundamental theorem was not satisfactorily and definitely proved before the last third of the nineteenth century. A precise definition of the irrational numbers and a thorough understanding of the differences between the rationals and the reals turned out to be a main ingredient needed to complete this proof. Dedekind's work provided, along with those of some others, the crucial ingredients to reach these insights.

Dedekind's construction of the reals out of the rationals, then, was motivated not just by a general desire to complete the overall, systematic picture of the world of numbers (a desire that is appealing in itself), but also by a pressing need arising from the hard core of contemporary advanced mathematical research (even though, it must be said, not all mathematicians considered themselves as pressed as Dedekind to address this issue, and most of them did not see any problem in the way that the fundamental theorem had been proved by that time). Dedekind based his construction of the irrationals on one general, abstract concept of the kind he strove to introduce at the foundations of any mathematical discipline. In this case, the said central concept was the concept of "cut." It is truly important to see, briefly, how it works.

One of the main concepts on which the entire edifice of calculus is built is the concept of "limit." It is fundamental for defining both the derivative and the integral, and it is the concept that allows, at bottom, the proof of the fundamental theorem in its classical formulation. At the inception of calculus, both Newton and Leibniz came up with concepts of their own that preceded the concept of limit as a foundation for the calculus, and which presented serious conceptual difficulties, each in its own way. "Fluxions," "fluents," "infinitesimals," and other similar concepts that are no longer part of the standard vocabulary of calculus had been at the core of Newton's and Leibniz's approaches, and they were far from being devoid of difficulties.

This did not prevent in any way the calculus from continuing its meteoric development over the eighteenth and nineteenth centuries. New techniques and new tools continued to appear and to be applied to a growing variety of problems, even in the absence of a solid conceptual foundation for the basic concepts. Only a handful of mathematicians were really concerned with this question of the foundations, and it was only in the context of teaching the calculus in institutes for advanced learning and research that more serious attention began to be paid to it. This was particularly the case with the establishment in France, in the aftermath of the Revolution of 1789, of institutions such as the Parisian École Normale Superieur and the École Polytechnique.

One of the leading figures who led the effort and in many senses achieved great success in clarifying the foundations of calculus was Augustin-Louis Cauchy (1789–1857). In the framework of his famous lectures at the École Polytechnique, he formulated in very precise terms the concept of limit, and satisfactorily based the entire calculus on it. (For the benefit of those readers whose background includes at least some undergraduate mathematical courses, I must point out here that it was in these lectures that Cauchy brought to the world the ε–δ language that unforgettably frightens so many beginners.)

A fundamental difference separating Newton's and Leibniz's original approaches to the questions of foundations from Cauchy's was that the former were essentially geometric, whereas the latter was purely arithmetic. Cauchy's concept of limit was based on the properties of the system of real numbers and its arithmetic, as Cauchy and his contemporaries conceived of them. These properties were generally considered to be self-evident and devoid of any problem requiring further clarification. Dedekind though otherwise. As a young teacher at Göttingen, beginning in 1854, he read with careful attention all textbooks on calculus that were available at the time, and he reached the conclusion that all existing proofs of the fundamental theorem were based on geometric analogies that were, in his view, insufficient and logically unacceptable. The laws of arithmetic of the real numbers had never received, in his opinion, adequate explanation and justification, and hence the fundamental theorem of calculus had never been adequately proved.

Dedekind proposed to provide the necessary justification based on a purely arithmetic construction of the irrationals starting from the rationals. This meant for him the need to avoid any kind of reliance on geometric theorems or even on geometric analogies. Rather, he strove to rely only on the laws of logic, and, in line with his own self-imposed methodological principles, on the properties of *collections* of numbers. He devised for this purpose a new, abstract general concept around which he would build his entire theory of real numbers. This was the concept of "cut." Dedekind began to develop the theory in the early years of his career, but, in a typical way, he continually delayed publication out of a desire to perfect and further develop it. Finally, in 1872, the theory was published in a booklet entitled *Stetigkeit und irrationale Zahlen* ("Continuity and Irrational Numbers").

10.6 Continuity and irrational numbers

The year when Dedekind's theory of cuts was published, 1872, was also the year when two other works appeared, suggesting alternative ways to construct the irrationals and to derive their properties from solidly defined foundations. Their aim was also to allow the completion of the proof of the fundamental theorem of calculus. One of these was by Georg Cantor (1845–1918), then a largely unknown mathematician. The other was published by Karl Weierstrass (1815–1897), then one of the leading Berlin mathematicians, and by all accounts the greatest living contributor to the development of the calculus (there was a third mathematician who came at roughly the same time with a theory of the real numbers, the French Mathematician Charles Méray (1835–1911), but his work had little resonance).

Cantor and Weierstrass proposed to build the irrationals with the help of ideas usually associated with the techniques of calculus as known at the time, and above all infinite sequences and series. Dedekind, in contrast, distanced himself from such an approach and constructed the irrationals completely independently from concepts related to the calculus and solely on the basis of the arithmetic of the rationals. While Cantor and Weierstrass focused on a very specific problem in relation to a specific field of mathematics (the calculus), and came up with a theory to address that problem, Dedekind's theory of cuts was certainly useful in addressing that same problem (and Dedekind certainly saw this as an important problem), but at the same time it was part of a much broader and ambitious program for clarifying the idea of number at large.

It may be useful to say at this stage some words about the state of the art, at Dedekind's time, of the proofs appearing in textbooks for the fundamental theorem of the infinitesimal calculus. There is a score of basic theorems that were variously used at crucial stages of existing proofs of the fundamental theorem of calculus. Thus, for instance, there is the "intermediate value theorem" (IVT), which, in one of its versions, can be formulated roughly as follows:

If a "continuous" function takes a negative value at a point a and a positive value at a different point b (with $a < b$), then there exists at least one intermediate point c (with $a < c < b$), such that $f(c) = 0$.

This situation is represented graphically in Figure 10.3.

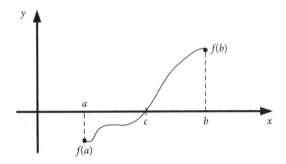

Figure 10.3 The intermediate value theorem (IVT): the continuous function f takes a negative value at a and a positive value at b. Therefore, there exists a number c such that $f(c) = 0$.

Taking this theorem as a starting point, one can go on and build an adequate proof, as already said, of the fundamental theorem (see Appendix 10.2). But among the reasons given in most books to justify the validity of the theorem itself, Dedekind never found a convincing argument that was based on anything other than some loose geometric reasoning. The same was the case with an entire family of additional theorems that appeared in some books. From each of them, the other theorems in the family (including the IVT) could be proved, and hence also the fundamental theorem could be proved. Each author had his own reason to choose this or that theorem as their starting point, but none of the various theorems, in Dedekind's view, was soundly justified.

Another well-known theorem in that family is that named after the French mathematician Michel Rolle (1652–1719), which states that if a continuous function takes identical values at two different points a and b, then there exists at least one point c between a and b at which the tangent to the graph of the function is horizontal. This situation is illustrated in Figure 10.4.

In formulating these two theorems, I have used the term "continuous functions." Roughly (and very imprecisely) speaking, this property is reflected in the graphs of the two theorems, namely, graphs that are drawn "in one stroke," without jumps or gaps. There is a very precise and generally accepted mathematical definition of continuity, based on the concept of "limit." (In the case of Rolle's theorem, one needs an additional condition, namely that the function be differentiable in the segment $[a, b]$; that is to

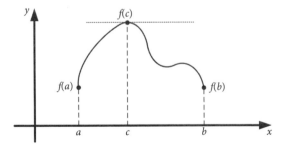

Figure 10.4 Rolle's theorem. The continuous function f takes identical values a and at b. Hence, there exists a number c such that the tangent to the graph of the function is horizontal at c.

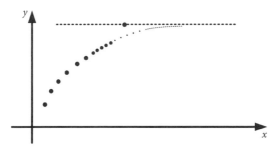

Figure 10.5 Every monotonically growing sequence of numbers for which there exists an upper bound has a limit.

say, the graph must be smooth and without any "cusp". But this condition also can be very precisely defined in terms of "limits.")

For Dedekind, the ability to define the properties of the real numbers independently of geometric considerations became a necessary condition for providing a legitimate proof of any of the theorems just mentioned and hence a solid foundations for the calculus at large. In his lecture notes, and later in his book, Dedekind focused on yet another theorem of the same family, which was widely used in existing texts. This theorem states that every monotonically growing sequence of rational numbers for which there is an upper bound converges to a well-defined limit (see Figure 10.5).

Dedekind's 1872 text on the irrationals ends with a proof of this theorem that is based on the theory of cuts, without any appeal to geometric intuitions or analogies. The proof itself is beyond the scope of this book, of course,[3] but I do want to present here very briefly some of the main ideas behind Dedekind's definition of cut, which are highly relevant to all of our discussion here.

As stated in the title of the book, what Dedekind sought to elucidate was the secret behind the idea of *continuity*. We know that the rational numbers satisfy a property nowadays called "density," namely, that between any two given rational numbers a and b there is always another rational number c (see Figure 1.4). It is clear that also the irrationals are a dense system of numbers, but the important question that Dedekind asked was whether there is some other property that the real numbers satisfy while the rationals do not. In order to find out, Dedekind focused on the straight line and asked what is the property that turns the line into a "continuous" mathematical entity. And the answer he gave to this question is surprisingly simple. Indeed, if we take any point P on the line, we will see that this point divides the line into two parts, A_1 to the right of P and A_2 to its left (see Figure 10.6). These two parts satisfy three simple properties:

(L1) A_1 and A_2 are disjoint sets (i.e., they have no common points).
(L2) If we take the union of A_1 and A_2 and add to it the point P, then we obtain the entire straight line.
(L3) Any point a_2 belonging to A_2 is always to the left of any point a_1 belonging to A_1.

---

[3] See (Dedekind 1963, pp. 24–30).

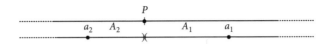

Figure 10.6 The straight line as a continuum.

Dedekind concluded that these three properties fully encapsulate the secret of the continuum, and he took a further, ingenious step that we find in all of his important works (this is also what he did when defining the ideals): he turned these *properties* into a *definition*, in this case the definition of a "cut." That is to say, given any system of numbers, and assuming that a relation of order is defined on it, a cut on this system is a pair (A_1, A_2) of subsets of elements in that system such that the following three conditions hold:

- (C1) A_1 and A_2 are disjoint.
- (C2) A_1 and A_2 taken together yield the entire system.
- (C3) Any number a_2 belonging to A_2 is always smaller than any number a_1 belonging to A_1.

So, how is it that this definition (which, on the face of it, implies no dramatic discovery, and in particular says nothing specific about the irrationals) helped Dedekind construct the reals out of the rationals? Well, in order to explain this, let us see how this idea works in the system $\mathbb{Q}$ of rational numbers. We take on $\mathbb{Q}$ a first cut, (A_1, A_2), defined as follows:

$$A_1 = \{x \in \mathbb{Q} \mid x > 2\}, \quad A_2 = \{x \in \mathbb{Q} \mid x \leq 2\}.$$

In words, A_1 is the set of all rational numbers greater than 2, whereas A_2 is the set of all rational numbers less than or equal to 2. It is very easy to check that the pair (A_1, A_2) satisfies conditions (C1)–(C3) above, and hence the pair is a cut of the rationals.

We take now a second cut of $\mathbb{Q}$, (B_1, B_2), defined as follows:

$$B_1 = \{x \in \mathbb{Q} \mid x > 0 \,\&\, x^2 > 2\}; \quad B_2 = \{x \in \mathbb{Q} \mid (x > 0 \,\&\, x^2 \leq 2 \text{ OR } x \leq 0\}.$$

This definition is less straightforward than the previous one, but, after giving it some further thought, one realizes that it just involves two sets of rationals that are easy to identify: B_1 is the set of all positive rational numbers with square greater than 2, whereas B_2 is the set of all positive rational numbers with square less than or equal to 2, taken together with all the negative rationals. Again, it is also easy to check here that the pair (B_1, B_2) satisfies conditions (C1)–(C3) and hence it is a cut of the rationals. But the different ways in which the two cuts, (A_1, A_2) and (B_1, B_2), are defined give rise to some questions. For one thing, why not define (B_1, B_2) in the following, equivalent, and seemingly simpler way, which imitates the definition of (A_1, A_2):

$$B'_1 = \{x \in \mathbb{Q} \mid x > \sqrt{2}\}, \quad B'_2 = \{x \in \mathbb{Q} \mid x \leq \sqrt{2}\}?$$

Well, the point is that although in appearance (B_1, B_2) and (B'_1, B'_2) define the same cuts, there is a fundamental difference between them. It is the following: the definition of (B'_1, B'_2) assumes the existence of a number $\sqrt{2}$, which is an irrational number. But remember that Dedekind's explicit intention was to construct the irrationals *out of the rationals*, and hence the number of $\sqrt{2}$ is actually "inexistent" before we have "constructed" it. "Constructed" here means, constructed, for instance, with the help of a cut of rationals. And this is indeed what we do with the help of (B_1, B_2), which is defined purely in terms of the rationals and of the relation of order defined on them. Dedekind did not want to assume the existence of the irrational number $\sqrt{2}$, but rather to make it arise, as it were, out of the existing rational numbers. His main insight, then, was that there are two fundamentally different kinds of cuts that one can define on the system of rationals:

- **Cuts of type 1** behave like (A_1, A_2), i.e., they are defined by a number, 2, which was part of $\mathbb{Q}$ to begin with.
- **Cuts of type 2** behave like (B_1, B_2), i.e., like (A_1, A_2), they satisfy properties (C1)–(C3), but, unlike it, they cannot be defined by straightforwardly separating $\mathbb{Q}$ into two by a number already existing in $\mathbb{Q}$.

The difference between these two kinds of cuts is illustrated in Figure 10.7, where we see how the point corresponding to (B_1, B_2), which was not originally in $\mathbb{Q}$, is created, *ex nihilo*, by the cut itself.

Having defined cuts, Dedekind proceeded to indicate how their addition and multiplication could be consistently defined. He showed that these two operations have well-defined inverses, namely, subtraction and division. Moreover, he also defined a relation of order among cuts. I do not give here the details of these definitions, which are not particularly difficult or interesting anyway.[4] The important point about them is that after all this process had been completed, Dedekind took the following simple step: he defined the system of real numbers $\mathbb{R}$ as the set of all cuts that one can define on the system $\mathbb{Q}$ of rational numbers, together with the four operations just defined and the relation of order. The existence of two types of cuts indicated above shows that some cuts do not correspond to already existing rational numbers. Hence, the set of all cuts involves some new entities, and these are the irrational numbers (like (B_1, B_2)).

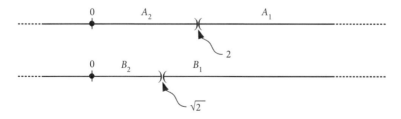

Figure 10.7 Two different kinds of cuts of rationals.

---

[4] See (Dedekind 1963, pp. 21–24).

These newly created irrational numbers, together with the already existing rationals, yield the real numbers. And the system thus created happens also to be an "ordered field." It is a "field" in the sense of Dedekind because the operations of addition and multiplication have inverses for all numbers. And it is "ordered" because the order defined on the numbers "behaves well" in relation with the operations. For example,

$$\text{if } a < b, \text{ and } c \text{ is any real number, then } a + c < b + c.$$

One would certainly expect these properties to hold in any system intended to represent $\mathbb{R}$.

To all of this, Dedekind added a last and very important step, by constructing the cuts of $\mathbb{R}$ in a way similar to that by which he had defined the cuts of $\mathbb{Q}$. The process is similar but the outcome quite different: it turned out that there are no cuts of reals of type 2. In other words, no cut of $\mathbb{R}$ creates an element not already existing in $\mathbb{R}$. And, in this way, finally, and strange as it may sound, Dedekind's concept of cut uncovered the secret of the continuum: a system of numbers is continuous if all its cuts are of type 1. i.e., if they *can* be defined with the help of numbers already existing in it. The system $\mathbb{Q}$, therefore, is not continuous, since, as we have seen, some of the cuts are of type 2. The system $\mathbb{Q}$, in contrast, as Dedekind showed, is continuous according to this definition.

The most surprising aspect of Dedekind's point of view is that, contrary to the natural expectation (and also contrary to the approach followed by Weierstrass and Cantor in their constructions of $\mathbb{R}$), continuity turns out not to be a property defined by considerations of proximity or *distance* (or, as mathematicians call them, "topological considerations"), but rather by considerations of *order* alone! Cuts are defined purely on the basis of order among the numbers in the system, and continuity, in Dedekind's approach, is defined purely on the basis of the cuts.

Dedekind's ideas were not easily accepted in the mathematical community, and initially they were not considered either interesting or important. This had also been the case with his introduction of the ideals, and now it happened with the cuts. An interesting example of this appears in a rather critical letter that Dedekind received from his friend Rudolf Lipschitz (1832–1903) shortly after the publication of *Continuity and Irrational Numbers*. Lipschitz had been among the few who immediately understood the importance of ideals when Dedekind began to publish his research on fields of algebraic numbers. Now, Lipschitz wrote that he could not understand the point of Dedekind's new work. In his view, Dedekind was using an idea, "cut," that was complex, obscure and hard to grasp, in order to explain another idea, "magnitude," that was simple and generally understood.

Dedekind disagreed, of course, and in order to stress the difficulties he had found in the current state of mathematical knowledge, he asked his friend for a sound justification of the commonly accepted operations with irrational numbers, such as in the case $\sqrt{2} \times \sqrt{3} = \sqrt{6}$. In Lipschitz's opinion, the ideas found in Euclid's *Elements* provided more than enough justification for this. He was referring, of course, to the contents of Book V, namely, Eudoxus' theory of proportions. We have already seen that ever since the works of Stevin in the seventeenth century, the idea of proportion was adopted, either explicitly or implicitly, as the basis for a new and broader concept of number seen as an abstract quantity. Among the quantities that were meant to be covered by this view were also the irrationals, which appeared as being represented by incommensurable quantities, as we saw, for example, in the case of Newton. Lipschitz, then, was partaking of a widely accepted view among his contemporaries (and indeed, well into the twentieth century, many continued to understand Dedekind's cuts as completely equivalent to Eudoxus' proportions. See Appendix 10.1).

But Dedekind was very clear in his opinion that even if we accept the assumption that irrational numbers can be identified with incommensurable quantities, there is no way to show that such quantities exhaust the *entire range* of *all* possible real numbers, and hence a coherent and comprehensive theory is needed to provide solid foundations for the system of those somewhat elusive entities. Moreover, Dedekind found no explicit reference in the *Elements* to the idea of the continuity of space (even though it seems reasonable to assume that Euclid and anyone using his books took for granted the continuity of space). And, for Dedekind, the definition of continuity was the vital and necessary task required for properly defining the real numbers.

Dedekind pointed out to Lipschitz that mathematicians typically admit the validity of the multiplication of roots, $\sqrt{2} \times \sqrt{3} = \sqrt{6}$, as correct without a sound justification. They use an algebraic identity, $(a \times b)^2 = a^2 \times b^2$, whose validity for the irrational numbers, he believed, had never been established—never, that is, before he introduced his cuts and defined real numbers and their arithmetic accordingly. It is true, he wrote to Lipschitz, that when we square both sides of the identity $\sqrt{2} \times \sqrt{3} = \sqrt{6}$, we obtain the result $2 \times 3 = 6$. But then, since the latter result is acknowledged as correct, mathematicians conclude that the initial identity, $\sqrt{2} \times \sqrt{3} = \sqrt{6}$, is by necessity itself correct. Dedekind found this reasoning faulty, for the simple reason that while the algebraic identity can be proved correct for rational numbers, no solid argument had been put forward for the conclusion that it must be equally true for the irrationals as well. Identifying irrational numbers with abstract quantities, Dedekind emphatically asserted, could in no way be considered a rigorous proof of that mathematical claim. Thus, Dedekind concluded, his theory of cuts was the first serious and successful attempt to provide such a proof. Moreover, he thought, his theory was the first successful account of the idea of continuity on a purely arithmetical basis and without the assistance of geometric or physical analogies. Dedekind's definition of the real numbers as cuts of rationals indeed signified a significant milestone in the history of numbers, and one of the climatic points of our story here. In order to complete the picture, we need to turn to his definition of the natural numbers, and this we will do in the next chapter.

Appendix 10.1 Dedekind's theory of cuts and Eudoxus' theory of proportions

It took some time before the ideas related to Dedekind's cuts began to attract the attention of his fellow mathematicians. As already stated, his friend Lipschitz even claimed that the concept involved no innovation at all. Others claimed (and perhaps this is also what Lipschitz had in mind) that Dedekind's cuts are equivalent to Eudoxus' proportions. The basis behind this claim is that if we are given any four integers a, b, c, d that are proportional in the sense of Eudoxus, $a:b :: c:d$, and if we write this in terms of fractions, $a/b = c/d$, then these two fractions define one and the same cut in the sense of Dedekind. In other words, according to this claim, cuts and rational fractions define the same system of numbers. In order to be able to judge by ourselves the validity of this claim, I will present now a proof of the said equivalence, which has some interest in itself. Let us take a fraction a/b and let us define two sets of rationals A_1 and A_2 as follows:

$$A_1 = \{\text{all rationals } n/m \text{ such that } n/m \leq a/b\},$$
$$A_2 = \{\text{all rationals } n/m \text{ such that } n/m > a/b\}.$$

It is easy to see that these two sets A_1 and A_2 constitute a cut of the rationals as defined by Dedekind, since the following conditions hold:

(1) A_1 and A_2 are disjoint.
(2) A_1 and A_2 taken together yield the entire set of rationals.
(3) Any rational number n/m belonging to A_2 is always smaller than any number n'/m' belonging to A_1.

One can likewise define a corresponding cut C_1, C_2 with the help of c/d as follows:

$$C_1 = \{\text{all rationals } n/m \text{ such that } n/m \leq c/d\},$$
$$C_2 = \{\text{all rationals } n/m \text{ such that } n/m > c/d\}.$$

And what needs to be proved now is that the two fractions are equal if and only if the corresponding cuts are equivalent, or, symbolically,

$$a/b = c/d \quad \Leftrightarrow \quad (A_1, A_2) = (C_1, C_2).$$

To prove this latter condition, let us take any rational number n/m belonging to the set A_1. We know, by definition of A_1, that $n/m \leq a/b$, and hence $nb \leq ma$. (We will assume, for the sake of simplicity, that in order to preserve the direction of the sign $\leq$, we take both m and b to be positive or both negative integers. If this were not the case, the proof could easily be corrected with some additional steps.) Now, if we take Eudoxus' definition in its symbolic formulation as presented in Appendix 3.2, then it follows from the proportion $a:b :: c:d$ that, given any two integers n and m, one has

$$ma >=< nb \quad \text{iff} \quad mc >=< nd.$$

In our case, we have $nb \leq ma$, and hence it follows from the above condition that $nd \leq mc$. Hence, by definition of C_1, it follows that any rational number n/m that is in A_1 is also always in C_1. In the language of set theory, one can say that we have proved that A_1 is a subset of C_1, or $A_1 \subseteq C_1$. In a similar way, one can prove the opposite inclusion, namely, $C_1 \subseteq A_1$. And these two inclusions taken together mean, simply, that $A_1 = C_1$. And then, in exactly the same way, one can also prove that $A_2 = C_2$. We have thus proved that the two cuts, (A_1, A_2) and (C_1, C_2), are one and the same, or, in other words, that two fractions that represent the same rational number generate the same cut.

But does this mathematical fact imply that Dedekind's definition of cuts did not involve a real conceptual innovation? Of course not. For one thing, even if it is true that every proportion yields an identity between two cuts, we need to remember that not every cut of rationals is generated by such a proportion. Indeed, we have already seen that there are cuts that are not generated by rationals, such as the cut

$$B_1 = \{x \in \mathbb{Q} \mid x > 0 \ \& \ x^2 > 2\}, \ B_2 = \{x \in \mathbb{Q} \mid (x > 0 \ \& \ x^2 \leq 2) \ \text{OR} \ x < 0\}.$$

But, beyond this important technical point, the conceptual difference that really interests us here, in the light of the historical account we have been looking at, is that for Eudoxus (as well as for all his followers up to the seventeenth century at least), neither ratio nor proportion should be seen as numbers in the first place. In particular, Eudoxus' theory provides no indication about how to define the arithmetic operations on ratios or proportions. Dedekind, in contrast, was very explicit on the aim of his theory as a way to provide, for the first time in history, a full foundation of the system of real numbers with all of its arithmetic. The same foundational idea should provide a natural answer to the question why this is a "continuous" system of numbers while the rationals are not. This is the main point that Lipschitz failed to see in his correspondence with Dedekind.

Appendix 10.2 IVT and the fundamental theorem of calculus

In this appendix, I provide a sketch of a possible proof of the fundamental theorem of calculus that is based on the use of the IVT. I use the IVT in the following, slightly more general, version: if a continuous function f takes two different values $f(p)$ and $f(q)$ for $p < q$ (and we can actually assume, without loss of generality, that $f(p) < f(q)$), and if K is any real number such that $f(p) < K < f(q)$, then there exists at least one intermediate point c ($p < c < q$) such that $f(c) = K$. The basic idea is represented in Figure 10.8. I also

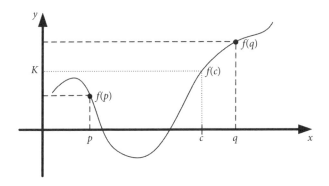

Figure 10.8 A more general version of the IVT.

invoke another, closely related, theorem, known as the extreme value theorem (EVT), which states that if a function f is continuous in a closed interval $[a, b]$, then f attains a maximal and a minimal value in that interval.

Now, the derivative of a function f at a point x is given by the expression

$$\frac{d}{dx} f(x) = \lim_{h \to 0} \frac{f(x+h) - f(x)}{h}.$$

The fundamental theorem of calculus states that the processes of differentiation and integration are inverses of each other for any function f satisfying certain basic conditions of derivability and integrability. Using the above notation, the theorem states that for any point x in the said interval $[a, b]$, we have

$$\frac{d}{dx} \left(\int_a^x f(t) \, dt \right) = f(x).$$

To prove this identity, we use the definition of the derivative and obtain

$$\frac{d}{dx} \left(\int_a^x f(t) \, dt \right) = \lim_{h \to 0} \frac{\int_a^{x+h} f(t) \, dt - \int_a^x f(t) \, dt}{h}.$$

Then, using the basic rules of integration, the right-hand side of this equation becomes

$$\frac{d}{dx} \left(\int_x^{x+h} f(t) \, dt \right).$$

Now, it follows from the EVT that for each value of h ($h > 0$) in the closed interval $[x, x+h]$, f attains a maximal value M and a minimal value m. Clearly then, over that same interval, the integral of f, representing the area under the graph of f between x and $x+h$, is bounded by two rectangles (as indicated in Figure 10.9):

$$h \cdot m \leq \int_x^{x+h} f(t) \, dt \leq h \cdot M.$$

It follows that

$$m \leq \frac{1}{h} \int_x^{x+h} f(t) \, dt \leq M.$$

But the values m and M are the values of the function for some numbers p and q in the interval $[x, x + h]$, say, $m = f(p)$, and $M = f(q)$. Hence,

$$f(p) \leq \frac{1}{h} \int_x^{x+h} f(t) \, dt \leq f(q).$$

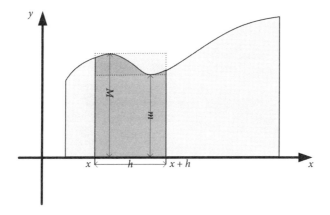

Figure 10.9 A proof of the fundamental theorem of calculus.

Now, it is right at this point that the IVT enters the picture, since it warrants the existence of a value c, between p and q, and hence belonging to the interval $[x, x + h]$, such that

$$f(c) = \frac{1}{h} \int_x^{x+h} f(t)\, dt.$$

Given the continuity of f, $\lim_{h \to 0} f(c) = f(x)$, and hence, clearly,

$$\frac{d}{dx}\left(\int_a^x f(t)\, dt\right) = \lim_{h \to 0} \frac{\int_a^{x+h} f(t)\, dt - \int_a^x f(t)\, dt}{h}$$

$$= \lim_{h \to 0} \frac{1}{h} \int_x^{x+h} f(t)\, dt = \lim_{h \to 0} f(c) = f(x),$$

which is what we wanted to prove.

APPENDIX 10.2

CHAPTER 11
Exact Definitions for the Natural Numbers: Dedekind, Peano and Frege

After formulating the theory of cuts and constructing with it the irrationals out of the rationals, Dedekind had completed the full picture of the various systems of numbers whereby each of them is constructed from the previous one in purely arithmetical terms, starting from the system of natural numbers:

$$\mathbb{N} \xrightarrow{(m,n)} \mathbb{Z} \xrightarrow{(p,q)} \mathbb{Q} \xrightarrow{(A_1, A_2)} \mathbb{R} \xrightarrow{(a,b)} \mathbb{C}.$$

This picture stresses in a precise way the basic role of the natural numbers as the foundation of all the universe of numbers. It also stresses that the successive creation of the various systems does not require any kind of reliance on geometric or physical analogies. Dedekind's ideas thus embody the manner in which, by the end of the nineteenth century, the entire debate on the nature of numbers had converged into a debate about the nature of natural numbers alone. And this debate was essentially different from all previous debates held on this issue, which we have been discussing throughout this book. A sharp departure from the previously prevailing, ages-old perspective became possible in the wake of specific contributions by mathematicians such as Hamilton and Dedekind to foundational questions. At the same time, it was also the consequence of more global changes in the mathematical discipline at large, such as those related to developments in the calculus and in geometry.

Dedekind's 1888 *What are numbers and what should they be?* was written precisely as an attempt to complete the entire job of understanding the architecture of the universe of numbers, by dealing with the question of the foundation of the system of natural numbers. As in his previous research on the foundations of the systems of algebraic and real numbers, here also Dedekind looked for a single, abstract and general concept from which the entire arithmetic of the natural numbers could be derived. Why is $1+1 = 2$ a true statement and what is the source of its validity? Why are the addition and multiplication of natural numbers commutative and associative operations? These are the kinds of questions for which Dedekind wanted to provide a mathematically sound answer. Indeed, it may sound strange that similar questions were not asked before (at any rate, they were never properly answered), and at the same time it may sound strange that Dedekind was asking them now.

But, of course, the point is not that Dedekind raised doubts about the *truth* of all known arithmetical results. What Dedekind was after was the *source* and possible justification of those truths. More specifically, Dedekind wanted to address such questions not by way of philosophical debate, as many of his predecessors did (even though he did approach the issue from a well-defined philosophical perspective), but rather to do so by following a mathematical approach that was similar to the one he had followed in establishing the basis for all other numerical systems (i.e., the real numbers with the help of cuts, and the algebraic numbers with the help of ideals). In other words, Dedekind insisted in defining the system of natural numbers and their rules of arithmetic by relying on a single general concept that would clarify this issue once and for all.

11.1 The principle of mathematical induction

Dedekind's work with the algebraic and the real numbers revolved around a central question to be elucidated in each of them (unique factorization and continuity, respectively). Also, in the case of the natural numbers, there was one central issue around which Dedekind focused his attention: the validity of the principle of mathematical induction. We need to devote some attention to this point before going on with the discussion. The principle of mathematical induction is an extremely important and efficient tool for proving general theorems in mathematics in which there appear indices or variables that are natural numbers. Take, for example, the well-known formula for the sum of all natural numbers from 1 to n:

$$1 + 2 + 3 + \ldots + n = \frac{n \cdot (n+1)}{2}.$$

This formula is typically proved by induction, which means that we need to follow a two-step procedure:

(I1) Prove that the formula is valid for $n = 1$. That is, prove that

$$1 = \frac{[1 \cdot (1+1)]}{2}.$$

(I2) Assume that the formula is valid for $n = k$, and from here deduce that it is valid for $n = k + 1$. That is, assume

$$1 + 2 + \ldots + k = \frac{[k \cdot (k+1)]}{2},$$

and from this derive

$$1 + 2 + \ldots + k + (k+1) = \frac{[(k+1)\cdot(k+1+1)]}{2}.$$

Now, every reader with a modicum of algebraic knowledge and skills can easily complete step (I2). The fundamental question I want to call attention to, however, is the following: Why is it that mathematicians universally accept that if we follow steps (I1) and (I2), then the theorem in question is considered to have been proved for all natural numbers? Are there any solid grounds on which we can give a sweeping justification to this way of reasoning with natural numbers? In Section 5.9, I mentioned Gersonides as a rather isolated, early example of someone using elaborate inductive arguments for proving general arithmetical statements, though in a less structured manner than that known nowadays. Inductive arguments started to be commonly used in Europe in the sixteenth century and gradually became generally accepted. But never before Dedekind was the question systematically asked (and much less properly addressed) of what are the grounds for accepting this kind of reasoning as mathematically legitimate.

Dedekind's approach to defining the system of naturals and to providing a justification for the use of the principle of induction was based on looking at this system as built around a principle of *order*. For Dedekind, the natural numbers were in the first place a system of *ordinals*, whereas their properties as *cardinals* were derivatives of the former. This was the first radical innovation in Dedekind's approach to the natural numbers. We have already seen that also in elucidating the question of the continuum, Dedekind's approach was based on looking at continuity as a property based on order rather than on distances between natural numbers. In the case of the natural numbers, he followed the very simple idea that the property defining the number 4, for example, is that it comes right after the number 3 and just before the number 5. The fact that 4 also represents, in some way, a quantity, Dedekind saw as deriving from this more basic, ordering essence of the naturals.

The single, general and abstract concept around which Dedekind suggested to construct his entire ordinal theory of the natural numbers was the concept of "chains." Rather than explaining this concept and how it fulfilled the aim that Dedekind had in mind for it, I will turn first to a brief description of a second, contemporary work that addressed the same issue from a different, yet closely related, perspective, namely, that provided by the so-called Peano postulates. Peano's point of view makes it much easier to understand the issues at stake, and it is also the one that has remained in use to this day. I will return to Dedekind and his work later on.

11.2 Peano's postulates

Giuseppe Peano (1858–1932) was a talented Italian mathematician who worked in the town of Torino at the turn of the twentieth century. Beyond the broad spectrum of mathematical topics that occupied him, he devoted great energy to developing and spreading a universal, artificial language of his invention, *Latino sine flexione*, along the tradition of Esperanto, which flourished among European intellectuals at the time. Peano's sensitivity for matters of language was also manifest in his mathematical work, and he made significant contributions to questions related to notation and formalization of mathematical argumentation.

In 1889, Peano published a pamphlet entitled *Arithmetices principia, nova methodo exposita* ("The principles of arithmetic, presented by a new method") containing the basic postulates of arithmetic that have been associated with his name ever since. Peano was explicit in declaring that Dedekind's *What are numbers . . .* was a crucial influence in consolidating the last stages of his own work. Indeed, the two mathematicians defined the natural numbers in ways that are equivalent, even if this equivalence is not evident on first sight. Peano's approach was easier to understand and more transparent, and it was the one that finally became of standard use.

Peano's definition of the natural numbers starts by assuming the existence of a set N whose elements are called "numbers." At the outset, nothing is known about these "numbers." Their properties will be determined by a set of abstractly defined postulates. Accordingly, at this point, we should identify neither N with $\mathbb{N}$ nor these "numbers" with the natural numbers about which we know—or think that we know— what our already existing, basic knowledge of arithmetic tells us. All the properties of numbers, I stress, are to be derived from the postulates alone. In the original version, Peano presented nine postulates, but these were later simplified and consolidated into just four. I will present now the postulates as it is standard to do so nowadays. I begin with the first three, which can be formulated as follows:

(P1) With every number n in N, we can associate another number in N, which is called the "successor" of n and which is denoted n'.
(P2) If $n \neq m$, then $n' \neq m'$ (i.e., different numbers have different successors).
(P3) There is one, and only one, number in N satisfying the property that it is not the successor of any number in N. This number is denoted 1.

If we now take the system of ordinal numbers as we knew it before we ever heard of Peano for the first time, it is easy to see that these ordinals do indeed satisfy the three postulates above (you may want to check that this is indeed the case). But the question immediately arises as to whether these three conditions *suffice* to define this system of ordinals alone or whether additional conditions may be necessary to attain a full characterization of them. On the face of it, one might think that other systems possibly exist that satisfy the same three conditions but essentially differ in other respects from the system of ordinals as we know it. In more technical terms, this question may be phrased by asking whether there are two or more non-equivalent models of the system. And this question is of particular importance when we think about the principle of induction: in any system such as defined by postulates (P1)–(P3), is the principle of induction justified? Can it be used on safe grounds as a basis for proof in arithmetic?

Both Dedekind and Peano, each in the framework of his own theory, understood that there are, indeed, systems that satisfy the three postulates but that are *substantially different* from that of the natural numbers, and they specifically differ when it comes to a situation where the principle of induction is invoked (for a detailed example, see Appendix 11.1). It turned out, then, that the only way to ensure the validity of the principle of induction in a system of ordinals satisfying (P1)–(P3), and to turn it into a system of postulates all of whose models are equivalent, is to add a fourth postulate embodying the principle itself. There is simply no way around this.

Peano's own formulation of this fourth postulate, the postulate of induction, is close to the following:

(P4) If A is a collection of elements of N that satisfy the two conditions
  (P4-a) 1 is in A
    (1 being the special element defined in P3),
  (P4-b) if n is A, then also n' is in A
    (i.e., if the element is in A, then its successor is also in A),
  then the collection A is actually the entire set N.

To see in what sense this postulate embodies the principle of induction the way we usually know it, let us assume that we have a certain formula that we want to prove to hold true for all natural numbers. Then, all we need to do is to take A to be the collection of all natural numbers that satisfy this formula. We then prove that 1 is in the collection A (i.e., we prove that 1 satisfies the formula), and then we prove that if a certain number n is in A, then its successor n' is also in A (i.e., we prove that if the formula is valid for n, then it is also valid for its successor). This being the case, then (P4) warrants that the collection A is indeed the entire collection of the natural numbers.

These four simple postulates (actually: three very simple, and a fourth one which is less straightforward) define in a sound way, then, the system of natural numbers known to us,

$$\mathbb{N} = \{1, 2, 3, 4, 5, 6, 7, 8, \ldots\},$$

and, in particular, they provide the foundation for a justified use of the principle of induction as a legitimate way of proving claims about number theory. And, moreover, this justification—this is a main point arising here—can only be provided axiomatically! As conceived by Peano and Dedekind, these postulates should define a system of the kind nowadays called categorical, namely, that no system that is essentially different from $\mathbb{N}$ satisfies these postulates. But this last claim immediately raises a concern, since it is very easy to suggest systems that seem to differ from $\mathbb{N}$, but do satisfy the postulates. Indeed, you can check that the postulates are satisfied in the following eight systems, $\mathbb{N}_1$–$\mathbb{N}_8$, by assuming in each sequence that the successor of each element is the one written immediately to its right, and that the special element defined in (P3) is the left-most element in each list:

$\mathbb{N}_1 = \{0, 1, 2, 3, 4, 5, 6, 7, 8, \ldots\}$,
$\mathbb{N}_2 = \{-2, -1, 0, 1, 2, 3, 4, 5, 6, 7, 8, \ldots\}$,
$\mathbb{N}_3 = \{4, 5, 6, 7, 8, 9, 10, 11 \ldots\}$,
$\mathbb{N}_4 = \{1, 10, 100, 1000, 10000, 100000, 1000000, \ldots\}$,
$\mathbb{N}_5 = \{-1, -2, -3, -4, -5, -6, -7, -8, \ldots\}$,
$\mathbb{N}_6 = \{1, 1/2, 1/3, 1/4, 1/5, 1/6, 1/7, 1/8, \ldots\}$,
$\mathbb{N}_7 = \{\&, ?, v, \aleph, 535, 0, \alpha, -3, \ldots\}$,
$\mathbb{N}_8 = \{!, @, X, \$, \%, 8, \&, *, \ldots\}$.

So, what are we actually saying here? If we can write down so many different systems that satisfy the postulates, do or do not, after all, the Peano postulates (P1)–(P4) achieve the aim of categorically defining the natural numbers? Well, the answer is simple, and it is crucial that we understand this point if we want to grasp the essential idea behind the theories of Dedekind and Peano. The symbols appearing in each of the above systems, $\mathbb{N}_1$–$\mathbb{N}_8$ (and, as a matter of fact, also in $\mathbb{N}$), have no *meaning* in themselves by virtue of which we place them in the sequence. They do not even represent an abstract quantity that allows us to order them according to their values. It is rather the other way round: the meaning that we can properly attach to any of those numbers *derives from their position* in the sequence.

In the system $\mathbb{N}$, the symbol 1 indicates the special element 1 that postulate (P3) defines for *N*. So, the number 1, under this view, does not embody the idea of a unit, or of a single apple, or of a centimeter, or of any other thing that has to do with quantity. Rather, it indicates an element with a well-defined position within a sequence that is ordered in a very specific way: it is that element in the system which is not a successor of any other element in the sequence. And exactly the same thing can be said about the symbol 0 in the sequence $\mathbb{N}_1$, the symbol 4 in the sequence $\mathbb{N}_3$, the symbol –1 in the sequence $\mathbb{N}_5$, or the symbol & in the sequence $\mathbb{N}_7$. For the same reason, all the symbols appearing in the sequences in the third place (i.e., respectively, 3, 2, 0, 6, 100, –3 , 1/3, v, X) represent exactly the same idea, namely, the element that is the successor of the successor of the element that is not a successor of any element. It is true that for centuries we have been following a convention in which that element is indicated by the symbol 3, but this is no more than that, a convention, i.e., a *contingent* result of a historical process. What is a *necessary* mathematical fact here is what concerns the *position* of the number in the sequence and whatever can be derived from that, but by no means the specific symbol traditionally used to represent it, in this case "3".

Whatever symbol we may choose to use, it is only the ordered structure of the natural numbers as a whole, that bestows its meaning on each of the symbols adopted according to its place in the sequence. This structure is fully, and indeed it is *defined* by the four postulates. As a matter of fact, each of the above sequences represents one and the same abstract idea that can be schematically represented as follows:

$$\mathbb{N} = \{T, T'', T''', T'''', T''''', T'''''', T''''''', T'''''''', \ldots\}.$$

In other words, the system of natural numbers is built in its entirety if one takes a number T (i.e., a number that is successor of none), and then takes the successor of T, the successor of the successor of T, and so on indefinitely. And, as in the other two theories of numbers previously developed by Dedekind (algebraic and irrational), it is the *entire structure* of the system of natural numbers that endows with meaning each of its individual members, rather than the other way round.

But the real significance of this way of defining the system of the natural numbers is that it allows us not just to grasp its essence as a sequence of ordinals and to realize that the only way to justify the principle of induction is by postulating it axiomatically. Rather, it also allows us to reconstruct the entire *arithmetic* of the natural numbers as we know it. Indeed, both Dedekind and Peano defined the addition of natural numbers in the system they defined, by means of the following two rules:

(**PA1**) For any number n, we have $n + 1 = n'$.

(**PA2**) For any two numbers m and n, we have $(m + n)' = m + n'$.

It is important to stress that also in this case we have a definition that is less trivial than it may first appear. Notice that addition is reduced to an operation that depends on the ordinal character of the natural numbers rather than on any concept of quantity. In particular, in rule (PA1), the symbol 1 is not taken to represent a unit in the quantitative sense, but rather that special element which is not the successor of any other number. Actually it works just the other way round: based on defining addition via (PA1) and (PA2) (and on defining multiplication similarly with the help of axioms as will be seen below), we can finally also give a *quantitative* meaning to the element 1, which the theory defined in the first place only as an ordinal. In other words, it is not that the successor n' is obtained by adding to n the quantity 1, but, rather, that when we add to n the number that is not successor of any number (i.e., 1), then we obtain n'.

Peano's axiomatic definition of addition, then, is a formal recipe to be followed without considering the meaning of the terms involved and certainly not the symbols used to denote them. Using terms that are common nowadays, we can see rules (PA1) and (PA2) as instructions that can be programmed in a machine and that the machine should follow every time that it encounters a situation of the kind $n + m$. The machine performs the instruction without having any idea of the meaning of the numbers involved, and certainly not of their meaning as quantities. If we are working with the usual symbols for $\mathbb{N}$, then, faced with the expression $6 + 1$, for example, the machine will act according to instruction (PA1), and will accordingly look in the list of ordinals for the successor of 6, considered as a sequence of meaningless symbols. Hence, the operation yields as a result the symbol 7. Following instruction (PA2) is slightly more complex, but it allows (in a way that most readers of this book will find easy to follow) a "recursive" calculation of an expression such as $2 + 3$, as follows:

- 3 is the successor of 2; hence $3 = 2'$. Hence $2 + 3 = 2 + 2'$.
- By (PA2), $2 + 2' = (2 + 2)'$. Hence, if we know what is the successor of $(2 + 2)$, we will have the result of $2 + 3$.
- But 2 is the successor of 1; hence $2 = 1'$. Hence $2 + 2 = 2 + 1'$.
- By (PA2), $2 + 1' = (2 + 1)'$. Hence, if we know what is the successor of $(2 + 1)$, we will have the result of $2 + 2$.
- And now, by (PA1), $2 + 1 = 2'$. Hence $2 + 1 = 3$ and $(2 + 1)' = 3' = 4$.
- But $(2 + 1)' = 2 + 1'$, and hence $2 + 1' = 4$, and hence $2 + 2 = 4$, and hence $2 + 2' = 4'$, and hence $2 + 3 = 5$.

In this way, it is possible to calculate any other addition of two natural numbers. Still, this definition is not intended to have us follow the process in order to make sure that, for example, $8 + 7 = 15$, or to be able to reconstruct the addition table. Rather, defining addition in this way was meant to provide a sound *justification* for the entire arithmetic of the natural numbers.

This arithmetic was well known before Peano and Dedekind, of course, but what they were both looking for was an axiomatic foundation that need not rely on any kind of geometric or physical analogies to support or explain it. Moreover, the postulates would also provide a solid ground for justifying the general rules underlying the arithmetic of natural numbers, such as $a + b = b + a$, which had previously been justified only on the basis of general and rather unconvincing arguments.

Going back to our examples above, we should notice that the instructions (PA1) and (PA2) work in exactly the same way on the other lists that satisfy the Peano postulates while using other symbols. For example, in $\mathbb{N}_1$, we would have $1 + 2 = 4$. Let us see why, step by step (and remembering to follow the instructions blindly while completely ignoring the meaning usually given to the signs involved):

- $1 + 2 = 1 + 1'$. Hence, by (PA2), $1 + 2 = (1 + 1)'$.
- $1 + 1 = 1 + 0'$. Hence, by (PA2), $1 + 1 = (1 + 0)'$.
- But, by PA1, $1 + 0 = 1'$. Hence $1 + 0 = 2$ (remember that in $\mathbb{N}_1$, 0 is the special number defined in (P3)).
- Thus, $1 + 1 = (1 + 0)' = 2' = 3$.
- Finally, $1 + 2 = (1 + 1)' = 3' = 4$.

You may want to try to work out in detail some additional examples in the other systems, such as the following:

$$\mathbb{N}_2: -1 + 0 = 2;$$
$$\mathbb{N}_3: 5 + 6 = 8;$$
$$\mathbb{N}_4: 10 + 100 = 10000;$$
$$\mathbb{N}_5: -2 + -3 = -5;$$
$$\mathbb{N}_6: 1/2 + 1/3 = 1/5;$$
$$\mathbb{N}_7: ? + v = 0;$$
$$\mathbb{N}_8: @ + X = \%.$$

These examples help stress the non-quantitative character of the operation. In N_3, for example, we say that $5 + 6 = 8$. This has nothing to do with the quantities that these symbols usually are taken to represent, but rather with the places in which they appear in the sequence $\mathbb{N}_3$: the symbol in the second place added to the symbol in the third place yield the symbol in the fifth place. More generally, $T'' + T''' = T'''''$. This is true for all the sequences above.

In order to complete the picture, it is necessary to define in a similar way the operation of multiplication of two natural numbers, and this is done by recourse to the following twofold definition:

(PM1) For any number n, we have $n \cdot 1 = n$.
(PM2) For any two numbers m and n, we have $(m \cdot n)' = m \cdot n + m$.

As in the case of addition, I stress that these are two formal rules that are not based on providing meaning to 1 as the quantity associated with the "unit" or on providing any quantitative meaning to the other symbols. We may find the value of the multiplication $1 \cdot 2$ in $\mathbb{N}_1$, for example, simply by following rules (PM1) and (PM2). Here, we must remember, of course, that the special element defined in (P3) and that in (PM1) appears as "1" is the element "0" because it is not the successor of any element in $\mathbb{N}_1$. We obtain the following result:

- $1 \cdot 2 = 1 \cdot 1'$. Hence, by (PM2), $1 \cdot 2 = 1 \cdot 1 + 1$.
- But $1 \cdot 1 = 1 \cdot 0'$. Hence, by (PM2), $1 \cdot 1 = 1 \cdot 0 + 1$.
- But, by (PM2), $1 \cdot 0 = 1$. Hence $1 \cdot 1 = 1 + 1$.
- Hence, $1 \cdot 2 = 1 \cdot 1 + 1 = 3 + 1 = 3 + 0' = (3+0)' = (3')' = 4' = 5$.

You are encouraged to try and perform some specific multiplications using symbols in the sequences $\mathbb{N}_1$–$\mathbb{N}_8$, while applying the formal rules for attaining the results. The results do depend, in fact, on the position of the symbols in the sequence, as in the example above; namely, the product of the third symbol in the sequence times the second symbol in the sequence always yields the sixth symbol in the sequence.

11.3 Dedekind's chains of natural numbers

As already mentioned, Dedekind formulated his own definition of the system of natural numbers in a manner that did not essentially differ from that of Peano, except for the way of presentation and the important fact that Dedekind's definition was part of his broader plan to provide a systematic account of the entire universe of numbers. In the final account, it was Peano's and not Dedekind's definition that became the standard and better known one, but as part of our historical account it is important to briefly comment on some of the specific methodological aspects of Dedekind's approach. These aspects are tightly connected to the sweeping process of adoption of the abstract idea of sets as an overall foundation for all of mathematics at the turn of the twentieth century, an idea that allowed mathematicians at the time to address, in a systematic and mathematically sound matter, the elusive and all-important idea of the infinite, as we will see below.

Dedekind built his theory of the natural numbers around the basic idea of "mapping" between two collections. Roughly, one can say that a mapping from the set A to the set B embodies the ability to relate, in some way, to each element in A one and only one clearly determined element in B. If, in addition, this mapping satisfies the property that any two different elements of A are mapped to two different elements of B, then one says that this is a one-to-one mapping (Figure 11.1).

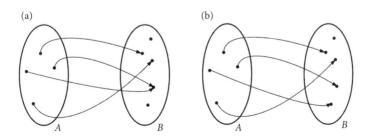

Figure 11.1 A mapping from A to B (a) and a one-to-one mapping from A to B (b).

In these terms, rather than saying, as Peano did in postulate (P1), that every number in N has a successor, Dedekind spoke of the existence of a mapping of N to N, which only implicitly means that we can associate a successor with each number. Likewise, rather than saying, as Peano did in postulate (P2), that two different numbers have different successors, Dedekind stated that the said mapping of N to N is one-to-one. Finally, the existence of a number that is not a successor of any other number, as in Peano's (P3), was formulated by Dedekind also in terms of the mapping: "there is one and only one element in N such that no element of N is mapped to it." Whenever a mapping with those three properties exists, then Dedekind would say that the set N is a "chain." "Chain" (*Kette*) is the abstract concept, defined in terms of a set with certain general properties, by means of which Dedekind attempted to address in this context the central constitutive question of the system of natural numbers (namely, the question of the legitimacy of induction). In this sense, chains play in Dedekind's theory of natural numbers a role similar to that played by ideals and cuts in his previous theories.

One important point to stress concerning the way in which these chains were used by Dedekind to formulate the principle of induction is that he showed that every system of elements that satisfies his postulates is an infinite system, and, moreover, that it is an infinite system of a very specific kind, which he called a "simply infinite system." But what does it mean that a system of numbers is infinite in a special way, to begin with? The very idea of giving true mathematical significance to such a statement was a groundbreaking innovation of Dedekind. He intended to emphasize that there is essentially only one kind of model satisfying the postulates (or, as we would say nowadays, that the set of axioms is categorical). But, in doing so, he also opened the way to a systematic, mathematically sound debate on infinity and infinite sets. In the history of mathematics, it is typically the name of Georg Cantor that is associated with the rise of the theory of infinite sets, and this association is, of course, not totally unjustified. However, the real story is a much more involved one, comprising a very fruitful, if sometimes tense, critical dialogue and collaboration between these two highly original mathematicians, Dedekind and Cantor. I will return to that part of the story, in the next and last chapter of this book.

At this point, I would like to stress yet another important point in Dedekind's approach to defining the natural numbers axiomatically, and that concerns his overall methodological and philosophical stance in all things mathematical. Dedekind conceived the universe of numbers as a successive extension of given numerical realms: from naturals to integers, from integers to rationals, and from rationals to reals. When dealing with the naturals, then, we are at the true underlying foundation of all of this universe of numbers. And at this point, it is necessary to look for a foundation of the system of naturals that lies outside the realm of numbers itself.

Dedekind saw this foundation as sustaining not just the universe of numbers but, indeed, of mathematics at large, since he saw geometry and the infinitesimal calculus as being built on arithmetic alone. His idea was, then, that this overall, underlying foundation should make use of notions that are even more general than numbers, that pertain to the operations of the mind, and without which human thinking would be inconceivable. He wanted to develop a concept of numbers that would be completely independent of the intuitions of space and time, and that would derive directly from the laws of pure thought. In other words, besides the precise technical aspects of the kind of arithmetic he was seeking to develop from the simplest ideas, Dedekind was working from a well-defined and elaborated philosophical agenda.

Dedekind's philosophical agenda, later to be known as "logicism," considers numbers and arithmetic to be direct products of the basic notions and insights of logic. It opposes, on the one hand, any empiricist perspective that sees mathematics as arising from the evidence of the senses, and, on the other hand, any position close to that of Kant, to whom the idea of "the pure intuitions of space and time" was fundamental (and we have already seen that Hamilton tried to follow a Kantian approach in defining the natural numbers). The idea of number derives, in Dedekind's approach.

11.4 Frege's definition of cardinal numbers

Side by side with Peano's and Dedekind's definition of natural numbers as a system of ordinals, some contemporary attempts focused on developing a view where cardinals are seen as the main concept, with ordinals appearing as only ancillary to them. These attempts are associated with the work of Gottlob Frege (1848–1925), in the last third of the nineteenth century, and with its criticism in the hands of Bertrand Russell (1872–1970). These are two important thinkers whose mathematical contributions, very much as was the case with Descartes, appear tightly interwoven with deep philosophical ideas that were also relevant for other fields of thought such as logic, language and metaphysics. This is not the place, of course, to describe in any detail the development of their ideas, but it is important to stress some of these inasmuch as they impinge on their understanding of natural numbers as cardinals. I devote the last part of this chapter to this point.

Frege started by trying to make sense of the basic idea that numbers are used to enumerate finite collections. What does it mean, for instance, that a collection has, say, 5 elements? The idea of "five," one can say, embodies a property that is common to all collections comprising that many elements. Indeed, it is the only property that is common to all of them. On the face of it, this may appear as a *petitio principi*, that is, as assuming beforehand what needs to be explained. Indeed, how can we define the number 5 on the basis of a property that depends on that number itself? But, when we take a closer look, we realize that no such problem arises here, because it is indeed possible to determine that two collections have the same number of elements, without having to count either of them. Think, for example, of a group of students entering a classroom to take chairs. We do not need to count either students or chairs to decide if there will be enough chairs for all students, or if, on the contrary, there will remain some chairs unused or students without a chair. Clearly, if no students remain unseated and no seats remain free, then the number of students and of chairs is the same. If some students remain unseated, we will know that there are more of them than of chairs, and of course also the other way round.

This is the basic sense in which Frege attempted to derive the notion of cardinal from the notion of one-to-one correspondence between two collections, and this is also the way in which we can establish that there may be a common property to two such collections without even knowing how to count them. Given two collections *A* and B such that a one-to-one correspondence can be established from A to B and from B to A, then we say that the collections are equivalent or "equipotent," or that they have the same power or cardinality. So, given for example the collections $\{a, b, c, d, e\}$ and $\{2, 4, 6, 11, 7\}$, we can certainly say that the idea of 5 represents a property that is common to both, and the idea of 5 arises from their equipotence, rather than the other way round. Frege, who shared with Dedekind the idea that logic is the necessary, sound basis for building all of arithmetic (even though they followed different paths when implementing this idea), put forward a very concrete program in order to build the naturals on the basis of the idea of equipotence of sets. I will describe some of these ideas now.

Consider a collection characterized by the fact that no element belongs to it. Such a collection is called the "empty set" and is usually denoted by the symbol $\varnothing$. On first reading, this idea of a collection without elements may sound strange to some readers. But, strictly speaking, there is no logical problem with such a definition, as can be seen when formulating it in the following way:

$$\varnothing = \{x | x \neq x\}.$$

The meaning of this formal expression is that in the collection $\varnothing$ we just gather together all those elements x characterized by the property that $x \neq x$. There is no logical impediment in defining the collection in this way, and yet it is clear that no element x can exist that fulfills the property. This being the case, we define the number zero, and we denote it, as usual, by 0, as the only property that is common to all collections that are equipotent with $\varnothing$. In this case, there is only one such collection, $\varnothing$ itself of course, but the definition is well formulated and it holds well. Moreover, as will now be seen, this is the way in which we will proceed with the next numbers in the sequence.

The next step defines, in purely logical terms (that is, without appealing to our senses or to some special kind of "intuition" or geometric analogy), a new collection $\{\varnothing\}$:

$$\{\varnothing\} = \{x | x = \varnothing\}.$$

Here we are speaking of the collection of all elements x defined by the property of being equal to $\varnothing$. Of course, this collection contains the element $\varnothing$ alone, and $\varnothing$ is an entity that was properly defined in the previous step, in purely logical terms as well. Notice that this is the only element in the collection, but we do not want to make appeal to the idea of "only one element," or anything of the sort that reminds us of the idea of 1, since the only things defined thus far in the procedure are $\varnothing$ and 0. But we can now define also the number 1: it is the property that is common to all collections that are equipotent with $\{\varnothing\}$. Notice that in the two steps taken thus far, we have made no appeal to any idea of quantity or of order. The only notions invoked are those of collection, or set, and equipotence, as well as rules of logic. Of course, we can put ourselves in the position of an "external" observer monitoring this process, and who already knows what is a natural number and what its arithmetic looks like. Such an observer who has previous knowledge of the arithmetic of the natural numbers would certainly realize the direction in which we are heading and why we are taking each of the successive steps in the process.

But this does not mean that this previous knowledge is ever part of the process itself. For instance, we know what all collections with the same cardinality as {∅} look like. Among them, we find {X}, {%}, and {8}, as well as many others. Our previous knowledge tells us that all of these collections have one element, but, in the process defined by Frege, all we are required to know is that we can establish the one-to-one correspondence between them and {∅}, abstractly and without having to invoke our previous knowledge. Thus, this process has allowed us to introduce, thus far, the numbers 0 and 1, based only on the rules of logic and the idea of a one-to-one correspondence.

It is now clear what the next step is in the construction. We can define a further set, using elements already constructed, namely, the set {∅, {∅}}, which comprises the elements ∅ and {∅}, and we accordingly define the number 2 as that property which is common to all collections that are equipotent with {∅, {∅}}. And clearly, in order to define the number 3, we will introduce a further set {∅, {∅}, {∅, {∅}}}. The process then continues indefinitely in the same way, thus creating all natural numbers as cardinals.

As with Dedekind and Peano, all of this effort would be futile if it did not also allow a definition of the arithmetic of the natural numbers and its order relation. This is easily done. If we want to know the value of, say, 2 + 3, we need to choose some set belonging to the class of collections defining 2 and a second set belonging to the class of collections defining 3. We impose a simple, rather unrestrictive, condition on the two sets chosen, namely, that they have no common elements. To give an example: we may choose $\{a, x\}$ and $\{1, \#, p\}$, and then create the set that is the union of those sets (namely, a set that contains all elements of those two sets taken together): $\{a, x, 1, \#, p\}$. The sum 2 + 3 is then the class of collections of which $\{a, x, 1, \#, p\}$ is a representative. We can then imagine looking up in the table of classes that we have constructed in the process above, and we will readily see that this class is the one denoted by 5.

We have not learnt anything new in terms of a result that was previously unknown, 2 + 3 = 5, but we have learnt that there is a procedure that can be justified in purely logical terms and that provides a solid ground for this result. This procedure is independent and is not a generalization of any kind of empirical evidence such as "if we have 2 apples in the basket and add to it 3 more apples, we obtain 5 apples." Moreover, using this definition of addition of cardinals, we can find a justification to other properties such a commutativity and associativity.

The product of two natural numbers can be defined in a similar way using a different operation with sets, known as the "Cartesian product." Based on this definition, the basic properties of the product (namely, commutativity and associativity) follow, once again, from the rules of logic alone. The Cartesian product of two sets A and B yields a third set $A \times B$ whose elements are ordered pairs (a, b), where the first element a is in A and the second element b is in B. If we want to know the value of, say 2·3, we need to choose some set belonging to the class of collections defining 2 and a second set belonging to the class of collections defining 3. Once again, to take an example, we may choose $\{a, x\}$ and $\{1, \#, p\}$ and then create a new set whose elements are all possible ordered pairs built with these elements: $\{(a, 1), (a, \#), (a, p), (x, 1), (x, \#), (x, p)\}$. The product 2·3 is the class of collections of which this latter set is a representative.

Finally, it is easy to also define order among the cardinals, using the idea of a subset of a given set. The number 2, for example, is smaller than the number 3, because if we take a set of the class representing 3, say {a, b, c}, then there is at least one subset of it, say {b, c}, that is in the class that represents 2. In this way, the idea of ordinal is easily derived from that of a cardinal. I skip all the details of how this is done, but most readers will not find any difficulty in doing so by themselves.

The definition of number on the basis of equipotence of sets allows us to reconstruct the entire system of the natural numbers with their arithmetic based solely on the rules of logic. One of the basic ideas implicitly used by Frege in his construction is the so-called axiom of comprehension or principle of comprehension, stating that any property that we can think of defines a set. For example, the property of being a prime number clearly defines a set (namely, the set of all prime numbers), and so does the property of being a country in Latin America. Dedekind also, in his own construction of the irrationals, relied on this seemingly innocuous assumption. It turned out very soon, however, that the principle is problematic and indeed that it leads to a logical paradox. This was indicated by Russell in a famous letter to Frege soon after reading the book in which Frege had presented his construction of the naturals. Russell also supported the view that arithmetic should be seen as arising purely from logic, and yet, using the paradox that has been since then associated with his name, he noticed a fatal flaw in Frege's system. The paradox is briefly discussed in the closing paragraphs of Chapter 12.

I would like to conclude this chapter by raising an additional, minor point that comes up when comparing the definitions of Peano and Dedekind with that of Frege. In the system of ordinals defined above with postulates (P1)–(P4), the first number in the sequence of ordinals is 1. In Frege's definition, we started with 0. This difference poses no real problem, and we can move from one to the other with minor modifications. Recall the case of the passage from naturals to integers according to Dedekind's approach: as part of it, Dedekind added the number 0 without much ado. In fact, in Peano's original work, and in many versions of it found in modern texts, the special element defined by postulate (P3) is identified as 0, rather than as 1. The postulates defined addition and product can be modified accordingly. In the end, this is a matter of convention, and the system can easily be modified according to whether one wants to begin with 1 or with 0. After making some adjustments in the rules of arithmetic, it makes little difference, for the broader frame of ideas involved here, whether or not we take 0 to be a natural number.

Appendix 11.1 The principle of induction and Peano's postulates

In this appendix, I present a system that satisfies Peano's first three postulates but does not satisfy the principle of induction. The existence of such a system implies that the principle of induction is logically independent of the first three postulates, and hence a system of postulates meant to define the natural numbers with all of their basic properties must explicitly include the principle as a further postulate, as Peano and Dedekind indeed realized.

Consider the system K defined as the union of two sets: $K = N \cup T$. Let the set N be the set of all natural numbers as we know it. Let T be defined as the set of all rational numbers less than 0 (we write them as $\frac{-p}{q}$). Now, for any element x in K, we define a "successor" x' as follows:

(K1) If x is an element in N then $x' = x + 1$.
(K2) If x is an element in T, say $\frac{-p}{q}$, then $x' = \frac{-2p}{q}$.

Under this rather artificial definition, we would have, for instance, $7' = 8$ or $\left(\frac{-5}{7}\right)' = \frac{-10}{7}$. It is completely straightforward to verify that K satisfies the three Peano postulates (P1)–(P3). Indeed,

(P1) Every element in K has a successor: because every element in K is either in N or in T, and in each case the successor is a well-defined element in K.
(P2) Any two different elements x and y in K have different successors x' and y': indeed, if both x and y are in N, then clearly $x + 1$ and $y + 1$ are different; if both x and y are in T, then clearly $2x$ and $2y$ are different; and, of course, if x is in N and y is in T, then, given the definition of successors here, x' is in N and y' is in T, and hence they are different.
(P3) There is only one element in K that is not successor of any other element: because in N there is one such element, 1, which is not a successor of anything, and there is no such element in T (notice that if we take any element of the form $x = \frac{-p}{q}$, then it is always the successor of another element in T, namely, $\frac{-p}{2q}$).

All that remains to be done now is to see that the system K does *not* satisfy the principle of induction (as defined in (P4) above). In order to do so, we take a subset A of K, and check that it satisfies the two properties that define (P4), but, nonetheless, it does not constitute the entire set K as stipulated by (P4). This will happen if we chose the set A to be N. Indeed, in this case, it is easy to see that

(a) The element 1, the element that is not successor to anything in N, is in A (this is obviously true: 1 is in A, because A is N).
(b) If an element x is in A, then its successor x' is also in A (again, this is obviously true for the same reason, namely, that A is N).

But, nonetheless, this subset A, which satisfies conditions (a) and (b), is not the entire K, contrary to the requirement of (P4). Hence, the system K as defined here satisfies (P1)–(P3), but does not satisfy (P4).

CHAPTER 12

Numbers, Sets and Infinity. A Conceptual Breakthrough at the Turn of the Twentieth Century

In Chapter 11, we discussed two attempts to address the question of the foundation of arithmetic as they appeared in the work of Dedekind and Frege. The idea of a set and the possibility of figuring out a one-to-one correspondence between two given sets were central to both of them. But it is with Dedekind, more than with anyone else, that we find a focused and systematic appeal to the idea of a set as a truly fruitful perspective for introducing abstract concepts in *mathematics in general*.

The new concepts that Dedekind was speaking about were by no means vague abstractions discussed for their own sake. Rather, he was looking for tools intended to address fundamental problems in advanced research in different disciplines, and especially (but not only) those associated with numbers. He introduced chains as the basic concept meant to provide the solid foundation for the arithmetic of the naturals, ideals as the basic tool for dealing with the question of unique factorization in the fields of algebraic numbers, and cuts as the basic tool for constructing the reals out of the rationals and for clarifying the riddle of continuity.

The line of thought promoted by Dedekind in introducing these kinds of concepts proved momentous in their overall influence on mathematical practice since the turn of the twentieth century, and this influence is related to an important general principle: whereas each of these three concepts involves focusing on certain sets of numbers defined by a few general properties, the problems they are meant to address are discussed in terms of *operations among the sets rather than among the individual numbers of which the sets are composed.*

This strong statement about Dedekind and sets may sound strange to some readers for whom the name of Georg Cantor is the one that they more commonly associate with the introduction of set-theoretical methods in mathematics and with the origins of the systematic handling of sets at the end of the nineteenth century. Of course, this association is justified in itself, but it very often diminishes or altogether ignores the fundamental contribution of Dedekind in this regard. The personal and professional relationship between Dedekind and Cantor was very complex, and its details are beyond the scope of this book.[1] They both focused their mathematical efforts on the idea of a set, but they came to it from different perspectives and with different motivations, and in this sense they mutually complemented each other's scopes of interests. Roughly speaking, one may say that while Cantor focused more narrowly on sets themselves as a topic of autonomous mathematical research and developed a systematic treatment of infinite sets, Dedekind saw the enormous potential of using the abstract idea of a set as a basic unifying language for many mainstream mathematical fields. The two met quite by chance for the first time in 1872, after both had independently started developing their respective ideas on sets and the infinite. They maintained an intensive correspondence for many years (with intermittent periods of total disconnection), which allowed them to follow each other's progress. In this chapter, I devote some attention to the idea of an infinite set, and to the way in which this idea allowed for a fundamental breakthrough in our conception of numbers at the turn of the twentieth century.

12.1 Dedekind, Cantor and the infinite

Dedekind's definition of the natural numbers as a "simply infinite system" appeared in 1880 in his book *What are numbers . . .* He began to develop this idea, however, much earlier than this, when he was a young teacher at Göttingen in the 1850s. That is also when he began to think about the question of the continuum and the irrational numbers. At that time, he identified the one-to-one correspondence as the key to understanding the idea of mathematical infinity, as he noticed that we can define a one-to-one correspondence between an infinite set of numbers and one of its proper subsets. The typical example of this is the correspondence between the natural numbers and the set of even numbers: $1 \rightarrow 2, 2 \rightarrow 4, 3 \rightarrow 6, \ldots$ Clearly, in a finite set, we can never establish a one-to-one correspondence between the set and a part of the set (try to figure this out with some examples).

Dedekind was certainly not the first to notice this difference when thinking about the mathematical infinite. In the seventeenth century, this is mentioned in various ways by mathematicians such as Galileo, Wallis and Bonaventura Cavalieri (1598–1647). The most interesting insights on this topic, however, appeared later, in the work of the Bohemian Catholic priest Bernard Bolzano (1781–1848), who also dealt in original ways with the question of the foundations of calculus prior to Cauchy. In a posthumously published book, *Paradoxes of the Infinite*, Bolzano showed explicitly how to establish a one-to-one correspondence between an infinite set of real numbers and a proper subset of it, as indicated in Figure 12.1.

---

[1] See (Ferreiros 1999, pp. 172–176).

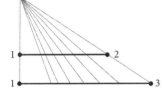

Figure 12.1 A one-to-one correspondence between a set of real numbers (between 1 and 3) and a proper subset of it (the numbers between 1 and 2). The dotted lines connect the points of the shorter line with those of the longer one, such that for each point in the upper one there corresponds one and only one in the bottom line, and vice versa.

But the reason why Bolzano brought up this example was in order to indicate that the idea of mathematical infinity is, indeed, a paradoxical one, precisely because it leads to this kind of situation that he deemed self-contradictory and hence unacceptable. In particular, the infinite appears to contradict the basic and seemingly obvious principle explicitly stated in Euclid's *Elements*, and unquestionably respected by generations of mathematicians, according to which "the whole is greater than the part." In Bolzano's view, the transgression of this principle in the example above, and in many other examples as well, just emphasized the known difficulties commonly associated with the idea of infinity in mathematics. Both Cantor and Dedekind were strongly impressed by the examples of Bolzano, but rather than seeing them as paradoxical and as providing a basis for banning the infinite from serious mathematical research, they saw in them a challenging gateway to a possible breakthrough in this fascinating and often mysterious question.

Dedekind, in particular, reacted to Bolzano's ideas in a way that was a hallmark of all of his mathematical innovations: what in Bolzano appears as a *property* of infinite sets (and, in this case, what Bolzano conceived as a paradoxical property) Dedekind took to be a general and abstract *definition*. In this way, the seemingly paradoxical was forgotten and from now on Dedekind had a straightforward definition of an infinite set:

An infinite set is a collection A that contains a proper subset S such that a one-to-one correspondence can be established between A and S.

As simple as that! And we can see that this definition works well in the simple example of the naturals, by taking the even numbers as the required proper subset and then establishing the following one-to-one correspondence:

$$
\begin{array}{cccccccc}
1 & 2 & 3 & 4 & 5 & \ldots & n & \ldots \\
\updownarrow & \updownarrow & \updownarrow & \updownarrow & \updownarrow & & \updownarrow & \\
2 & 4 & 6 & 8 & 10 & \ldots & 2n & \ldots
\end{array}
$$

As requested, each number in the upper row is associated with one, and only one, in the bottom row, and vice versa. Dedekind focused on this kind of property when he defined the natural numbers as a simply infinite system. Cantor, on the other hand, focused more closely on the question of the possibility of having systems of numbers of *different* cardinalities. In the first place, the question arises concerning the cardinalities of $\mathbb{N}$ and $\mathbb{Z}$. At first sight, one might conclude that these two sets have the same cardinality simply because both of them are, well, infinite. But if we try to be more precise and base the concept of cardinality on that of one-to-one correspondence, then an apparent difficulty arises, given that, in the standard way of ordering it, $\mathbb{Z}$ displays two "infinite tails," one to the left, and the other to the right:

$$\mathbb{Z} = \{ \ldots -4, -3, -2, -1, 0, 1, 2, 3, 4, \ldots \}.$$

This apparent difficulty, however, is easily overcome, once we adopt a different ordering in which only one infinite tail appears. Indeed, it is easy to conceive the one-to-one correspondence of naturals and integers in the following terms:

1	2	3	4	5	6	7	...
↕	↕	↕	↕	↕	↕	↕	
0	1	−1	2	−2	3	−3	...

And what happens when we move to the next step and ask ourselves about the cardinalities of $\mathbb{N}$ and $\mathbb{Q}$? The previous example of $\mathbb{Z}$ shows that to establish a one-to-one relationship of any set of numbers with $\mathbb{N}$, one may think of a reordering of the entire set in such a way that we only see one infinite tail, and this ordering will in and of itself define the correspondence. In the case of $\mathbb{Z}$, it was not particularly difficult to find this reordering. In the case of $\mathbb{Q}$, it is obviously a more complicated task. Both Dedekind and Cantor found such a reordering after some thought, and it yields a nice, surprising trick that is often repeated in other, similar circumstances. To figure out this reordering, we must first of all arrange the rational numbers in an array of infinite rows and columns, as displayed in Figure 12.2. It is easy to see that this arrangement indeed comprises all elements of $\mathbb{Q}$, since every number of the form $\frac{p}{q}$ is found with certainty in row q, in the column under number p. Notice that some numbers do appear in more than one place, because of the non-simplified fractions such as −2, −4/2, 6/−3, etc. But, as we will see, this will not be an obstacle in defining the desired one-to-one correspondence. At any rate, in this arrangement, we have not yet established the desired reordering with only one infinite tail. Indeed, we still have an infinite number of infinite tails, both to the right and to the bottom, so that we still seem to be far away from accomplishing our aim. But here is where the original insight comes in, and it is based on drawing diagonals in the direction right-top to left-bottom, starting from the right-uppermost corner. These diagonals grow gradually, and eventually they cover the entire infinite arrangement (Figure 12.3). The

0	1	−1	2	−2	3	−3	4	−4	5	−5	...
½	−½	2/2	−2/2	3/2	−3/2	4/2	−4/2	5/2	−5/2		...
⅓	−⅓	2/3	−2/3	3/3	−3/3	4/3	−4/3	5/3	−5/3		...
¼	−¼	2/4	−2/4	3/4	−3/4	4/4	−4/4	5/4	−5/4		...
⅕	−⅕	2/5	−2/5	3/5	−3/5	4/5	−4/5	5/5	−5/5		...
⋮	⋮	⋮	⋮	⋮	⋮	⋮	⋮				

Figure 12.2 The rational numbers arranged in an infinite array of rows and columns.

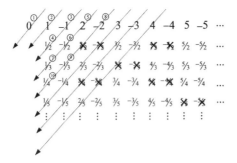

Figure 12.3 A one-to-one correspondence between the rational and natural numbers.

one-to-one correspondence is now built along the diagonals, skipping those numbers that appear repeatedly along the way (such as $\frac{2}{2}$, which appears first as 1).

The diagonal arrows Figure 12.3 help us define the correspondence between the two sets, and this is what we obtain:

1	2	3	4	5	6	7	8	9	10	...
↕	↕	↕	↕	↕	↕	↕	↕	↕	↕	
0	1	−1	$\frac{1}{2}$	2	$\frac{-1}{2}$	$\frac{1}{3}$	−2	$\frac{-1}{3}$	$\frac{1}{4}$	...

The ability to prove in this mathematically precise way that $\mathbb{N}$ and $\mathbb{Q}$ are two sets with the same cardinality, even though the former is a proper subset of the latter, was the first non-trivial contribution of the new approach to the idea of an infinite set. Still, the notion itself—that these three systems have "the same number of elements"—is perhaps not very surprising, precisely because, in the naïve idea of the mathematical infinite, we are speaking of three sets, $\mathbb{N}$, $\mathbb{Z}$ and $\mathbb{Q}$, that are *infinite* in the first place, and hence why should they be other than equivalent in their "sizes" (whatever meaning we give to this term for infinite sets prior to Cantor and Dedekind)?

But, if we go along with this naïve point of view, we could likewise expect that the set of real numbers, $\mathbb{R}$, will also be of the same size as the others. Well, it was here that the first real surprise arose in the works of Dedekind and Cantor, since we can now prove that $\mathbb{N}$ and $\mathbb{R}$ *cannot be equipotent*. As a consequence of this, we realize for the first time in the history of mathematics that there are, indeed, *several kinds of infinities*! No wonder that many mathematicians were instinctively opposed to accept these new ideas when Cantor first publicized them.

12.2 Infinities of various sizes

In their early correspondence, Dedekind confessed to Cantor that he was not able to decide on the question whether or not $\mathbb{N}$ and $\mathbb{R}$ are equipotent. Cantor, on his side, came up in 1873 with a proof that they are not. But this proof was rather involved and unconvincing. It was only in 1890 that Cantor was able to complete one of his most famous proofs, the so-called Cantor diagonal proof. This proof had an enormous influence on the mathematics of the twentieth century, not just because of the surprising claim it helped establish, but also because it introduced a new kind of reasoning that was repeated in proofs of many other important results developed in various branches of mathematics, logic and computer science. I want to present Cantor's proof now. It is not particularly lengthy, and understanding its details does not require any specific mathematical background. If you find my explanation somewhat difficult to follow, I urge you to carry on bravely nonetheless to the end, if only for the sake of coming close to grasping the main ideas behind this famous argument.

Let us begin with some preliminary considerations. I will not exactly prove that $\mathbb{N}$ and $\mathbb{R}$ have different cardinalities. It will be enough to prove that only a small part of $\mathbb{R}$ cannot be equipotent with $\mathbb{N}$. I will focus on the set $(0,1)$, namely, the collection comprising *all* real numbers between 0 and 1 (but not including 0 and 1). The proof begins with the assumption that there is a one-to-one correspondence between $\mathbb{N}$ and $(0,1)$, and this will lead to a contradiction. Now, recall that a real number α between 0 and 1 can always be written as a decimal fraction starting with 0 before the decimal point. We can represent any such number symbolically as follows:

$$\alpha = 0.\alpha^1\alpha^2\alpha^3\alpha^4\alpha^5\alpha^6\alpha^7\ldots$$

This notation may be confusing on first reading, but it is actually very straightforward: each symbol α^i stands here for a digit between 0 and 9; the superscript i is to be taken as an *index* and not an exponent as we usually do when writing powers. So, for instance, if we take $\alpha = \frac{31}{99} = 0.3131\ldots$, then the first digit after the decimal point, 3, corresponds to α^1; the second digit, 1, corresponds to α^2, and likewise we have $\alpha^3 = 3$, $\alpha^4 = 1$ and so on. This is an example where the number α is *rational*, and hence the digits in the decimal expansion are *periodic*, i.e., they repeat themselves indefinitely after some stage. A similar example, would be with $\alpha = \frac{23}{56} = 0.41071428571428571\ldots$ Here we have the sequence of digits 714285 as the period appearing in the expansion. But since we are considering all real numbers in $(0,1)$, our number α could be *irrational* as well, in which case no such period would appear in the tail of the expansion. Think about the number π, for example. It is an irrational number whose decimal expansion 3,1415... does not become periodic no matter how far we go into it.

A second preliminary remark brings us back to an issue already mentioned in Chapter 1: we can write a number in two seemingly different forms, such as in the case of 0.4999... and 0.5. For the purposes of our proof, and to avoid misunderstandings, we will consider only one of these forms as valid, for example, 0.4999... Finally, to make things even simpler, we will use a binary notation, so that all of the digits appearing in the expansion $\alpha = 0.\alpha^1\alpha^2\alpha^3\alpha^4\alpha^5\alpha^6\alpha^7\ldots$ are either 0 or 1.

As already stated, we begin by assuming that a one-to-one correspondence between $(0,1)$ and $\mathbb{N}$ has been defined. We saw in the previous examples that if such a correspondence existed, then we could write all the numbers in $(0,1)$ as a sequence $\alpha_1, \alpha_2, \alpha_3, \alpha_4, \alpha_5, \ldots, \alpha_n, \ldots$ Notice that now I am using subscripts instead of the superscripts introduced above. Each subscript indicates the place of a real number between 0 and 1 in the putative sequence in which we would have ordered all of them. The superscripts, on the other hand, are used in the expansion $\alpha = 0.\alpha^1\alpha^2\alpha^3\alpha^4\alpha^5\alpha^6\alpha^7\ldots$ to indicate the place of a digit in the decimal expansion of a certain given number α. It is important to keep this distinction in mind throughout the proof.

$$\alpha_1 = 0.\ \alpha_1^1\ \alpha_1^2\ \alpha_1^3\ \alpha_1^4\ \alpha_1^5\ \alpha_1^6\ \alpha_1^7\ \alpha_1^8\ \alpha_1^9\ \ldots$$

$$\alpha_2 = 0.\ \alpha_2^1\ \alpha_2^2\ \alpha_2^3\ \alpha_2^4\ \alpha_2^5\ \alpha_{21}^6\ \alpha_2^7\ \alpha_2^8\ \alpha_2^9\ \ldots$$

$$\alpha_3 = 0.\ \alpha_3^1\ \alpha_3^2\ \alpha_3^3\ \alpha_3^4\ \alpha_3^5\ \alpha_3^6\ \alpha_3^7\ \alpha_3^8\ \alpha_3^9\ \ldots$$

$$\alpha_4 = 0.\ \alpha_4^1\ \alpha_4^2\ \alpha_4^3\ \alpha_4^4\ \alpha_4^5\ \alpha_4^6\ \alpha_4^7\ \alpha_4^8\ \alpha_4^9\ \ldots$$

$$\alpha_5 = 0.\ \alpha_5^1\ \alpha_5^2\ \alpha_5^3\ \alpha_5^4\ \alpha_5^5\ \alpha_5^6\ \alpha_5^7\ \alpha_5^8\ \alpha_5^9\ \ldots$$

$$\vdots$$

$$\alpha_n = 0.\ \alpha_n^1\ \alpha_n^2\ \alpha_n^3\ \alpha_n^4\ \alpha_n^5\ \alpha_n^6\ \alpha_n^7\ \alpha_n^8\ \alpha_n^9\ \ldots$$

$$\vdots$$

Figure 12.4 A one-to-one correspondence between (0,1) and $\mathbb{N}$.

Now, by combining the subscripts and the superscripts together, we can write, as in the table depicted in Figure 12.4, all the real numbers between 0 and 1, arranged one after the other according to their putative order $\alpha_1, \alpha_2, \alpha_3, \ldots$, while to the right of each of them we see its full expansion as decimal fractions. Any reader looking at this table for the first time may be daunted by the abundance of indices, but with a little practice it is very easy to read. For instance, the symbol α_3^4 stands for the fourth digit in the decimal expansion of the number α_3, and, in turn, α_3 is the number that occupies the third place in our putative, exhaustive list. As already mentioned, α_3^4 can be either 0 or 1, since we have decided to write all the numbers as binary expressions, without thereby losing any generality in the argument. To make it more concrete, if in our list the number in the seventh place (that is, the number that the putative correspondence assigns to the number 7) were $\alpha_7 = 0.101001000100001\ldots$, then, in the seventh row of the table, we would have

$$\alpha_7^1 = 1,\ \alpha_7^2 = 0,\ \alpha_7^3 = 1,\ \alpha_7^4 = 0,\ \alpha_7^5 = 0,\ \alpha_7^6 = 1,\ \alpha_7^7 = 0,\ \alpha_7^8 = 0, \ldots$$

Now comes the decisive step of Cantor's proof and the one that gives it its famous name. I will construct a special real number β, which is itself a real number between 0 and 1. The above list, by definition, is supposed to be exhaustive. This means that β, by virtue of being a number between 0 and 1, should also be one of the numbers in the list, for some index r, $\beta = \alpha_r$. But the way I will construct this special number β will imply that, no matter what index r I take, it is *never* possible that $\beta = \alpha_r$. This may appear strange at first sight, but a straightforward argument shows that this is indeed the case. Now, like any other number in the interval (0,1), β also can be thought of as a string of 0s and 1s that represent the number in a binary expansion, $\beta = 0.\beta^1\beta^2\beta^3\beta^4\beta^5\beta^6\beta^7\ldots$ Again, here each β^i is either 0 or 1. To construct our special number β, we can proceed as follows:

- In Figure 12.4, check to see what is α_1^1. It can be either 0 or 1.
- If α_1^1 is 0, then set β_1 to be 1; if α_1^1 is 1, then set β_1 to be 0. In other words, we make sure that $\beta_1 \neq \alpha_1^1$, and hence $\beta \neq \alpha_1$, because they differ in the *first* digit of their respective expansions.

- Do the same for β_2 and α_2^2: if α_2^2 is 0, then set β_2 to be 1; if α_2^2 is 1, then set β_2 to be 0. In other words, we make sure that $\beta_2 \neq \alpha_2^2$, and hence $\beta \neq \alpha_2$, because they differ in the *second* digit of their respective expansions.
- Continue with the same process for each index i. The result is that for all indices i, we know for sure that $\beta_i \neq \alpha_i^i$, and hence $\beta \neq \alpha_i$, because they differ in the ith digit of their respective expansions.

One way to visualize the meaning of Cantor's diagonal procedure is shown in Figure 12.5. This is a modification of the table in Figure 12.4, with the elements in the diagonal now appearing in boldface. In the decimal expansion of β, each of the digits β_i is different from each of the corresponding digits in the diagonal, and this is why, as stated above, there can be no index r for which $\beta = \alpha_r$. So, we began by assuming that the list is exhaustive, and then the outcome of the diagonalization is that we have found an element of the interval (0,1) that by necessity cannot be part of that list! Well, this is the contradiction that we were looking for. This leads us to conclude that there can be no such one-to-one correspondence between (0,1) and $\mathbb{N}$, as assumed.

Now, even though the proof is sound and incontestable, a skeptical reader might still claim that the number β is no more than a minor "mathematical accident." Since we are dealing with infinities, the skeptic might say, even if we thought that the list $\alpha_1, \alpha_2, \alpha_3, \alpha_4, \alpha_5, \ldots \alpha_n, \ldots$ was exhaustive and we nevertheless found one instance of a number that should have been there but is not, then this situation can be easily corrected simply by adding β at the head of the list. The corrected, putative exhaustive list should simply now read as $\beta, \alpha_1, \alpha_2, \alpha_3, \alpha_4, \alpha_5, \ldots \alpha_n, \ldots$, and not as we had originally thought. That's all. But this seemingly ingenious objection to the proof will not hold water: indeed, we may take this new list and once again apply Cantor's diagonal argument to it, thus yielding another number β_1, in the interval (0,1), that is not part of the new list. The decimal expansion of this number is $\beta_1 = 0.\beta_1^1 \beta_1^2 \beta_1^3 \beta_1^4 \beta_1^5 \beta_1^6 \beta_1^7 \beta_1^8 \beta_1^9 \ldots$, and the digits of this expansion are taken as differing from those in the diagonal of the new list that includes β, as indicated in Figure 12.6.

$$\alpha_1 = 0.\ \boldsymbol{\alpha_1^1}\ \alpha_1^2\ \alpha_1^3\ \alpha_1^4\ \alpha_1^5\ \alpha_1^6\ \alpha_1^7\ \alpha_1^8\ \alpha_1^9 \ldots$$

$$\alpha_2 = 0.\ \alpha_2^1\ \boldsymbol{\alpha_2^2}\ \alpha_2^3\ \alpha_2^4\ \alpha_2^5\ \alpha_2^6\ \alpha_2^7\ \alpha_2^8\ \alpha_2^9 \ldots$$

$$\alpha_3 = 0.\ \alpha_3^1\ \alpha_3^2\ \boldsymbol{\alpha_3^3}\ \alpha_3^4\ \alpha_3^5\ \alpha_3^6\ \alpha_3^7\ \alpha_3^8\ \alpha_3^9 \ldots$$

$$\alpha_4 = 0.\ \alpha_4^1\ \alpha_4^2\ \alpha_4^3\ \boldsymbol{\alpha_4^4}\ \alpha_4^5\ \alpha_4^6\ \alpha_4^7\ \alpha_4^8\ \alpha_4^9 \ldots$$

$$\alpha_5 = 0.\ \alpha_5^1\ \alpha_5^2\ \alpha_5^3\ \alpha_5^4\ \boldsymbol{\alpha_5^5}\ \alpha_5^6\ \alpha_5^7\ \alpha_5^8\ \alpha_5^9 \ldots$$

$$\vdots$$

$$\alpha_n = 0.\ \alpha_n^1\ \alpha_n^2\ \alpha_n^3\ \alpha_n^4\ \alpha_n^5\ \alpha_n^6\ \alpha_n^7\ \alpha_n^8\ \alpha_n^9 \ldots$$

$$\vdots$$

Figure 12.5 The diagonal in the putative one-to-one correspondence between (0,1) and $\mathbb{N}$.

$$\beta = 0. \; \boldsymbol{\beta^1} \; \beta^2 \; \beta^3 \; \beta^4 \; \beta^5 \; \beta^6 \; \beta^7 \; \beta^8 \; \beta^9 \ldots$$

$$\alpha_1 = 0. \; \alpha_1^1 \; \boldsymbol{\alpha_1^2} \; \alpha_1^3 \; \alpha_1^4 \; \alpha_1^5 \; \alpha_1^6 \; \alpha_1^7 \; \alpha_1^8 \; \alpha_1^9 \ldots$$

$$\alpha_2 = 0. \; \alpha_2^1 \; \alpha_2^2 \; \boldsymbol{\alpha_2^3} \; \alpha_2^4 \; \alpha_2^5 \; \alpha_2^6 \; \alpha_2^7 \; \alpha_2^8 \; \alpha_2^9 \ldots$$

$$\alpha_3 = 0. \; \alpha_3^1 \; \alpha_3^2 \; \alpha_3^3 \; \boldsymbol{\alpha_3^4} \; \alpha_3^5 \; \alpha_3^6 \; \alpha_3^7 \; \alpha_3^8 \; \alpha_3^9 \ldots$$

$$\alpha_4 = 0. \; \alpha_4^1 \; \alpha_4^2 \; \alpha_4^3 \; \alpha_4^4 \; \boldsymbol{\alpha_4^5} \; \alpha_4^6 \; \alpha_4^7 \; \alpha_4^8 \; \alpha_4^9 \ldots$$

$$\alpha_5 = 0. \; \alpha_5^1 \; \alpha_5^2 \; \alpha_5^3 \; \alpha_5^4 \; \alpha_5^5 \; \boldsymbol{\alpha_5^6} \; \alpha_5^7 \; \alpha_5^8 \; \alpha_5^9 \ldots$$

$$\vdots$$

$$\alpha_n = 0. \; \alpha_n^1 \; \alpha_n^2 \; \alpha_n^3 \; \alpha_n^4 \; \alpha_n^5 \; \alpha_n^6 \; \alpha_n^7 \; \alpha_n^8 \; \alpha_n^9 \ldots$$

$$\vdots$$

Figure 12.6 The diagonal of the extended, putative one-to-one correspondence between (0,1) and $\mathbb{N}$.

So, we construct β_1 by taking $\beta_1^1 \neq \beta^1$, $\beta_1^2 \neq \alpha_1^2$, $\beta_1^3 \neq \alpha_2^3$, $\beta_1^4 \neq \alpha_3^4$ and so on. And, as above, we conclude that for all indices i, $\beta_1^i \neq \alpha_{i-1}^i$, and hence $\beta_1 \neq \alpha_i$, because they differ in the $(i + 1)$th digit of their respective expansions.

This process of adding the newly found number to the existing list and then finding another one not included in the new list can be extended indefinitely. Moreover, notice that the putative one-to-one correspondence was not defined in any particular way, and hence Cantor's diagonal argument dismisses the possibility of *all possible* one-to-one correspondences, rather than of a particularly defined one. Hence, it definitely proves that the putative list can never be made to be truly exhaustive, as required by the assumption. We conclude, then, that these two infinite sets, $\mathbb{N}$ and $\mathbb{R}$, have different cardinalities.

Cantor's surprising discovery marked a truly momentous breakthrough with long-standing consequences for the foundations of mathematics. Among other things, it immediately broadened to an unprecedented extent and in truly unexpected directions the very idea of number. The new world of transfinite cardinals turned out to be extremely rich in mathematical ideas, and to this day it provides a highly active field of advanced mathematical research. And as in the gradual construction of new systems from existing ones as part of the hierarchy of (finite) numbers, the passage to the transfinite realm is interesting because it involves the creation of a full arithmetic of the newly created cardinals (which is quite different from that of the finite ones). We will of necessity have to make do with only a brief glance into all this world.

A first remarkable result is that, there are not only two infinite sizes of sets, those of $\mathbb{N}$ and $\mathbb{R}$, but in actual fact there are infinitely many different sizes. This can be seen with the help of another important idea introduced by Cantor, in relation to the cardinality of the "power set" of any given set. Indeed, if we are given any set, say $A = \{a, b, c\}$, the power set of A, denoted $P(A)$, is a set whose elements are all the possible subsets of A. It is customary to consider the set A as a subset of itself, and always to consider the empty set $\emptyset$ as a subset of any given set. Hence, it is easy to see that in this example the set $P(A)$ has 8 elements:

$$P(A) = \{\{a,b,c\},\ \{a,b\},\ \{a,c\},\ \{b,c\},\ \{a\},\ \{b\},\ \{c\},\ \emptyset\}.$$

That is, given a set A with 3 elements, its power set $P(A)$ has 8 elements, $8 = 2^3$. More generally, it is easy to see for a finite set A that if A has n elements, then the set $P(A)$ has 2^n elements (you are invited to check that this is the case, for instance, with $n = 2$ and $n = 4$). In particular, if we denote the cardinality of a set X by $\#X$, then it is clear that in the finite case $\#P(A) > \#A$. The question of course immediately arises as to whether this inequality continues to hold true for transfinite cardinals. If A is infinite, then clearly $P(A)$ also is infinite, but there is no a priori reason to decide if in this case it is still always true that $\#P(A) > \#A$. Well, one of the interesting results proved by Cantor (actually this is often called the Cantor theorem) is that, indeed, the same inequality continues to hold for transfinite cardinals. In particular, a consequence of the Cantor theorem is that given any transfinite cardinality, there is always a set with cardinality greater than that of the given set, and consequently there is an infinite, increasing sequence of different transfinite cardinalities, such as in the case:

$$\#\mathbb{N}\ <\ \#P(\mathbb{N})\ <\ \#P(P(\mathbb{N}))\ <\ \ldots$$

Now, clearly, the smallest transfinite cardinality is that of the set of naturals $\mathbb{N}$. In addition, $\mathbb{Z}$ and $\mathbb{Q}$ have this same cardinality. A set whose cardinality is the same as that of $\mathbb{N}$ is called "countably infinite," or countable. Cantor introduced a new symbol to indicate the cardinalities of transfinite numbers, using the first letter of the Hebrew alphabet, "aleph" ($\aleph$), which is also the first letter of the Hebrew word for infinity "einsoph." The cardinality of $\mathbb{N}$ (and of all countably infinite sets) is denoted $\aleph_0$ (aleph-nought). The arithmetic of the natural numbers as defined by Frege may naturally be extended to that of the transfinite cardinalities, whereby we obtain some rules that differ from those of the finite case. For example, $\aleph_0 + 1 = \aleph_0$, because if we add a single element to a set that is countable, we obtain a set that is also countable, as suggested by the following correspondence:

$$\begin{array}{cccccccccc} 1 & 2 & 3 & 4 & 5 & 6 & 7 & 8 & 9 & 10 & \ldots \\ \updownarrow & \updownarrow & \updownarrow & \updownarrow & \updownarrow & \updownarrow & \updownarrow & \updownarrow & \updownarrow & \updownarrow & \\ 0 & 1 & 2 & 3 & 4 & 5 & 6 & 7 & 8 & 9 & \ldots \end{array}$$

More generally speaking, and for the same reason, for any finite cardinal n, we have $\aleph_0 + n = \aleph_0$, and indeed also $\aleph_0 + \aleph_0 = \aleph_0$. Moreover, if we have a countable collection of countable sets, then it is easy to see that the set obtained as the union of all sets in this collection is also countable (a detailed proof of this appears in Appendix 12.1). As a consequence of this, we have the following additional results:

$$\aleph_0 + \aleph_0 + \aleph_0 + \ldots + \aleph_0 = \aleph_0 \quad \text{for a finite number of factors,}$$

$$\aleph_0 + \aleph_0 + \aleph_0 + \ldots = \aleph_0 \quad \text{for a countably infinite number of factors.}$$

David Hilbert used to explain the special situation arising with the use of transfinite cardinalities with the help of the "hotel metaphor," which can be stated as follows: if a person comes to a hotel that is fully occupied and requests a room, the receptionist will have no choice—in a normal hotel—but to reject the request. But if the hotel has a countably infinite number of rooms, the receptionist can easily get a free room for the new guest, by moving each existing guest from the room where he is now hosted, say room n, to the next one in the list, say $n + 1$. When all guests have entered their new rooms, room number 1 is free for use by the new guest, and no other guest has been left without a room. Also, if k new guests arrive at a certain moment, the receptionist in the fully booked, countable hotel can free rooms for all of them by moving each guest from room n, where she is now, to room $n + k$. And, indeed, what is really interesting is that even if an infinite, but countable, number of new guests arrive, the receptionist will have no trouble in freeing a room for each of them: he simply needs to move the guest currently hosted at room n to room $2n$. This will set free all rooms with odd numbers for the countable set of new guests to occupy them.

Once aware of the arithmetic of the transfinite cardinalities and of the infinite sequence of transfinite cardinalities, many intriguing questions arise, and these have been investigated by mathematicians ever since the days of Cantor and Dedekind. Cantor himself was the first to pose a truly important question in this regard, while reasoning along the following lines. If $\aleph_0$ is the cardinality of $\mathbb{N}$, it seems natural to indicate the cardinality of $P(\mathbb{N})$ as $2^{\aleph_0}$. Notice that this is just a convenient symbol and it does not indicate that we actually raise 2 to the power of $\aleph_0$, since that would be a meaningless operation. Further, we denote the cardinality of $\mathbb{R}$ by $\mathfrak{c}$ (which stands for the *continuum*). Now, both $\mathfrak{c}$ and $2^{\aleph_0}$ are cardinalities known to be, for different reasons, greater than $\aleph_0$. Cantor proved that $\mathfrak{c} = 2^{\aleph_0}$, or, in other words, that $\#P(\mathbb{N}) = \#\mathbb{R}$. The proof is not too difficult, but is beyond the scope of this book and I will not present it here. But the point is that, this being the case, Cantor brought up the issue of whether the sequence of alephs can be fully ordered, and what this order would look like.

In the context of this attempt to order the alephs, Cantor suggested a very strong conjecture about $\aleph_1$, namely, about that transfinite cardinal that appears right after $\aleph_0$ in the ordering sequence of the alephs. The conjecture (of whose correctness he was convinced, even though he had no proof of it) is the following:

There is no set X of numbers such that $\mathbb{N}$ is contained in X while at the same time X is contained in $\mathbb{R}$, and such that $\aleph_0 < \#X < \mathfrak{c}$.

To put it differently, Cantor conjectured that the cardinality of *any non-countable infinite set of real numbers* is $\mathfrak{c}$, i.e., the cardinality of the *entire set of real numbers*. Or, stated directly in the language of transfinite cardinalities, $\aleph_1 = 2^{\aleph_0}$. That is, in the sequence of transfinite cardinalities, the cardinality immediately following $\aleph_0$ is $2^{\aleph_0}$.

With time, this conjecture became known as Cantor's Continuum Hypothesis (CH). The attempts to prove or disprove it led to many important insights in the theory of sets, beginning at the turn of the twentieth century and continuing up until the 1960s. Remarkably enough, as an outcome of the work of prominent mathematicians such as Kurt Gödel (1906–1978) and Paul Cohen (1934–2007), it became clear that the CH is a statement that is *independent* of the theory of sets as typically conceived nowadays. The meaning of this is that mathematicians can build mathematical worlds based on the theory of sets in which CH is true, but they can likewise build alternative mathematical worlds, also based on the theory of sets, but where the negation of CH is true. Philosophers of mathematics continue to debate to this day the significance of this startling result.

One specific set found between $\mathbb{N}$ and $\mathbb{R}$, which deserves consideration from the point of view of its cardinality, is the set of algebraic numbers, which I denote here by $\mathbb{A}$. This set was already mentioned in Chapter 1, as part of the description of the entire hierarchy of number systems (see Figure 1.1). It comprises all real numbers that are solutions of polynomial equations with rational coefficients. Clearly, this set comprises all the rational numbers, for the simple reason that any given rational number, say $\frac{p}{q}$, is a solution of the polynomial equation $x - \frac{p}{q} = 0$. But the set also includes other numbers such as $\sqrt{2}$, which is a solution of the equation $x^2 - 2 = 0$. Informally speaking, we can say that $\mathbb{A}$ is the set of all rational numbers with the addition of all irrational numbers that are roots of various degrees of rational numbers, or sums or ratios of such numbers, and some others as well. Since $\mathbb{Q}$ is a proper subset of $\mathbb{A}$, this latter set became a primary focus of interest concerning the question of the sequence of the alephs. Early in their dealings with transfinite cardinalities, both Dedekind and Cantor investigated carefully the cardinality of $\mathbb{A}$, and they were able to prove that this set is *countable* (the proof appears in Appendix 12.1). This lends immediate support to Cantor's conjecture.

Cantor published his proof in 1873, and he did not mention the name of Dedekind in his publication, even though they had explicitly discussed the topic, and some of the important ideas in the proof came from Dedekind. This was one of those instances when their correspondence was interrupted for several years. Two additional reasons for such interruptions were (a) that the details of the theory of transfinite numbers was of less interest to Dedekind than to Cantor and (b) that Cantor suffered from several mental breakdowns that led to prolonged interruptions to his own work. At any rate, it was through his work on the cardinalities of various systems of numbers, and particularly that of $\mathbb{P}$, that Cantor was led to formulate the Continuum Hypothesis.

Now, since the cardinality of $\mathbb{P}$ is $\aleph_0$ and the cardinality of $\mathbb{R}$ is $\mathfrak{c}$, with $\aleph_0 < \mathfrak{c}$, we can conclude that there are, of necessity, real numbers that are not algebraic. Although this conclusion seems simple and straightforward, it is important to give some additional thought to the very kind of argument for mathematical existence that it involves. Recall that in Chapter 1, I mentioned the numbers e and π as two examples of numbers that are not algebraic (I called such numbers "transcendental"). I also said that proving that either e or π is not the solution of any possible polynomial equation with rational coefficients is a very demanding mathematical task. So, for the purposes of our account here, it can be said that we have not yet seen any real, specific example of a number that we know with certainty is *not* the solution of a polynomial equation with rational coefficients.

On the other hand, we have just "counted," as it were, the elements of two transfinite entities, $\mathbb{A}$ and $\mathbb{R}$ (which, by their very nature, cannot be counted in any physical sense of the word), and from the fact that this count yields different results, $\aleph_0$ and $\mathfrak{c}$, we have drawn conclusions about the *existence* of certain mathematical entities of which we have not directly seen any instance (i.e., transcendental numbers). Moreover, even though we have not seen any such individual number, we know for certain some of the properties of the collection of which they are part, for example, that the set of these transcendental numbers has a cardinality that is larger than that of the algebraic numbers (indeed, $\mathbb{R}$ is the union of the set of transcendental numbers and $\mathbb{A}$, and since $\#\mathbb{A} = \aleph_0$ and $\#\mathbb{R} = \mathfrak{c}$, it follows that the set of transcendental numbers cannot be countable). I will return to the implications of this interesting argument below.

12.3 Cantor's transfinite ordinals

Cantor discussed transfinite sets not only from the point of view of their cardinality, but also from the perspective of their properties as ordinals, which is no less interesting and surprising on first inspection. As already hinted in previous chapters, whenever we are considering finite sets, it makes no real difference to speak about them as cardinalities or as ordinals. If we are given two sets of four elements, they are clearly equipotent, since it is easy to establish a one-to-one correspondence between them. At the same time, they are also equivalent from the point of view of any order relation that we may want to define on them. If we take two different orderings of four elements, say *a, b, c, d* and *b, c, d, a*, then it is obvious that, seen, abstractly they are one and the same. Thus, the element appearing in the second place in both arrangements (*b* and *c* respectively) is preceded by one element and is followed by two elements. Similarly, any other property related to their order appears in both arrangements simultaneously, and hence we can say that there is essentially only one way to order these four elements.

But when we come to the transfinite case, the situation changes dramatically. We have already seen that if we take the set $\mathbb{N}$ and add the number 0 before the number 1, then we obtain a set that is equipotent to the original, in the sense that we can establish a one-to-one correspondence between them:

$$\begin{array}{ccccccccccc} 1 & 2 & 3 & 4 & 5 & 6 & 7 & 8 & 9 & 10 & \ldots \\ \updownarrow & \updownarrow & \updownarrow & \updownarrow & \updownarrow & \updownarrow & \updownarrow & \updownarrow & \updownarrow & \updownarrow & \\ 0 & 1 & 2 & 3 & 4 & 5 & 6 & 7 & 8 & 9 & \ldots \end{array}$$

The new set obtained in this way turns out to be equivalent also from the point of view of their ordinality. Indeed, in both sets, we have a successor for every element, and also there is only one element that is not a successor of any other element (in the first case it is 1, in the second case it is 0). But now we can think of an alternative way to add this extra element 0 in such a way that an essentially different order structure will arise. This alternative ordering may look artificial or strange, and even misleading, but you should be aware at this point that sometimes mathematicians can follow seemingly unorthodox paths if only to examine certain ideas from every possible perspective.

The only limitations to be observed in such cases are those that pertain to the introduction of possible logical contradictions in our system, and, inasmuch as we do not incur such contradictions, we are then free to proceed as we like in the exploration of ideas. In this spirit, consider for example the following way to add the extra element 0 to the known sequence of natural numbers:

$$1, 2, 3, 4, 5, 6, 7, 8, 9, 10, \ldots 0.$$

In this arrangement the new element 0 is defined as *greater* than all other numbers, or, in a more neutral terminology, as appearing *after* all other numbers in the sequence. Notice that now we have not one but *two* elements, 1 and 0, that are not successors of any other number in the sequence (consequently, the natural numbers arranged in this way do not satisfy all of the Peano postulates). In addition, of these two, only one has successors, while the other has none. From an abstract perspective, the following arrangement is also equivalent to the previous one:

$$2, 3, 4, 5, 6, 7, 8, 9, 10, \ldots 1.$$

Here it is 1 that is defined as the number that comes after all other numbers in the sequence, but again, seen abstractly we have two numbers, 1 and 2, that are not successors of any other number. And here 2 has a successor, while 1 has none. As we saw in Section 11.2, when dealing with the Peano postulates and their possible models, it is not any specific value of the elements, or the symbols that stand for them, that count, but rather the abstract properties that define the order within the sequence. Based on this idea of various possible orderings for a set with a given cardinality, Cantor investigated the transfinite ordinals. He had used the symbol $\aleph_0$ to denote the cardinality of $\mathbb{N}$, and now he also introduced a new symbol, ω, to denote the ordinal embodied in the standard ordering of the natural numbers: 1, 2, 3, 4, ... He called it "the first transfinite ordinal." Cantor further used this idea to define in a canonical way new ordinals of countable sets in the direction already hinted at above. An example of this is the ordinal that I defined above as

$$1, 2, 3, 4, 5, 6, 7, 8, 9, 10, \ldots 0.$$

Cantor defined it as

$$1, 2, 3, 4, 5, 6, 7, 8, 9, 10, \ldots \omega.$$

This ordinal is clearly different from ω, for the reasons explained above (two numbers in it, 1 and ω, are not successors of any number). Cantor denoted this new ordinal by $\omega + 1$. And from here he continued in a natural way by defining another countable ordinal $\omega + 2$ as follows:

$$1, 2, 3, 4, 5, 6, 7, 8, 9, 10, \ldots \omega, \omega + 1.$$

It is easy to see how this is generalized to other ordinals $\omega+n$, for any natural number n. But the process does not stop here. A further step comes with the ordinal called 2ω, or $\omega + \omega$, that represents the following arrangement:

$$1, 3, 5, 7, 9, 11, \ldots 2, 4, 6, 8, 10, \ldots$$

Notice that in this arrangement there are *two* numbers that are not successors of any number, and that there are two subsequences, each of which is equivalent to ω. We can then imagine additional ordinals such as $2\omega + 1, 2\omega + 2, \ldots, 2\omega + n, \ldots, 3\omega, 3\omega + 1, \ldots, m\omega + n, \ldots$ (with m and n being any natural numbers). But, even after all of these, we can still add a new ordinal, which is defined as being greater than all previous ordinals of the type $m\omega + n$. We call this new ordinal ω^2, and, following it, we can also define $\omega^2 + 1$, and many others such as $\omega^2 + 7\omega + 5$ and $5\omega^3 + 4\omega^2 + 32$, and as much as the imagination can conceive in this direction. And it turns out that Cantor's imagination was very fruitful in this regard! He suggested to order, for example, a part of the natural numbers according to the ordinal ω, immediately thereafter another part according to ω^2, then another part according to ω^3, and so on with ω^n for each natural number n. The rather intricate ordinal obtained in this way is called ω^ω. From here, we can go on and define an ordinal ω^{ω^2} and then continue further on in the direction of ω^{ω^ω}. We can proceed indefinitely in this same direction. One further step would be to order a part of the natural numbers according to ω, then another part according to ω^ω, yet another part according to ω^{ω^ω} and so on indefinitely. In this way, we obtain a very complex and interesting ordinal called ϵ_0, and there is no need to add that we can continue building further ordinals of $\mathbb{N}$ by similar kinds of combinations.

Cantor further noticed that all the above ordinals taken together create a set that is interesting in itself. On the one hand, we can ask about the cardinality of this set of all ordinals. It turns out that it is greater than $\aleph_0$. Without going into too many details about it, I will just say that Cantor proved this fact using again a diagonal argument of the kind described above. And what about the ordinal of the set of *all* ordinals? Well, if we denote this ordinal by the Greek letter Ω, then we can, as in all previous cases, build the ordinal that comes right after Ω. It seems reasonable to denote this new ordinal by $\Omega + 1$. It can be thought of as appearing in the sequence of all ordinals, by definition, right after Ω. But, on the other hand, given that Ω is the ordinal of the sequence of all ordinals, then, again by definition, it cannot be smaller than $\Omega + 1$, given that the latter is just another ordinal. Stating this symbolically, we realize that there is an apparent logical paradox involved here, as we get

$$\Omega \;<\; \Omega + 1 \;<\; \Omega.$$

The Italian mathematician Cesare Burali-Forti (1861–1931), one of Peano's students, was the first to call attention, in an article published in 1897, to this perplexing situation (which eventually became known as a "paradox" associated with his name). He became aware of it while studying in detail Cantor's new theories. Cantor's work, to be sure, was received with great mistrust among many of the leading mathematicians of the time. It was slow to be seriously acknowledged, but it gradually gained ground among younger mathematicians, beginning from the turn of the twentieth century. It eventually became the unifying language of a broad range of mainstream mathematical domains. Burali-Forti was among the few early enthusiasts, but he was not alone in both his interest in the theory and the difficulties he encountered. Hilbert was also an early promoter, and he realized from very early on that the new theory of transfinite numbers contained conceptual difficulties that required further clarification. But since these mathematicians also saw in Cantor's theory the gateway to an extremely rich and appealing new mathematical reality, they did not regard these difficulties as insurmountable obstacles. Rather, they viewed the difficulties as challenges to be met, such as are very often found in many incipient mathematical disciplines.

12.4 Troubles in paradise

For the critics of Cantor's theory of transfinite numbers, difficulties such as those indicated by Hilbert or Burali-Forti came as no surprise. Attacks on the theory focused on the way in which it handled infinite sets as actual, *completed collections*. This was in opposition to a long held view of the infinite as a *process*, rather than as an actual reality. The traditional view of the "potential infinite" laid stress on the ability to indefinitely add ever new natural numbers to any given collection of them. The existence of the *complete*, infinite set of natural numbers was seen, from this perspective, as mathematical nonsense. The renowned French mathematician Jules Henri Poincaré (1854–1912) described the theory as a "disease" from which mathematics would someday be cured. But the most vocal opposer of Cantor and of his innovative ideas was Leopold Kronecker.

Kronecker, as I have already mentioned, was one of the leading figures of German mathematics in the last third of the nineteenth century. Together with Dedekind, he had played a key role in the creation of the new theory of fields of algebraic numbers, and more generally he was considered a world authority on all matters related to the theory of numbers. The basis of Kronecker's approach to numbers is subsumed in the often-quoted motto: "God created the integers, all else is the work of man." He thought that an overall, abstract theory of real numbers as implied by Cantor's or Dedekind's approach was unnecessary in mathematics. Rather, one could legitimately speak about specific numbers (for example π) only inasmuch as there was a specific, well-defined procedure to construct them.

Besides the substantive, mathematical debate embodied in Kronecker's controversies with Cantor, there was also a strong personal dimension to them. Kronecker, who had been Cantor's teacher in Berlin, strongly attacked his former student, and dubbed him a charlatan and "corrupter of youth." Kronecker's efforts were instrumental in blocking Cantor's attempt in 1879 to obtain a much desired appointment in Berlin. Cantor was deeply disappointed by Kronecker's reaction, and also his long friendship with Dedekind reached a dead-end at the time. He had suffered throughout his life from severe nervous breakdowns, which now became increasingly acute and frequent. The last years of his life were spent in mental institutions. Outside the texts published in professional mathematical journals, Cantor willingly mixed his ideas with various kinds of mystical and theological arguments. His choice of the Hebrew letter *aleph* for his sets had a clear kabbalistic undertone. He became convinced, at some point, that transfinite numbers had come to him as a message from God. He corresponded with theologians who showed interest in his theories and in their possible philosophical implications. He also addressed one of his letters and some of his pamphlets to Pope Leo XIII.

We can make more precise sense of the mathematical reasons behind the initial opposition to Cantor's new ideas on the infinite by going back to the issue of the cardinality of the set $\mathbb{A}$ of algebraic numbers. Recall that after proving that the cardinalities of $\mathbb{A}$ and $\mathbb{R}$ are different, I stated that we can deduce the *existence* of transcendental numbers (and also some properties of the collection of such numbers). We can do so, I emphasized, without having shown even a single instance of such a number. An argument of this kind is usually called a proof of mathematical existence based on contradiction, as opposed to a *constructive* argument of the kind that Kronecker favored at the time. The contradiction in this particular case is the one found at the core of Cantor's proof that $\mathbb{N}$ and $\mathbb{R}$ are not equipotent. That proof was based on assuming the existence of a one-to-one correspondence between the two sets and then showing that this assumption leads to a contradiction. Subsequently, based on the different cardinalities of $\mathbb{A}$ and $\mathbb{R}$ we could deduce the existence of transcendental numbers.

For a mathematician like Kronecker, with his strongly constructivist views, this argument was totally unacceptable. The very idea of a given one-to-one correspondence between the two sets $\mathbb{N}$ and $\mathbb{R}$ would make no sense to him, in the first place, because he would not admit the existence of those two collections as completed entities. Rather, he would speak about concrete processes that give rise to specific numbers, which in turn can be classified as being either algebraic or transcendental. And notice that for Kronecker there would be no problem with arguments by contradiction *per se* (as, for e.g., in the proof of the irrationality of $\sqrt{2}$ in Appendix 3.1, where we proved that two integers with a certain property *cannot exist*). His problem was with arguments by contradiction meant to prove that some kind of mathematical entity *does exist*.

So, once Cantor began to publish his results on cardinalities, there was on the one hand the strong opposition of some leading mathematicians, like Kronecker and Poincaré, and, on the other hand, the growing interest on his ideas, especially among younger mathematicians. At the same time, however, even enthusiastic promoters of the theory were aware of some specific difficulties, of the kind suggested by the Burali-Forti argument. But, as already stressed, these difficulties were not initially considered to be real obstacles that would come to put a full stop to the further development of the theory. The definitive insight, however, that these difficulties were not just transient and that there was some deeper problem that required focused consideration came in 1903, when Russell published his highly influential book *The Principles of Mathematics*. The book, to be sure, played a crucial role in helping disseminate the new ideas of the theory of sets, as a significant step toward establishing a solid logical foundation for mathematics at large. But, at the same time, this was the text where Russell publicized the paradox now associated with his name, which calls to attention a basic logical difficulty inherent in Cantor's theory.

As we saw in Chapter 11 Section 11.4, Frege's definition of cardinals, and indeed all of his approach to the foundations of arithmetic, was based on the implicit, sweeping adoption of the "principle of comprehension," that is, the assumption that any property that one can think of determines a well-defined set. Dedekind, also, in his own work, implicitly assumed this principle, which appeared self-evident to contemporary mathematicians and logicians. Russell now came up and showed that this assumption involved a deep logical fallacy that damages the entire edifice that Frege worked so hard to construct. Russell's argument is simple and bold, and it is perhaps this very simplicity that forced some mathematicians working on the foundations of mathematics to rethink some of the basic notions that were until then considered to be totally unproblematic.

In order to understand the paradox, let us begin by pointing out that in the typical case, if we look at the list of elements of a set, we will not find the set itself as part of this list, and indeed it even seems unnatural to think about such a possibility. If we are given, for example, the set $X = \{1, b, \&, X\}$, it may seem strange at first sight that X

is one of the elements of the set X, but there is no immediate mathematical reason to reject this possibility. After all, we have a clear list of the elements of the set X, and in all cases we can give a clear-cut answer to the question whether or not a certain element belongs to X. For example, 1 belongs to X, 2 does not belong to X, *a* does not belong to X, Y does not belong to X, X belongs to X, and so on. Specifically, if we think this situation from the point of view of the principle of comprehension, we find no immediate reason to discard the possibility that a set will be one of the members of itself. So for instance, if we consider the set A of all sets, then it is clear that A is one of the elements of the set A, since A is a set. Moreover, we can separate the set A into two clearly distinguished classes, as follows:

G = {all sets that *do not* appear as one of the elements of the set itself},
B = {all sets that *do* appear as one of the elements of the set itself}.

Of course, most of the sets we usually handle are part of the class G, while sets like X or A, the set of all sets, are part of B. And it is also clear, at any rate, that any possible given set will be either in G or in B, and that there is no set that is simultaneously in B and in G. Now, the Russell paradox arises when we ask ourselves the following question about the set G: is G in the class G or in the class B? Let us consider, with Russell, what happens in either case:

- Let us assume "G is an element in G." Then, according to the definition of G, this would mean that G cannot be in the class G (since in the class G we will never find a set that is an element in the set itself). But this contradicts what we are assuming, namely, "G is an element in G." Hence, the assumption that "G is an element in G" is untenable.
- Let us assume "G is an element in B." Then, according to the definition of B, this would mean that G is in the class G (since in the class B we find all sets that are an element in the set itself). But this contradicts what we are assuming, namely "G is an element in B." Hence, the assumption that "G is an element in B" is untenable.

And here is the paradox: G is, as any other set, an element of the set A, and hence it should be either in G or in B. But it turns out that each of these situations leads to its negation, and hence none of the alternatives is logically possible. Hence, G cannot be in A! Russell himself coined an often-cited metaphor to illustrate what lies at the heart of the paradox. In a certain village, he said, the barber is the man who shaves all those men who do not shave themselves. Who, then, shaves the barber?

What is actually wrong here? What has led to this logically impossible situation? It turns out that the problem lies in the sweeping adoption of the principle of comprehension. This principle has allowed us to assume that, for instance, "being a set" is a property that gives rise to a well-defined set. Some properties, Russell concluded, do not give rise to sets. This is, by the way, the cautionary reason for using above the term "class," rather than set, in defining G and B. Although when defining them we do not immediately see that they are truly problematic, what the paradox shows is that, ultimately, one cannot treat them as sets, without further ado.

Need we limit the scope of the principle of comprehension? Which are exactly those properties that cannot give rise to sets? These are interesting questions that remained open after the publication of the Russell paradox, and it turned out that answering them is not an easy task. In particular, let us not assume that the paradox can be solved simply by banning all sets of the kind accounted for by class B above. This ban would certainly leave out the set A and hence the direct reason for the Russell paradox, but at the same time it would also leave out large classes of sets that are not logically damaging in the same sense and that indeed there is no reason to ban from mathematical use. The meaning and the possible ways to deal with the Russell paradox continue to be at the heart of many of the most crucial discussions in the philosophy of mathematics.

Two other important, strongly interconnected ideas related to the new theory of sets were also at the heart of intense debates at the turn of the twentieth century: the Well-Ordering Theorem (WO) and the Axiom of Choice (AC). I want to comment briefly on them.

A set A is said to be well-ordered if every non-empty subset of A has a *least element* under the given ordering. The set $\mathbb{N}$ of natural numbers is the most basic example of an infinite set that is well-ordered. In its usual ordering, every subset of $\mathbb{N}$ has indeed a minimal element. By contrast, the set $\mathbb{Z}$ of integers is *not* well-ordered in its usual ordering,

$$\mathbb{Z} = \{\ldots, -4, -3, -2, -1, 0, 1, 2, 3, 4, \ldots\}.$$

Think of the set of multiples of 2, for example. It grows indefinitely in both directions and cannot be said to have a *least* element. Still, it is easy to find a different ordering under which $\mathbb{Z}$ becomes a well-ordered set. Actually, we already know how to do this:

$$\mathbb{Z} = \{0, 1, -1, 2, -2, 3, -3, 4, -4, \ldots\}.$$

According to this same logic, it is clear that any countable set can be well-ordered, and this with the help of the one-to-one correspondence with $\mathbb{N}$ that makes the set countable. So, in the case of the rational numbers $\mathbb{Q}$, we have already seen how to order this set so that the correspondence can be established. This alternative ordering is as follows:

$$\mathbb{Q} = \left\{0, 1, -1, \frac{1}{2}, 2, \frac{-1}{2}, \frac{1}{3}, -2, \frac{-1}{3}, \frac{1}{4}, 3, \frac{2}{3}, \frac{-1}{4}, \frac{1}{5}, \ldots\right\}.$$

The truly interesting question concerns the possibility of well-ordering *any* set, and, in particular, sets that are not countable such as $\mathbb{R}$.

The question of well-ordering came up very early on in the work of Cantor. For example, an important, related result that Cantor formulated in 1878 concerns the comparability of cardinals. It reads as follows:

Whenever M and N do not have the same cardinality, then either M is equipotent to a part of N (i.e., the power of M is less than that of N) or N is equipotent to a part of M (i.e., the power of N is less than that of M).

This result appears to be intuitively self-evident, and indeed Cantor stated it without proof. It would gradually become clear, however, that it requires a proof, and this proof assumes WO. By 1884, Cantor had realized the centrality of well-ordering for his entire theory, but he did not see it as a principle that itself required a proof. In his view, this was a "basic law of thought with far-reaching consequences, especially remarkable for its general validity."[2] Not that he had an idea, for example, of what was exactly that alternative ordering of the set of real numbers that turns it into a well-ordered one. At any rate, it gradually became clear to him, that the principle was not so evident as it originally had appeared him to be, and that it actually required a proof.

The well-ordering principle came prominently to the forefront of mathematical attention in 1900. At the International Congress of Mathematicians held in Paris, Hilbert gave a lecture that became one the most memorable milestones in the history of modern mathematics. He presented a list of twenty-three open problems in various fields of mathematics that in his opinion should set the agenda of research for the mathematical community at large in the century that was about to begin. The influence of this list was enormous. Of course, I will not be able to go into the details of that story here.[3] I just want to focus on the great impact that Hilbert's list had on the development of Cantor's theory. The very first open problem in his list was the proof of Cantor's Continuum Hypothesis. There could be no more direct statement on the side of Hilbert of how central he considered Cantor's theory to be for the entire edifice of mathematics. But, moreover, he actually suggested, specifically, the need to prove WO for the case of the real numbers as a possible first step in the proof of CH.

Hilbert's speech motivated many of his students and collaborators, and many who were not part of the Hilbert circle as well, to begin studying the intricacies of a theory that had remained thus far largely in the margins of mainstream mathematical research. Still, the status of the well-ordering principle within the theory of infinite cardinalities remained unsettled for the next few years. An important turning point came in 1904, when Ernst Zermelo (1871–1953) finally published a proof. In 1897, Zermelo had arrived at Göttingen as an expert on statistical mechanics and mathematical physics, but soon shifted to set theory under the influence of Hilbert. He devoted his best efforts at the time to prove WO because, in his view, this principle had to be considered "the true fundament of the whole theory of number."[4]

The main idea underlying Zermelo's 1904 proof of WO is the use of the Axiom of Choice. In informal terms, what AC states is that given any collection of non-empty sets, there is always another set C such that C contains exactly one element (a "representative") from each of the sets in the collection. If the collection is finite, the validity of the axiom is self-evident. For instance, suppose that we are given the following four sets:

$$A_1 = \{a, b, c\}, \quad A_2 = \{x, a, p, 5\},$$

$$A_3 = \{t, c, 3, U\}, \quad A_4 = \{5, t, d\}.$$

---

[2] Quoted in (Ferreirós 1996, p. 277, note 2).
[3] See (Corry 2004; Gray 2001).
[4] Quoted in (Ebbinghaus and Peckhaus 2007, p. 63).

Two different examples of possible choice sets for this collection are the following:

$$C = \{b, p, U, 5\} \quad \text{or} \quad C = \{c, a, t, d\}.$$

Sometimes, even when the collection is infinite, a choice of representatives is totally unproblematic. This happens when there is some obvious selection rule available for making the choice. To see, however, that there may be situations in which the existence of the choice set C is not that obvious, we can use a nice illustration suggested by Russell, referring to the different cases of pairs of shoes or of socks. Given an infinite collection of pair of shoes, it is easy to construct the choice set C by taking, for example, the left shoe of each of the pairs. But given an infinite collection of pairs of socks, how do we know exactly how to build the choice set C? How can we know what exactly C looks like, what is there representing each pair and what is not? The Axiom of Choice is meant to come to our help here, by *stipulating*, without further explanations or arguments, that even when we cannot explicitly formulate the selection rule, we can simply be sure that the choice set C for the given collection just exists.

At the time of Zermelo's proof, the status of AC was no less controversial than that of WO itself. Some mathematicians thought that AC was self-evident and could be freely used wherever needed. Some thought that it was wrong or even devoid of meaning, and hence unacceptable in mathematics in general. Some thought that WO was more self-evident than AC and that it made no sense to prove the former with the help of the latter. Among those who rejected AC, there were some who had been using it (or some other result equivalent to AC) inadvertently in their own work, and in a crucial manner at that. One of the main merits of Zermelo's proof of WO using AC was that it sparked lively and fruitful debates about these two statements, about their mutual logical relationship, about their relation to some other similar mathematical statements, and about the way in which they had been used—explicitly or implicitly—in important proofs in various branches of mathematics.

It soon became clear, in the first place, that WO and AC are logically equivalent. Each of them can be derived from the other, and hence neither is more or less self-evident than the other. It also became clear that all known proofs of the fundamental theorem of calculus required and made use, at bottom, of AC or of some equivalent result. For this reason alone, AC could not just be ignored as a piece of esoteric mathematics. Zermelo was also led to formulate a detailed system of axioms for sets that helped clarify in a systematic manner what should be the basic assumptions at play behind a consistent implementation of the Cantorian theory. The Zermelo–Fraenkel system of axioms for set theory, ZF (so called because Zermelo's original system was corrected and slightly modified in the 1920s by Abraham Halevy Fraenkel (1891–1965)), soon became one of the most widely accepted conceptual frameworks for conducting research on sets.

Within ZF, for instance, it seemed possible to find a way to bypass the Russell paradox. In this system, the scope of validity of the principle of comprehension was not universal and had limitations. In this way, one could discern between sets, properly speaking, and "classes" (the latter being collections to which the principle of comprehension does not apply). Stated in these terms, the collection A of all sets, for example, is itself a class and not a set, and in this way the Russell paradox could be bypassed. Another important result was that AC could be consistently added to ZF as an independent axiom. More interestingly, this meant that, alternatively, ¬AC (that is, the negation of AC) can likewise be added consistently to ZF. We can accordingly imagine a mathematical world based on set theory as defined by ZF + AC and another, alternative one defined by ZF + ¬AC. I have already said something similar above about CH, and now we can say it more precisely: we can imagine a mathematical world based on set theory as defined by ZF + CH, and another defined by ZF + ¬CH.

I will not go any further into the story of the development of set theory and all the important ideas elaborated around it in the twentieth century.[5] That would really take us well beyond the topic of this book. But I do want to conclude this chapter by calling attention to the direct connection between the main ideas underlying the constructivist criticism by Kronecker of Cantorian set theory and the controversy around WO and AC. Kronecker, as I have stressed above, opposed the idea of completed infinities as unnecessary in mathematics and as potentially leading to logical inconsistencies. Few ideas embody so boldly as AC the kind of reasoning that Kronecker saw as pernicious for the discipline. The very essence of AC is to *postulate* the existence of a completed infinity without having to specify any procedure that explicitly defines or singles out its individual elements. The axiom tells us that the collection just exists but says nothing about its elements. AC is the opposite to what a constructivist mathematician of any kind would like to see as providing a respectable foundation for his field of knowledge. And yet, as Zermelo's work showed, accepting AC as a valid principle seemed a necessity for which there was no alternative at the time.

Over the early decades of the twentieth century, foundational debates in mathematics, to the extent that they attracted the attention of leading mathematicians, focused on the ideas of set theory, the logical paradoxes, and the status of AC and its equivalents. Controversies reached a peak in the 1920s, in the form of an open confrontation between Hilbert and Luitzen E. J. Brouwer (1881–1966). Brouwer was a strong supporter of mathematical constructivism. His own constructivist doctrine, known as "mathematical intuitionism," became the most visible alternative to the mainstream views about the Cantorian infinite throughout the twentieth century. In his strong opposition to the use of ideas like AC in mathematics, Brouwer was willing to pay the full price, both personal and in terms of the scientific substance. He proposed alternative ways to elaborate the foundations of calculus, without having to rely on AC or on any other non-constructive principle. As a result, he obtained certain fundamental theorems that differed, in significant ways, from those known and practiced as a matter of course by the entire community of mathematicians. Hilbert reacted quite aggressively to Brouwer's proposals. He did not hesitate to use his moral and institutional authority to marginalize Brouwer from the main venues of the profession. In the framework of his confrontation with Brouwer, Hilbert came up with one of his most famous and oft-cited statements. It has come to represent what for many was the core of the foundations of mathematics over the twentieth century, based on an unrestricted adoption of infinite sets seen as completed collections: "No one," wrote

---

[5] See (Ferreirós 1996; Hesseling 2003; Moore 2013).

Hilbert in 1926 in reaction to Brouwer's brand of intuitionism, a shall expel us from the paradise that Cantor has created for us."

Appendix 12.1 Proof that the set of algebraic numbers is countable

Cantor in 1873 proved that the set of algebraic numbers is countable. I give here a sketch of his proof. Recall that this set, which we denote here by $\mathbb{A}$, comprises all real numbers that are solutions of polynomials with rational coefficients. Notice, in the first place, that if q is a rational number, then q is surely algebraic, since q is the solution of the following polynomial equation with rational coefficients: $x - q = 0$. But, of course, there are also some irrational numbers that are algebraic. The simplest example is that of $\sqrt{2}$, which is the solution of the equation $x^2 - 2 = 0$. Now, in order to prove our statement about the cardinality of $\mathbb{A}$, we begin from the following, more general, result:

Lemma 12.1. *Given a countable collection, $A_1, A_2, A_3, \ldots, A_i, \ldots$, where each of the sets A_j is countable, if we define the set A to be the union of all sets in this collection, then A is a countable set.*

The proof of this general result is very similar to the proof, given in Section 12.1, that $\mathbb{Q}$ is a countable set. As usual, all we need to do is find a way to arrange all elements in A such that there is a first element, a second element, a third one, etc. So, we begin the proof by writing each of the sets in the collection as a sequence of this kind, given that, by hypothesis, these sets are themselves countable. For instance, we can write the elements of the set A_n, arranged as a sequence, as follows:

$$A_n : a_1^n, a_2^n, a_3^n, a_4^n, a_5^n, \ldots, a_i^n, \ldots$$

I am using here a format that is similar to the one used above for discussing the cardinality of $\mathbb{R}$: the symbol a_i^j, represents the ith element of the set A_j. We are now in a position to arrange all the members of the entire collection in an array of rows as displayed in Figure 12.7. Here each row i comprises the elements of A_i arranged as a sequence. Now, exactly as we did with the rationals, we start drawing diagonals from top-right to bottom-left, the diagonals increasingly growing as we move to the right, and thus eventually covering the entire collection, as displayed in Figure 12.8.

$a_1^1, a_2^1, a_3^1, a_4^1, a_5^1, \ldots a_i^1, \ldots$

$a_1^2, a_2^2, a_3^2, a_4^2, a_5^2, \ldots a_i^2, \ldots$

$a_1^3, a_2^3, a_3^3, a_4^3, a_5^3, \ldots a_i^3, \ldots$

$a_1^4, a_2^4, a_3^4, a_4^4, a_5^4, \ldots a_i^4, \ldots$

$a_1^5, a_2^5, a_3^5, a_4^5, a_5^5, \ldots a_i^5, \ldots$

$\vdots \quad \vdots \quad \vdots \quad \vdots \quad \vdots \quad \quad \vdots$

Figure 12.7 Members of a set in an infinite array of rows and columns.

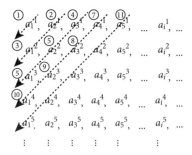

Figure 12.8 Diagonals in an infinite array of rows and columns.

In other words, now taking the numbers assigned to each element in the array, we see that all the elements in the array, which are all the elements in the union set A, can be ordered as follows:

$$A_n : a_1^1, a_2^1, a_1^2, a_3^1, a_2^2, a_1^3, a_4^1, a_3^2, a_2^3, a_1^4, a_5^1, \ldots$$

This essentially completes the proof, since, as requested, all the elements of the union set appear here in a sequence. Still, the following two clarifications are important:

- It is possible that a certain element appears in more than one of the sets of the collection. In this case, it would appear more than once in the array of rows and columns above. This is, however, not an obstacle to asserting that the sequence constructed with the help of the diagonals is exhaustive. If it so happened that, for example, the three elements a_5^1, a_2^2, a_1^3 appearing in three different sets were actually one and the same, then, in establishing the sequence of elements, we could simply skip them, so that the beginning of the ordering of elements in A would look as follows:

$$a_1^1, a_2^1, a_1^2, a_3^1, a_2^2, a_4^1, a_3^2, a_2^3, a_1^4, \ldots$$

- If one or more of the sets in the countable collection (and indeed even if all of them) were finite rather than countably infinite, then the same proof would work without problems. A more precise statement of the lemma would be

  Given a countable collection, $A_1, A_2, A_3, \ldots A_i, \ldots$, where each of the sets A_j is *either countable or finite*, if we define the set A to be the union of all sets in this collection, then A is a countable set.

We can now return to the set $\mathbb{A}$ of algebraic numbers and complete the proof that this is a countable set, based on Lemma 12.1. Indeed, let us denote by Q_1 the collection of all polynomials with rational coefficients and degree 1, that is, polynomials of the form $a_1x + a_0$, with a_1 and a_0 rational. For a fixed rational number q, let Q_{q1} be the collection of all polynomials of the form $qx + a_0$, with a_0 rational. Clearly, Q_{q1} is a countable set, since every rational number a_0 defines one and only one polynomial in Q_{q1}, and vice versa. Thus, there are as many polynomials in Q_{q1} as there are rational numbers. But now Q_1 is the union of all sets Q_{q1} when q runs over all rational numbers. Hence Q_1 is the union of a countable collection of countable sets, and hence, by Lemma 12.1, Q_1 is itself countable. Now let P_1 denote the set of all numbers that are roots of polynomials in Q_1. Since the polynomials in Q_1 are of degree one, there is one root for every polynomial in Q_1, and we can therefore conclude that Q_1 is a countable set.

We now move on to consider the set Q_2 of all polynomials of the form $a_2x^2 + a_1x + a_0$, with a_2, a_1 and a_0 rational. As before, for a fixed rational q, let Q_{q2} be the collection of all polynomials of the form $qx^2 + a_1x + a_0$, with a_1 and a_0 rational. Clearly, Q_{q2} is a countable set, since every polynomial $a_1x + a_0$ with a_1 and a_0 rational defines one and only one polynomial in Q_{q1}, and vice versa. So, there are as many polynomials in Q_{q2} as there are polynomials $a_1x + a_0$. But, as we have just seen, the set of all those polynomials is countable, and hence Q_{q2} is countable for each rational q. But Q_2 is the union of all sets Q_{q2} when q runs over all rational numbers. Hence Q_2 is the union of a countable collection of countable sets, and hence, again by Lemma 12.1, Q_2 is itself countable. If we now let P_2 denote the set of all numbers that are roots of polynomials in Q_2, then, since the polynomials in Q_2 are of degree two, there are at most two real roots for every polynomial in Q_2, and we can therefore conclude that P_2 is a countable set.

It is easy to see how this kind of reasoning can be extended to the sets Q_n and P_n, for any degree n. But, by definition, $\mathbb{A}$ is the union of all sets P_n, and hence $\mathbb{A}$ is the union of a countable collection of countable sets, and hence, by a final application of Lemma 12.1, we conclude that $\mathbb{A}$ is itself countable.

CHAPTER 13

Epilogue: Numbers in Historical Perspective

We began this long journey into our brief history of numbers with the Pythagoreans and their peculiar view of the role of numbers as the building blocks underlying all phenomena in the universe. In spite of their relatively high degree of technical sophistication, their overall program soon run into serious difficulties with the discovery of incommensurable magnitudes. As we saw, even by the seventeenth century, a period of time that signified a true watershed in so many scientific realms, the concept of number was, in many important respects, closer to that of the ancient Greeks than to the one that would be consolidated by the end of the nineteenth century. The elaborate picture of the entire hierarchy of number systems that arose in the wake of the works of mathematicians such as Dedekind, Cantor, Frege and Peano and which I have discussed in the preceding chapters is, in essence, one that is still accepted today. The various systems and the relationships among them are clearly defined in rigorous mathematical terms. The specific features of each of these systems, including the fields of algebraic numbers, are well known, as are their relationships with polynomials, their coefficients and their roots.

All of this has been exhaustively investigated over more than a century, and this knowledge has been thoroughly organized in the form of well-elaborated, advanced mathematical theories. Still, research on numbers and their properties continues to be a thriving field of activity, with ever new open questions and methodologies being developed by younger generations of talented mathematicians. Powerful computations interact in fascinating ways with innovative conceptual breakthroughs to yield important and sometimes unexpected results.

But the kind of questions about the legitimacy of using this or that kind of numbers, which were so central in the developments discussed throughout this book, are seen nowadays as meaningless, and indeed even as silly. Without the historical contexts discussed in the foregoing chapters, we can only consider as preposterous statements like "Newton denied the legitimate existence of negative numbers." What sense does it make to say, otherwise, that a mathematician of the caliber of Newton would not understand something as simple as the idea of a negative number? I hope that readers have come to understand the true meaning of these debates in their historical context.

And yet, as I have suggested when explaining the Russell paradox, the outcome of the long and winding historical process leading to the modern idea of number and the seemingly harmonious picture that I have presented at the end were themselves not devoid of problems and challenges. Indeed, mathematicians interested in foundational questions were quick to recognize them. Debates about the legitimacy of using irrationals or imaginary numbers receded into the background, but gave way to new debates about the legitimacy of using completed infinities and relying without restrictions on principles like the Axiom of Choice. Followers of Kronecker, Brouwer and the intuitionists favored, on both practical and philosophical grounds, banning them altogether from mathematical practice.

The mainstream foundational perspective throughout the twentieth century, however, remained the one promoted by Hilbert and his plea for taking full advantage of the completed infinities found in Cantor's paradise. The attempts to address the challenges that arose within this view led, from the early twentieth century on and to a large extent under the impact of Hilbert's own groundbreaking research in this area, to the rise of new mathematical disciplines typically assembled under the umbrella term "meta-mathematics." These disciplines explore various kinds of questions that originally arose from the works discussed in the preceding chapters, and that soon expanded into new, and rather surprising, directions as well. The hierarchical view of the entire system of numbers allows most of the foundational concerns in mathematics to be funneled toward technical and rather specific questions about sets, logic, and axiomatic systems. New and deep questions arose about the precise capabilities of the logical systems that provide the underlying infrastructure that supports the deductive edifice of the various mathematical theories. The famous incompleteness theorems associated with the names of Kurt Gödel (1906–1978) and Alan Turing (1912–1954) opened new avenues of research where the deductive capabilities of any axiomatic system could be analyzed in detail and their limitations could be very precisely determined.

Research on fields related to the foundations of mathematics has provided full-time employment to a large number of talented and diligent mathematicians since the beginning of the twentieth century, and will continue to do so. The fascinating mathematical results that they have produced have also provided job security to many philosophers who have added their share of interpretation and clarification (and sometimes obscurity as well). But it is important to stress that advanced research in the various mathematical domains, classical as well as those newly established in the twentieth century, continued to make progress and to flourish while expressing little concern about (and often with little knowledge of) the important work done by fellow meta-mathematicians in foundational questions.

At the same time, important developments over the last few decades—and prominently so in the early twenty-first century—are leading to new kinds of questions that might have the potential to turn the tide once again. The increasing predominance of the use of electronic computers in mathematics is a phenomenon that is bound to have long-ranging consequences also on foundational conceptions of what are numbers and what should they be. The legitimacy of Cantorian actual infinities is being questioned once again along the lines of Kroneckerian constructivism, but from surprisingly new, and perhaps stronger, perspectives.

This situation is, of course, no different from what we have seen repeatedly throughout this book. Even before the work of Dedekind et al., and certainly after that, there has typically been an interesting interplay between the development of mathematics in its various branches and the development of the foundations of mathematics at large (the concept of number being included here), but there has hardly been a necessary binding dependence of the former on the latter. From the time of the ancient Greeks and up to our own days, the internal dynamics of research—sometimes in interactions with external factors such as research in physics and more recently also biology, economics, or computer science—was typically the main driving force behind the development of new ideas in mathematics, even when these developments seemed to contradict basic underlying assumptions (implicit or explicit) about numbers. These assumptions were continually challenged and reconsidered, and also eventually modified under the pressure of developments in research in the cutting-edge disciplines of advanced mathematics.

A forceful description of this fundamental tension inherent in the development of mathematics was expressed in 1905 by Hilbert in one of his lectures in Göttingen. It is very appropriate to conclude this book by quoting him. Hilbert, as I have stressed in the preceding chapters, left his mark on many fields of advanced mathematical research, but also in foundational, meta-mathematical domains. He was as aware as anyone can be of the dialectial relationship between these two layers of knowledge. Prone to express his ideas in the form of illuminating metaphors, he conveyed this important message to his students in the following words:[1]

The edifice of science is not raised like a dwelling, in which the foundations are first firmly laid and only then one proceeds to construct and to enlarge the rooms. Science prefers to secure as soon as possible comfortable spaces to wander around and only subsequently, when signs appear here and there that the loose foundations are not able to sustain the expansion of the rooms, it sets about supporting and fortifying them. This is not a weakness, but rather the right and healthy path of development.

Throughout the chapters of this book, we have visited some of the main historical crossroads where significant gaps appeared between the foundations provided by the prevailing ideas about numbers and the new directions of thought toward which advanced research was pushing. I hope to have conveyed a clear and convincing account of the ways in which those prevailing ideas were repeatedly revised, so that the shaky edifice of mathematics could be fortified and set once again on solid ground. But I also hope to have made it clear that its rooms and corridors continue to be enlarged in ever new and unsuspected ways, and that the process of clarification and fortification of the pillars, for all of its successes and breakthroughs, is, in fact, a never-ending one.

---

[1] Quoted in (Corry 2004 [1996], p. 162).

REFERENCES AND SUGGESTIONS FOR FURTHER READING

The history of mathematics is a very active and dynamic field of academic research. Historians continually add to its basic repertoire innovative perspectives of enquiry, previously unexplored archival sources, and new interpretive approaches to existing material. They critically analyze and often contest existing interpretations, while engaging in ongoing, learned debates, some of which remain unsettled. Readers whose curiosity and intellectual interest has been aroused by this book can further develop and deepen their knowledge by reading a wide range of professional journals and books devoted to focused studies of the mathematical cultures, the ideas, and the mathematicians whose works we have only been able to discuss in a somewhat cursory way here. With this aim in mind, I have put together a sample of titles, selected from among the much larger collection of existing, relevant publications. I have included works that are more or less directly connected with the issues discussed in the various chapters. Typically, the question of the development of the concept of number, which was at the focus of our discussion here, appears in the texts cited below only as part of broader discussions on other topics related to the historical development of mathematics in general.

History of mathematics: general sources

It goes without saying that the Internet is filled with informative and well-designed websites that deal with different aspects of the history of mathematics. One of the most popular, broad and reliable of these is the *Mac Tutor History of Mathematics Archive*, created and maintained by John J. O'Connor and Edmund F. Robertson at the University of St. Andrews, Scotland. This is a website definitely worth visiting for anyone interested in th history of mathematics. Among other things, it offers a comprehensive collection of biographies of the most prominent, as well as of some of the less prominent, mathematicians in history. It can be reached at <http://www-history.mcs.st-and.ac.uk>.

The best single piece of advice for anyone interested in the history of mathematics and, no less than that, for anyone working in mathematical research and teaching is, simply, "read the masters." An unmediated acquaintance with the texts where the great ideas were originally presented and explored is the best way to appreciate not just the intrinsic beauty of those ideas, but also the difficulties inherent in trying to come to terms with them in their initial stages of emergence.

Also in this regard, the Internet provides an invaluable tool, since large amounts of scanned texts of all periods and in all mathematical disciplines are continually made available for anyone interested. I mention here just three of the existing websites where many such sources can be found:

- DML: Digital Mathematics Library, <http://www.mathematik.uni-bielefeld.de/~rehmann/DML/dml_links.html>
- Livres Numériques Mathématiques–Digital Math Books, <http://sites.mathdoc.fr/LiNuM>
- *Namas Te*: Sources on Natural Sciences in Antiquity, <http://wilbourhall.org>.

English translations of original sources are also available in collections that will be of interest to readers of this book. They cover various historical periods of time and various geographical contexts. The following is a list some of them:

- Ewald, William (ed.) (1996), *From Kant to Hilbert: A Source Book in the Foundations of Mathematics*, Oxford: Clarendon Press.
- Fauvel, John and Jeremy Gray (eds.) (1988), *The History of Mathematics. A Reader*, London: Macmillan.
- Katz, Victor (ed.) (2007), *The Mathematics of Egypt, Mesopotamia, China, India, and Islam: A Sourcebook*, Princeton, NJ: Princeton University Press.
- Katz, Victor et al (ed.) (2016), *The Mathematics of Medieval Europe and North Africa: A Sourcebook*, Princeton, NJ: Princeton University Press.
- Mancosu, Paolo (ed.) (1998), *From Brouwer to Hilbert. The Debate on the Foundations of Mathematics in the 1920s*, Oxford: Oxford University Press.
- Smith, David .E. (ed.) (1929), *A Source Book in Mathematics*, New York: Dover.
- Stedall, Jacqueline (ed.) (2008), *Mathematics Emerging: A Sourcebook (1540–1900)*, New York: Oxford University Press.
- van Heijenoort, Jean (1967), *From Frege To Gödel, A Source Book in Mathematical Logic, 1879–1931*, Cambridge, MA: Harvard University Press.

Also of interest for readers of this book are several works devoted to general overviews of the history of mathematics. The following are among the most well-known and frequently used ones:

- Grattan-Guinness, Ivor (1994), *Companion Encyclopedia of the History and Philosophy of the Mathematical Sciences* (2 vols.), London: Routledge.
- Katz, Victor J. (2008), *A History of Mathematics - An Introduction*, 3rd edn., Reading, MA: Addison-Wesley.
- Katz, Victor J. and Karen Hunger Parshall (2014), *Taming the Unknown: A History of Algebra from Antiquity to the Early Twentieth Century*, Princeton, NJ: Princeton University Press.
- Jahnke, Hans Niels (ed.) (2003), *A History of Analysis*, Providence, RI: American Mathematical Society.
- Kleiner, Israel (2006), *A History of Abstract Algebra*, Boston: Birkhäuser.
- Martinez, Alberto (2014), *Negative Math: How Mathematical Rules Can Be Positively Bent*, Princeton, NJ: Princeton University Press.
- Robson, Eleanor and Jacqueline Stedall (eds.) (2009), *The Oxford Handbook of the History of Mathematics*, Oxford: Oxford University Press.
- Stedall, Jacqueline (ed.) (2012), *The History of Mathematics: A Very Short Introduction*, Oxford: Oxford University Press.
- Stillwell, John (2010), *Mathematics and Its History*, 3rd edn., New York: Springer.

I should also mention some books that focus on the history of the concept of number or that deal with the history of specific kinds of numbers. Their overall approach and their scope differ in many respects from the one I have followed in this book.

- Benoit, Paul et al (eds.) (1992), *Historie de fractions / Fractions d'histoire*, Basel: Birkhäuser.
- Berggren, J. Len, Jonathan M. Borwein, and Peter Borwein (eds.) (2004), *Pi: A Source Book*, 3rd edn., New York: Springer.
- Clawson, Calvin C. (1994), *The Mathematical Traveler: Exploring the Grand History of Numbers*, Cambridge, MA: Perseus.
- Crossley, John N. (1987), *The Emergence of Number*, Singapore: World Scientific.
- du Sautoy, Marcus (2003), *The Music of the Primes: Searching to Solve the Greatest Mystery in Mathematics*, New York: HarperCollins.
- Flegg, Graham (2002), *Numbers: Their History and Meaning*, Mineola, NY: Dover.
- Higgins, Peter M. (2011), *Numbers: A Very Short Introduction*, Oxford: Oxford University Press.
- Mazur, Barry (2003), *Imagining Numbers (Particularly the Square Root of Minus Fifteen)*, New York: Farrar, Straus and Giroux.
- Sondheimer, Ernst and Alan Rogerson (2006), *Numbers and Infinity: A Historical Account of Mathematical Concepts*, Mineola, NY: Dover.

Suggestions for further reading by chapter

Readers interested in more specialized texts on any of the topics covered in the various chapters of this book will find suggestions in what follows. I have endeavored to include updated research literature—both articles and books—alongside references to primary sources in translation. In some cases, I have also included items in languages other than English. This is not intended, in any way, as an exhaustive report on the existing, relevant literature. Rather, it is meant as a representative sample. Of course, sources from which I quoted passages in the book, do appear in this list.

Chapters 1 and 2

- Chrisomalis, Stephen (2010), *Numerical Notation: A Comparative History*, Cambridge: Cambridge University Press.
- Porter, Theodore M. (1996), *Trust in Numbers*, Princeton, NJ: Princeton University Press.
- Serfati, Michel (2010), *La révolution symbolique: La constitution de l'écriture symbolique mathématique*, Paris: Editions PETRA.

Chapters 3 and 4

- Christianidis, Jean (2004), "Did the Greeks have the notion of common fraction? Did they use it?," in Christianidis, J. (ed.), *Classics in the History of Greek Mathematics*, Dordrecht: Kluwer, pp. 331–336.
- Christianidis, Jean (2007), "The way of Diophantus: some clarifications on Diophantus' method of solution," *Historia Mathematica* 34, 289–305.

- Christianidis, Jean and Jeffrey Oaks (2013), "Practicing algebra in late antiquity: the problem-solving of Diophantus of Alexandria", *Historia Mathematica* 40, 127–163.
- Dijksterhuis, Eduard Jan (1987), *Archimedes* (translated by C. Dikshoorn; with a new bibliographic essay by Wilbur R. Knorr), Princeton, NJ: Princeton University Press.
- Fowler, David (1987), *The Mathematics of Plato's Academy. A New Reconstruction*, New York: Oxford University Press.
- Fried, Michael and Sabetai Unguru (2001), *Apollonius of Perga's Conica: Text, Context, Subtext*, Leiden: Brill.
- Heath, Thomas L. (1956), *The Thirteen Books of Euclid's Elements*, 2nd edn., New York: Dover.
- Klein, Jakob (1968), *Greek Mathematical Thought and the Origin of Algebra* (translated by Eva Brann), Cambridge, MA: MIT Press.
- Knorr, Wilbur R. (1975), *The Evolution of the Euclidean Elements*, Dordrecht: Reidel.
- Knorr, Wilbur R. (1985), *The Ancient Tradition of Geometrical Problems*, Boston: Birkhäuser.
- Mueller, Ian (1981), *Philosophy of Mathematics and Deductive Structure in Euclid's Elements*, Cambridge, MA: MIT Press.
- Netz, Reviel (1999), *The Shaping of Deduction in Greek Mathematics*, Cambridge: Cambridge University Press.
- Schappacher, Norbert (2005), "Diophantus of Alexandria: a Text and its History," <http://www-irma.u-strasbg.fr/~schappa/NSch/Publications_files/Dioph.pdf>.
- Tannery, Paul (1893–95), *Diophanti Alexandrini Opera omnia cum graecis commentariis* (2 vols.), Leipzig: Teubner. (Reprint, Stuttgart: Teubner 1974.)
- Unguru, Sabetai (1975), "On the need to rewrite the history of Greek mathematics", *Archive for History of Exact Sciences* 15(1): 67–144.
- Van Brummelen, Glen (2009), *The Mathematics of the Heavens and the Earth: The Early History of Trigonometry*, Princeton, NJ: Princeton University Press.
- Ver Eecke, Paul (1959), *Diophante d'Alexandrie, les six livres arithmétiques et le livre des nombres polygones*, Paris: Blanchard.

Chapter 5

- Berggren, Len (1986), *Episodes in the Mathematics of Medieval Islam*, Berlin: Springer (2nd edn., 2003).
- Brentjes, Sonja (2014), "Teaching the mathematical sciences in Islamic societies: eighth–seventeenth centuries," in Karp, Alexander and Gert Schubring (eds.), *Handbook on the History of Mathematics Education*, New York: Springer, pp. 85–107.
- Djebbar, Ahmed (2005), *L'algèbre arabe: genèse d'un art*, Paris: ADAPT.
- Freudenthal, Gad, (ed.) (1992), *Studies on Gersonides: A Fourteenth-Century Jewish Philosopher-Scientist*, Leiden: Brill.
- Gutas, Dimitry (1998), *Greek Thought, Arabic Culture: The Graeco-Arabic Translation Movement in Baghdad and Early 'Abbāsid Society*, London: Routledge.
- Lange, Gerson (1909), *Sefer Maassei Chosheb. Die Praxis des Rechners. Ein hebräisch-arithmetisches Werk des Levi ben Gerschom aus dem Jahre 1321*, Frankfurt am Main: Louis Golde.
- Levy, Tony (1997), "The establishment of the mathematical bookshelf of the medieval Hebrew scholar: translations and translators," *Science in Context* 10: 431–451.

- Linden, Sebastian (2012), *Die Algebra des Omar Chayyam*, München: Edition Avicenna.
- Netz, Reviel (2004), *The Transformation of Mathematics in the Early Mediterranean World: From Problems to Equations*, Cambridge: Cambridge University Press.
- Oaks, Jeffrey (2011a), "Al-Khayyām's Scientific Revision of Algebra," *Suhayl: Journal for the History of the Exact and Natural Sciences in Islamic Civilisation* 10: 47–75.
- Oaks, Jeffrey (2011b), "Geometry and proof in Abū Kāmil's algebra," *Actes du 10ème Colloque Maghrébim sur l'Histoire des Mathématiques Arabes*. Tunis: L'Association Tunisienne des Sciences Mathématiques, pp. 234–256.
- Oaks, Jeffrey and Haitham M. Alkhateeb (2007), "Simplifying equations in Arabic algebra," *Historia Mathematica* 34(1): 45–61.
- Rashed, Roshdi (1994), *The Development of Arabic Mathematics: between Arithmetic and Algebra* (translated by A. F. W. Armstrong), Dordrecht: Kluwer.
- Rashed, Roshdi (2009), *Al-Khwārizmī: The Beginnings of Algebra*, London: Saqi.
- Rashed, Roshdi and Bijan Vahabzadeh (1999), *Al-Khayyām Mathématicien*, Paris: Blanchard.
- Saidan, Ahmad Salim (1978), *The arithmetic of al-Uqlīdisī: The story of Hindu-Arabic arithmetic as told in Kitāb al-Fuṣūl fī al Ḥisāb al-Hindī, by Abū al-Ḥasan Aḥmad ibn Ibrāhīm al-Uqlīdisī, written in Damascus in the year 341 (A.D. 952/3)*, Dordrecht: Reidel.

Chapter 6

- Corry, Leo (2013), "Geometry and arithmetic in the medieval traditions of Euclid's *Elements*: a view from Book II", *Archive for History of Exact Sciences* 67(6): 637–705.
- Folkerts, Menso (2006), *The Development of Mathematics in Medieval Europe. The Arabs, Euclid, Regiomontanus*, Aldershot: Ashgate.
- Franci, Raffaella (2010): "The history of algebra in Italy in the 14th and 15th centuries: some remarks on recent historiograpphy", *Actes d'Història dela Ciència i de la Tècnica*, N.E. 3(2): 175–194.
- Gavagna, Veronica (2012), "La soluzione per radicali delle equazioni di terzo e quarto grado e la nascita dei numeri complessi: Del Ferro, Tartaglia, Cardano, Ferrari, Bombelli," <http://web.math.unifi.it/archimede/note_storia/gavagna-complessi.pdf>.
- Gavagna, Veronica (2014), "Radices sophisticae, racines imaginaires: the Origin of complex numbers in the late Renaissance", in Rossella Lupacchini and Annarita Angelini (eds.), *The Art of Science. Perspectival Symmetries Between the Renaissance and Quantum Physics*, Cham: Springer, pp. 165–190.
- Heeffer, Albrecht (2008), "On the nature and origin of algebraic symbolism", in Bart Van Kerkhove (ed.), *New Perspectives on Mathematical Practices. Essays in Philosophy and History of Mathematics*, Singapore: World Scientific, pp. 1–27.
- Heeffer, Albrecht and M. van Dyck (eds.) (2010), *Philosophical Aspects of Symbolic Reasoning in Early Modern Mathematics*, London: College Publications.
- Høyrup, Jens (2007), *Jacopo da Firenze's Tractatus Algorismi and Early Italian Abbacus Culture*, Basel: Birkhäuser.
- Rommevaux, Sabine et al. (eds.) (2012), *Pluralité de l'algèbre à la Renaissance*, Paris: Champion.
- Rose, Paul Lawrence (1975), *The Italian Renaissance of Mathematics. Studies on Humanists and Mathematicians from Petrarch to Galileo*, Genève: Librairie Droz.

- Sesiano, Jaques (1985), "The appearance of negative solutions in mediaeval mathematics," *Archive for History of Exact Sciences* 32 (2): 105–150.
- Stedall, Jackie (2011), *From Cardano's Great Art to Lagrange's Reflections: Filling A Gap in the History of Algebra*, Zürich: EMS Publishing House.
- Wagner, Roy (2010), "The natures of numbers in and around Bombelli's *L'algebra*," *Archive for History of Exact Sciences* 64 (5): 485–523.

Chapters 7 and 8

- Bos, Henk (2001), *Redefining Geometrical Exactness*, New York: Springer.
- Descartes, René (1637), *The Geometry of René Descartes* (translated by David Eugene Smith and Marcia L. Lantham), New York: Dover (1954).
- Feingold, Mordechai (ed.) (1999), *Before Newton—The Life and Times of Isaac Barrow*, Cambridge: Cambridge University Press.
- Guicciardini, Niccolò (2003), *Reading the Principia: The Debate on Newton's Mathematical Methods for Natural Philosophy from 1687 to 1736*, Cambridge: Cambridge University Press.
- Guicciardini, Niccolò (2009), *Isaac Newton on Mathematical Certainty and Method*, Cambridge, MA: MIT Press.
- Hill, Katherine (1996–97), "Neither ancient nor modern: Wallis and Barrow on the composition of continua," *Notes and Records of the Royal Society* 50, 165–178; 51, 13–22.
- Mahoney, Michael Sean (1994), *The Mathematical Career of Pierre de Fermat, 1601–1665*, Princeton, NJ: Princeton University Press.
- Malet, Antoni (2006), "Renaissance notions of number and magnitude," *Historia Mathematica* 33 (1), 63–81.
- Mancosu, Paolo (1996), *Philosophy of Mathematics and Mathematical Practice in the Seventeenth Century*, New York: Oxford University Press.
- Neal, Katherine (2002), *From Discrete to Continuous: The Broadening of the Number Concepts in Early Modern England*, Dordrecht: Kluwer.
- Pierce, R.J. (1977), "A brief history of logarithms," *The Two-Year College Mathematics Journal* 8 (1), 22–26.
- Pycior, Helena (2006), *Symbols, Impossible Numbers, and Geometric Entanglement. British Algebra through the Commentaries On Newton's Universal Arithmetick*, Cambridge: Cambridge University Press.
- Whiteside, Derek Thomas (1970), "The mathematical principles underlying Newton's *Principia Mathematica*," *Journal for the History of Astronomy* 1, 116–138.

Chapters 9, 10, 11, and 12

- Crowe, Michael J. (1994), *A History of Vector Analysis: The Evolution of the Idea of a Vectorial System*, New York: Dover.
- Corry, Leo (1996), *Modern Algebra and the Rise of Mathematical Structures*, Basel and Boston: Birkhäuser (2nd edn., 2004).
- Corry, Leo (2004), *Hilbert and the Axiomatization of Physics (1898–1918): From "Grundlagen der Geometrie" to "Grundlagen der Physik,"* Dordrecht: Kluwer.
- Dedekind, Richard (1963), *Essays on the Theory of Numbers*, New York: Dover.

- Ebbinghaus, Heinz-Dieter and Volker Peckhaus (2007), *Ernst Zermelo. An Approach to His Life and Work*, New York: Springer.
- Euler, Leonhard (2006), *Elements of Algebra*, St. Albans, Hertfordshire: Tarquin Reprints.
- Ferreirós, José (1996), *Labyrinth of Thought. A History of Set Theory and Its Role in Modern Mathematics*, Basel: Birkhäuser (2nd edn., 2007).
- Goldstein, Catherine, Norbert Schappacher, and Joachim Schwermer (eds.) (2007), *The Shaping of Arithmetic after C. F. Gauss's Disquisitiones Arithmeticae*, New York: Springer.
- Grattan-Guinness, Ivor (2000), *The Search for Mathematical Roots, 1870–1940: Logics, Set Theories and the Foundations of Mathematics from Cantor through Russell to Gödel*, Princeton, NJ: Princeton University Press.
- Gray, Jeremy J. (2001), *The Hilbert Challenge*, Oxford: Oxford University Press.
- Gray, Jeremy J. (2008), *Plato's Ghost: The Modernist Transformation of Mathematics*, Princeton, NJ: Princeton University Press.
- Hesseling, Dennis E. (2003), *Gnomes in the Fog: The reception of Brouwer's Intuitionism in the 1920s*, Basel: Birkhäuser.
- Lützen, Jesper (ed.) (2001), *Around Caspar Wessel and the Geometric Representation of Complex Numbers*, Copenhagen: The Royal Danish Academy of Sciences and Letters.
- Moore, Gregory H. (2013), *Zermelo's Axiom of Choice: Its Origins, Development, and Influence*, New York: Dover.
- Schubring, Gert (2005), *Conflicts Between Generalization, Rigor, and Intuition, Number Concepts Underlying the Development of Analysis in 17th–19th Century France and Germany*, New York: Springer.

NAME INDEX

A

Abū Kāmil ibn Aslam (ca. 850–ca. 930) 103, 105, 106, 118, 125, 129
Adelard of Bath (ca. 1080–ca. 1150) 126
al-Aḥdab, Yitzhak Ben Shlomo (ca. 1350–ca. 1429) 117
al-Kāshī, Ghiyāth al-Dīn (1380–1429) 110
al-Karajī, Abū Bakr (ca. 953–ca. 1029) 110, 118
al-Khayyām, 'Umar (1048–ca. 1131) 111–114, 116, 121–123, 129, 139, 201
al-Khwārizmī, Muḥammad ibn Mūsā (ca. 780–ca. 850) 90–106, 108–112, 115–118, 121, 125, 128, 131, 140–143, 151, 201
al-Kindī, Abu Yūsuf ibn 'Isḥāq (ca. 800–ca. 873) 90
al-Ma'mūn, Abū Ja'far (ca. 783–833) 89
al-Manṣūr, Abū Ja'far (714–775) 89
al-Rashīd, Hārūn (ca. 766–809) 89
al-Samaw'al, Ibn Yaḥyā al-Maghribī (ca. 1125–1174) 109, 118
al-Uqlīdisī, Abū al-Ḥasan Aḥmad ibn Ibrāhīm (ca. 920–ca. 980) 107, 109, 167
Alkhateeb, Haitham M. 98
Apollonius of Perga (ca. 262–ca. 190 BCE) 29, 32, 51, 90, 97, 121–123, 125, 183, 191
Archimedes of Syracuse (ca. 287–ca. 212 BCE) 29, 32, 48–51, 70, 81, 90, 125, 158
Argand, Jean-Robert (1768–1822) 212, 213, 218
Aristarchus of Samos (310–230 BCE) 29
Aristotle (384–322 BCE) 52, 54

B

Bar-Ḥiyya ha-Nasi, Abraham (1070–1136) 117, 125
Barrow, Isaac (1630–1677) 188–192, 198, 201, 203–205
Berggren, Len 50, 68, 163
Bolzano, Bernard (1781–1848) 266, 267
Boole, George (1815–1864) 218
Bortolotti, Ettore (1866–1947) 148
Bos, Henk 180, 181
Briggs, Henry (1561–1630) 167–171
Brouwer, Luitzen E. J. (1881–1966) 286
Buée, Adrien-Quentin (1748–1826) 212, 213
Burali-Forti, Cesare (1861–1931) 279–281

C

Campanus de Novara (1220–1296) 126
Cantor, Georg (1845–1918) 16, 237, 242, 258, 265–281, 287, 291
Cardano, Girolamo (1501–1576) 138, 139, 141–148, 156–158, 162, 163, 180, 194
Cauchy, Augustin-Louis (1789–1857) 236, 266
Cavalieri, Bonaventura (1598–1647) 266
Cayley, Arthur (1821–1895) 221
Chuquet, Nicolas (ca. 1445–ca. 1500) 134, 167
Clavius, Christopher (1538–1612) 59, 149, 150, 156, 163, 202
Clifford, William Kingdon (1845–1879) 221
Cohen, Paul (1934–2007) 275
Copernicus, Nicolaus (1473–1543) 156
Corry, Leo (1996–2013) 201, 214, 229, 284, 293

D

d'Alembert, Jean le Rond (1717–1783) 208, 209, 212
da Vinci, Leonardo (1452–1519) 137
Dardi, Maestro (14th century) 137, 145
De Morgan, Augustus (1806–1871) 218

Dedekind, Richard
(1831–1916) 9, 43,
220, 223–225, 228–245,
249–255, 257–263,
265–269, 275, 276, 281,
291–293
del Ferro, Scipione
(1465–1526) 139
Descartes, René
(1596–1650) 14,
50–52, 82, 142, 159,
161–163, 175–184,
187–193, 196, 198, 208,
211, 259
Diderot, Denis
(1713–1784) 208
Dijksterhuis, Eduard Jan
(1892–1965) 50
Diophantus of Alexandria
(ca. 201–ca. 285) 32,
71–80, 83–85, 98, 111,
127, 147, 156–159

E

Ebbinghaus, Heinz-Dieter 284
Euclid of Alexandria (fl. 300
BCE) 32, 43–45, 50,
51, 55, 58, 59, 63–66,
70, 78, 90, 93, 103, 105,
117, 119–121, 125, 126,
136, 149, 150, 163–165,
188, 191, 214, 227, 242,
243, 267
Eudoxus of Cnidus (408–355
BCE) 42, 43, 209
Euler, Leonhard
(1707–1783) 169,
210–213, 218

F

Fauvel, John 69, 123
Ferrari, Ludovico
(1522–1565) 139
Ferreirós, José 284, 286
Fibonacci, Leonardo Pisano
(ca. 1170–ca. 1240) 117,
128–130, 136, 150, 167
Fraenkel, Abraham Halevy
(1891–1965) 285
Français, Jacques
(1775–1833) 212, 213

Frege, Gottlob (1848–1925)
259–263, 265, 274,
281, 291
Frobenius, Georg Ferdinand
(1849–1917) 219, 220

G

Gauss, Carl Friedrich
(1777–1855) 14, 152,
210, 211, 214, 215, 217,
223–226, 230
Gavagna, Veronica 140
Gerard de Cremona
(ca. 1114–ca. 1187) 117
Gersonides: Levy Ben
Gerson (Ralbag)
(1288–1344) 117–120,
125, 251
Girard, Albert
(1595–1632) 180
Gödel, Kurt
(1906–1978) 275, 292
Grassmann, Hermann Günther
(1809–1877) 221
Gray, Jeremy J. 69, 123, 284
Gregory, Duncan Farquharson
(1813–1844) 218

H

Halley, Edmond
(1656–1742) 191
Hamilton, William Rowan
(1805–1865) 216–224,
231–234, 249, 259
Harriot, Thomas
(1560–1621) 135,
182
Heath, Sir Thomas L.
(1861–1940) 38, 44,
51–57, 198, 200, 201
Heeffer, Albrecht 137
Hermite, Charles
(1822–1901) 13
Heron of Alexandria (ca.
10–ca. 70) 80, 81
Hesseling, Dennis 286
Hilbert, David (1862–
1943) 230, 275, 279, 280,
284, 293
Hippasus of Metapontum (fl.
5th century BCE) 41
Hippocrates of Khios (ca.
470–ca. 410 BCE) 68, 70

Hobbes, Thomas
(1588–1679) 188
Hypatia of Alexandria (ca.
370–415) 31

I

Iamblichus (245–325) 41
ibn al Bannā' al Marrākushī
(1256–1321) 110, 117
ibn Qurra, Thābit
(826–901) 90
Ibn-Ezra, Abraham
(1092–1167) 117

K

Kant, Immanuel
(1724–1804) 219,
220, 259
Katz, Victor 101, 132
Klein, Jacob (1899–1978) 41,
42, 47, 76
Kleiner, Israel 147
Kronecker, Leopold (1823–
1891) 228, 280, 281,
286
Kummer, Eduard Ernst
(1810–1893) 227–229

L

Lange, Gershom 119
Lange, Gerson 118
Leibniz, Gottfried Wilhelm
(1646–1716) 149, 207,
208, 235, 236
Lejeune-Dirichlet, Peter
(1805–1859) 225
Lindemann, Ferdinand von
(1852–1939) 12, 13
Lipschitz, Rudolf
(1831–1916) 242,
243, 245

M

Mahoney, Michael S. 159
Martinez, Alberto 211
Menaechmus (ca. 380–ca. 320
BCE) 68, 69
Méray, Karl (1835–1911) 237
Moore, Gregory H. 286

N

Napier, John (1550–1617) 167, 169–173
Neal, Katherine (2002) 202, 204
Newton, Isaac (1642–1727) 57, 155, 175, 188, 190–195, 207, 208, 211, 235, 236, 243, 291
Noether, Emmy Amalie (1882–1935) 230

O

Oaks, Jeffrey 98, 105, 106
Oresme, Nicole (ca. 1323–1382) 131
Oughtred, William (1575–1660) 182, 190

P

Pacioli, Luca (1445–1517) 137
Pappus of Alexandria (ca. 290–ca. 350) 70, 71, 81, 83, 129, 159, 179
Parshall, Karen 132
Peacock, George (1791–1858) 218
Peano, Giuseppe (1858–1932) 251–259, 261–263, 278, 279, 291
Peckhaus, Volker 284
Poincaré, Jules Henri (1854–1912) 280, 281
Ptolemy, Claudius (85–165) 30, 32, 111, 113, 125
Pythagoras of Samos (ca. 596–ca. 475 BCE) and the Pythagoreans 32–41, 44, 52, 54, 64, 67, 78, 291

Q

Qusṭā ibn Lūqā (820–912) 72, 75, 111

R

Rashed, Roshdi 92, 98, 114, 115
Recorde, Robert (1510–1558) 135
Riemann, Bernhard 1826–1866) 224
Robert of Chester (fl. c. 1150) 126
Roche, Estienne de La (1470–1530) 135
Rolle, Michel (1652–1719) 238
Rudolff, Christoff (1499–1545) 131, 135
Russell, Bertrand (1872–1970) 259, 262, 281–283, 292

S

Saidan, Ahmad S. 108, 109
Senruset I (1956–1911 BCE) 25
Smith, David E, (1860–1944) 141
Stevin, Simon (1548–1620) 28, 149, 163–167, 169–171, 184, 242
Stifel, Michael (1487–1567) 135, 137, 138, 158, 167, 168
Sylvester, James Joseph (1814–1897) 221

T

Tūsī, Sharaf al-Dīn (ca. 1135–1213) 110
Tait, Peter Guthrie (1831–1901) 221, 222
Tannery, Jules (1848–1910) 73
Tartaglia, Niccolò Fontana (1500–1557) 139
Thales of Miletus (ca. 624–ca. 546 BCE) 31
Theon of Smyrna (ca. 70–ca. 135 BCE) 28
Turing, Alan Mathison (1912–1954) 292

V

Vahabadzeh, Bijan 114, 115
ver Eecke, Paul 73, 76
Vesalius, Andreas (1514–1564) 156
Viète, François (1540–1603) 77, 157–164, 166, 171, 176, 178, 182, 183, 188, 190, 192
Vlacq, Adriaan (1600–1667) 170

W

Wallis, John (1616–1703) 182–187, 189, 190, 192, 193, 201–204, 209, 212
Weierstrass, Karl (1815–1897) 237, 242
Weisz, Rachel 31
Wessel, Caspar (1745–1818) 212, 213
Whiston, William (1667–1752) 191

Z

Zermelo, Ernst (1871–1953) 284, 285

SUBJECT INDEX

A

Agora (film) 31
abbacus tradition 129–131, 136–140, 142, 145–147, 156–158, 161, 164
AC, Axiom of Choice 283, 284, 286
Acadians 28
al-jabr wa'l-muqābala 77, 91, 98–100, 111, 114, 115, 130
al-Khwārizmī's six kinds of equations 91, 143
al-kitāb al-muḥtaṣar fī ḥisāb al-jabr wa'l-muqābala. Treatise by al-Khwārizmī 90
algebra
  Arabic 91, 94, 127, 128
  British symbolic 217, 218
  symbolic 99, 111
algebra, fundamental theorem of 14, 180, 193, 194, 211, 214, 225
algebraically closed 14
Algorismus proportionum. Book by Oresme 131
Almagest. Book by Claudius Ptolemy 30, 111
Arabic translations 32, 72, 75, 88, 90
Archimedes, Principle 59
arithmetic, fundamental theorem of 5, 214, 225, 227, 230
Arithmetica. Book by Diophantus 71–77, 83, 111
Arithmetica Infinitorum. Book by Wallis 183
Arithmetica Integra. Book by Stifel 135
arithmos (Diophantus) 75–78, 84, 85, 170
arithmos (Diophantus) 74, 77
arithmoston (Diophantus) 75
Ars Magna, sive de Regulis Algebraicis. Book by Cardano 143, 146, 147, 156
Ars Magna. Book by Cardano 138
Artis Analyticae Praxis. Book by Harriot 182
astrology 138
astronomy 28–30, 32, 45, 47, 89, 90, 101, 111, 113, 126, 129, 138, 156, 166–168, 170–172, 191, 218

B

Babylonian mathematics 19, 27, 28, 30
Baghdad 89, 90
Bayt al-Ḥikma 89, 90
Biblical exegesis 117
Bologna 125
British currency 23
British Parliament 191
Brougham Bridge, Dublin 221
Byzantine sources 32, 128

C

calculus, fundamental theorem of 235–237, 245, 246
Cambridge 188, 190, 193
  Lucasian Chair 188, 191
  Trinity College 191
cardinal 15, 16, 251, 259–262, 267–279, 281, 287
cartography 166, 167
CH, Cantor's Continuum Hypothesis 275, 276, 284, 286
China vii, 24, 87
Christians 87, 116
circle 38, 47, 48, 68, 101, 121–124, 177, 178, 185, 186, 197, 225, 226
  squaring problem 49, 68
Clavis Mathematicae. Book by Oughtred 182
completion of the square 94, 96, 120
comprehension, principle of 281, 282, 285
conic sections 68–70, 82, 112, 121, 129, 183
Constantinople 127
constructivism 281, 286
continuity 9, 237–239, 242, 243, 247, 250, 251, 265
continuous magnitudes 42, 59, 80, 163, 189
cossic tradition 134–138, 142, 148, 150, 156, 158, 160, 161, 164
Crotona 32
cube 67–69, 72, 74, 80, 110–112, 124, 129, 131, 139–143, 148, 160, 179, 184
  duplication problem 67
cuneiform texts 27
cut (Dedekind) 236–245
cyclotomy 226

D

Damascus 107
De humani corporis fabrica.
    Book by Vesalius 156
De Regula Aliza. Book by
    Cardano 145, 147
decimal metric system 23
decimal system 17–20, 22,
    23, 28, 93, 117, 129, 130,
    166, 167
denominator 26, 35, 41, 52,
    78, 85, 123
density 9, 239
Descartes' rules of signs 180
diagonal of a square 40, 41,
    52–54, 189
Die Coss Book by Rudolff 131
Discours de la methode. Book
    by Descartes 175
discrete magnitudes 80, 150,
    162, 192
Disquisitiones Arithmeticae.
    Book by Gauss 210, 214
double *reductio ad
    absurdum* 58, 59
dynamis (Diophantus)
    74–78, 84
dynamoston (Diophantus) 75,
    78, 84

E

École Normale Superieur 236
École Polytechnique 236
Egyptian hieroglyphic
    writing 24–26
Egyptian mathematics 24–26,
    36, 45, 128
electronic computers 19, 20
ellipse 183
*Encyclopédie, ou dictionnaire
    raisonné des sciences, des
    arts et des métiers* 208
English Civil War
    (1642–1651) 182
equation
    algebraic 5–7, 12, 66, 69, 71,
        72, 77
    cubic 112, 141–144
    differential 4
    Diophantine 71
    fourth-degree 147
    polynomial 11–14, 177,
        180, 181, 193, 194, 211,
        229, 276, 287
    quadratic 10, 12, 67,
        91–94, 117, 120, 135,
        142–144, 179, 182,
        183, 186, 187, 196, 198
    quartic 140
Euclid's *Elements* 38, 42–44,
    50–52, 55, 57–59, 63, 64,
    66, 78, 93, 97, 103, 117,
    118, 121, 126, 127, 133,
    136, 139, 142, 144, 149,
    150, 156, 163, 164, 188,
    191, 195, 197–203, 227,
    242, 243
Eudoxus' theory of propor-
    tions 43, 45, 55–59, 64,
    66, 68, 123, 149, 164, 208,
    242–245
Euler formula 210
exhaustion, method of 59

F

fluxions and fluents 236
fractions 6, 7, 9, 12, 21, 25–27,
    35, 36, 41, 45, 46, 52, 78,
    79, 81, 84, 85, 93, 94,
    101, 102, 128, 129, 135,
    149, 156, 164–167, 169,
    170, 179, 183, 184, 189,
    193, 209, 231, 232, 234,
    244, 268
    common 6, 26, 35, 45–47,
        102, 134, 135, 150
    decimal 6, 9, 21, 27,
        35, 107–110, 134,
        164–167, 169–171,
        184, 189, 270, 271
    Egyptian 26
    sexagesimal 101
    unit 25–27, 45, 78
French Revolution 23, 236

G

geography 32, 87, 90, 125
geometric algebra 54, 66, 67,
    198, 200
geometry
    analytic 68, 69, 82, 121, 163,
        176, 179, 182, 187
    non-Euclidean 215, 218
gigabyte 23
Göttingen 224, 225, 230, 231,
    236, 266, 284, 293
golden ratio 37, 38, 66, 67
Great Pyramid 37
Greece, ancient 5, 24, 31, 35,
    50, 87, 127
Greek mathematics 26, 28, 31,
    32, 38, 42, 45, 54, 63, 78,
    80, 81, 83, 84, 88

H

Hebrew numeration system 20
*Ḥibbūr ha-meshīhah we-ha-
    tishboret.* Treatise by
    Bar-Ḥiyya 117
higher reciprocity 215
Hindu–Arabic numerals 89,
    90, 101, 107, 117, 128, 151
Hollywod 31
humanism 127
hyperbola 68–70,
    121–123, 183
hypotenuse 33

I

ICM 1900, International Con-
    gress of Mathematicians,
    Paris 284
incommensurability 39–43,
    52–54, 64, 126, 136, 149,
    164, 192, 243, 291
India vii, 24
induction, principle of 118,
    250–254, 258, 262, 263
infinitesimal calculus 4, 60,
    155, 157, 183, 191,
    195, 207, 210, 211, 215,
    218, 235–239, 247, 249,
    258, 266
intuitionism 286, 287
Islamicate mathematics 77, 83,
    87, 88, 91, 94, 100, 110,
    112, 116, 117, 121, 125
isosceles triangle 38

J

Japan vii, 24
Jews 87, 116, 125

K

Karnak Temple, Egypt 24
Kitāb al-Fuṣūl fī al Ḥisāb al-Hindī. Treatise by al-Uqlīdisī 107
Korea vii, 24
kyboskybos (Diophantus) 78
kybos (Diophantus) 74, 75

L

L'invention en algebra. Book by Girard 180
La Disme (De Thiende). Book by Stevin 164
La géométrie. Book by Descartes 175, 178, 192, 196
L'arithmétique. Book by Stevin 164
Latin America vii, 24, 262
Liber Abbaci. Book by Fibonacci 128, 129, 167
Liber Embadorum. Latin translation of a treatise by Abraham Bar-Ḥiyya 117
linear methods of solution (Pappus) 71
locus 81, 82, 179
logarithm 13, 167–173, 210
logicism 259
logistics 32
Louvre Museum, Paris 24

M

Ma'aseh Hoshev. Treatise by Gersonides 117, 118
Maghreb 87, 88, 110
marked ruler 38
mathematical physics 284
Mathesis Universalis. Book by Wallis 202
Mayan mathematics 19
mean
  arithmetic 9, 36
  double geometric 68
  geometric 36, 68, 185, 186, 212
  harmonic (or sub-contrary) 36, 37
On the Measurement of the Circle. Treatise by Archimedes 49, 68
Menon. Book by Plato 54

Meshed, Iran 72
Mexico 19
Milan 138
monades (Diophantus) 74, 76, 84

N

Notre Dame 37
numbers
  algebraic 228–230, 250, 265, 276, 277, 287, 288, 291
  algebraic integers 229, 230
  amicable 33
  complex 10–14, 22, 148, 164, 180, 186, 207, 209–222, 224–226, 228, 231, 232, 234
  cubic 75, 78
  cyclotomic integers 226–228
  even 16, 34, 52, 266, 267
  figurate 34, 54, 73, 74
  Gaussian integers 214, 225
  hypercomplex 221
  integers 5–8, 12, 14, 21, 41, 43, 44, 52, 56, 79, 101, 102, 128, 136, 138, 143, 152, 165, 166, 193, 210, 214–216, 219, 225–230, 232–234, 244, 258, 262, 268
  irrational 4, 7–10, 12, 13, 41, 42, 79, 136–138, 168, 184, 189, 193, 235–237, 239–243, 249, 262, 266, 276, 287
  natural 3–5, 7, 14–16, 26, 33, 41, 42, 63–65, 93, 94, 101, 118, 134, 184, 189, 212, 220, 230, 233–235, 243, 249–263, 265–269, 274, 277–279
  negative 5, 6, 79, 93, 112, 135–138, 143–147, 162, 168, 181–187, 189, 193, 208, 209, 212, 233, 234, 291
  negative, square roots of 11, 144, 145, 147, 148, 164, 181, 184, 186, 193
  oblong 34
  odd 34, 53, 275

  ordinal 260
  perfect 33, 73
  positive 8, 112, 143, 145
  prime 4, 5, 64, 65, 73, 119, 214, 227, 228, 230
  rational 7, 9, 14, 136, 138, 156, 168, 184, 189, 231–245, 249, 258, 262, 265, 287
  real 8, 9, 43, 186, 216, 217, 219, 220, 222, 224, 231–243, 245, 250, 258, 265–271, 275, 276, 287
  square 121
  transcendental 13, 68, 276, 277
  triangular 34
numerator 25, 35, 41, 45, 46, 78, 85, 123

O

On Polygonal Numbers. Treatise by Diophantus 74
ordinal 15, 16, 251, 252, 254, 255, 259, 262, 277–279
Oxford 125, 169
  Savilian Chair of Geometry 182, 202

P

papyrus 24–26, 32, 46
parabola 68, 69, 176, 183
parallelogram 48, 50, 51, 213
Paris 24, 125
Parthenon 37
philosophy 138
π 7, 10, 12, 13, 21, 47–49, 68, 158, 183, 210, 270, 276
plane methods of solution (Pappus) 70, 129
positional system 17–22, 24, 25, 27, 29, 89, 90, 93, 94, 101, 111, 117, 130, 151, 166
Principia Mathematica. Book by Newton 191, 195
Principles of Mathematics. Book by Russell 281
Prior Analytics. Treatise by Aristotle 52
prism 82

probability 138
proportion 35–37, 41–45, 55–59, 64, 66, 68, 69, 122–124, 126, 149, 161, 164, 170, 171, 177, 183, 189, 194, 208, 209, 212, 242–245
Provence 117
Pythagorean pentagram 38, 39, 41, 67

Q

quaternions 219–223
Quesiti et Inventioni Diverse. Book by Tartaglia 140

R

ratio 33, 35–37, 40–45, 47, 52–58, 64, 83, 138, 149, 156, 165, 170, 183, 184, 189, 192, 193, 208, 245, 276
  composite 83
rectangle 51, 52, 66, 82, 83, 95, 121, 122, 144, 145, 162, 176, 177, 185, 246
Renaissance 31, 35, 70, 71, 76, 77, 83, 88, 127, 131, 138, 142, 148, 150, 151, 156, 160
representation
  binary 19–22, 270, 271
  hexadecimal 19
  hexagesimal 20
  octal 19
  sexagesimal 27, 129
  sexagesimal 30, 102, 111, 129, 166

right-angled triangle 33, 40
Roman Empire 125
Roman numeration system 21
Russell's paradox 281–283, 285, 286

S

Sand Reckoner. Treatise by Archimedes 29
Sanskrit sources 89
Sefer ha-Ehad. Treatise by Ibn-Ezra 117
Sefer ha-Mispar. Treatise by Ibn-Ezra 117
sign multiplication 76, 136, 145, 147
solid methods of solution (Pappus) 70, 129
Spain 87
Stetigkeit und irrationale Zahlen. Book by Dedekind 9, 237
straight line 8, 9, 38, 48, 51, 59, 82, 83, 121, 122, 171, 172, 176, 184, 186, 209, 239
straightedge-and-compass constructions 38, 67, 68, 70, 71, 121, 123
Sumerians 28
Summa de arithmetica geometria proportioni et proportionalita. Book by Pacioli 137
Synagogue (Collection). Book by Pappus 70, 81

T

Talkhīṣ 'amal al-ḥisāb. Treatise by Ibn al Bannā' al Marrākushī 110, 117
tetragônoi (Diophantus) 73
texts
  astronomical 29, 30, 32, 45, 47, 89, 101, 111, 172
  commercial 45–47, 90, 144
The Whetstone of Witte. Book by Recorde 135
theology 126, 138, 182
Toledo, Spain 125
Treatise on Algebra. Book by Wallis 184
triangle 38, 48, 81, 196, 197
trigonometry 129, 170, 172, 173

U

Universal Arithmetick. Book by Newton 191, 195, 208

W

Was sind und was sollen die Zahlen?. Book by Dedekind 224
WO, Well-Ordering Principle 283, 284, 286

Z

zero 20, 25, 27, 30, 93, 94, 101, 112, 128, 143, 180, 186, 209, 232, 260
ZF, Zermelo–Fraenkel axioms for sets 285
Zoroastrians 87